Lecture Notes in Mathematics 1496

C. Foias B. Francis J. W. Helton
H. Kwakernaak J. B. Pearson

H_∞-Control Theory

Lectures given at the 2nd Session of the
Centro Internazionale Matematico Estivo (C.I.M.E.)
held in Como, Italy, June 18-26, 1990

Editors: E. Mosca, L. Pandolfi

Springer-Verlag
Berlin Heidelberg New York
London Paris Tokyo
Hong Kong Barcelona
Budapest

Authors

Ciprian Foias
Department of Mathematics
Indiana University
Bloomington, IN 47405, USA

Bruce Francis
Department of Electrical Engineering
University of Toronto
Toronto, Canada M5S 1A4

J. William Helton
University of California, San Diego
La Jolla, CA 92093-9112, USA

Huibert Kwakernaak
Eindhoven University of Technology
Department of Mathematics and Computer Science
P. O. Box 513
5600 MB Eindhoven, The Netherlands

J. Boyd Pearson
Rice University
Department of Electrical and Computer Engineering
P. O. Box 1892
Houston, TX 77251-1892, USA

Editors

Edoardo Mosca
Dipartimento di Sistemi e Informatica
Università di Firenze
Via di Santa Marta 3
I-50139 Firenze, Italy

Luciano Pandolfi
Dipartimento di Matematica
Politecnico di Torino
Corso Duca degli Abruzzi, 24
I-10129 Torino, Italy

Mathematics Subject Classification (1991): 93B36, 93B35, 93C57

ISBN 3-540-54949-8 Springer-Verlag Berlin Heidelberg New York
ISBN 0-387-54949-8 Springer-Verlag New York Berlin Heidelberg

Typesetting: Camera ready by author

46/3140-543210 - Printed on acid-free paper

PREFACE

The fundamental problem of control engineering is to provide robust and satisfactory performance to plants affected by several impairments such as parameter and structural uncertainties as well as stochastic disturbances. Many tools have been developed to answer these needs.

Historically, they date back to the prewar fundamental studies of Bode and Nyquist. During the sixties and seventies, in an attempt to address systematically the issue for multivariable plants, optimal control, deterministic and stochastic, as well as adaptive control, were used to tackle the problem. Despite the tremendous related research effort, at the beginning of the eighties it appeared that some of the most popular "modern" control theoretical tools were inherently lacking the necessary robustness requirements. H_∞ - control theory began in the early eighties as an attempt to put over rigorous bases the classical automatic control ideas, yielding robust feedback design tools, and to extend them to multivariable systems. It soon appeared that H_∞ - control theory was at the intersection of several important roads: state-space approach to control theory and classical interpolation problems; harmonic analysis and operator theory; minimax linear quadratic stochastic control and integral equations. Many results were found in a short time by both control engineers and mathematicians. This CIME session was aimed at presenting the underlying fundamental ideas, problems and results of H^∞ - control theory to a mixed audience of both engineers and mathematicians, so as to stimulate a cross fertilization and further advances in this interesting research area.

Three Seminars by leading mathematicians active in this field were also offered to the participants with the aim of enlighting related topics and methods. The texts of the seminars are also contained in the volume.

Edoardo Mosca and Luciano Pandolfi

TABLE OF CONTENTS

COMMUTANT LIFTING TECHNIQUES FOR COMPUTING OPTIMAL H^∞ CONTROLLERS

Ciprian Foias
Department of Mathematics, Indiana University
Bloomington, Indiana 47405, U.S.A.

Introduction

During the 1980's, a general approach was formulated for the design of control systems to meet frequency-domain performance specifications which came to be known as H^∞-optimization theory ([Z1]; see also [Fr1] and the references therein). In these lectures, we will discuss an operator theoretic method (so-called "skew Toeplitz" theory) for the solution of such H^∞-problems for distributed plants introduced by A. Tannenbaum, G. Zames, H. Bercovici, and the author in [FTZ1],[FTZ2],[FT1],[FTZ3], [BFT1], [FT2],[FT4]. This method is based on reducing the given optimization problem to computing the kernel of a certain "skew Toeplitz" operator. This reduction is carried out via the commutant lifting theorem which will be discussed below. A key point to be made is that the complexity of the computations in solving a given optimization problem only depends on the MacMillan degree of the weighting filters (which are always chosen to be finite dimensional) and not that of the plant (which may be infinite dimensional). Hence we can handle distributed parameter systems. More details about skew Toeplitz theory may be found in [BFT1],[FT3], and [FF1]. The method can also be coupled to some nonlinear systems [FT5], [FT6]. In this case even if the systems are not distributed, the mathematics is similar to that of the distributed ones. This will be briefly presented at the end of these lectures.

Acknowledgement. I thank Allen Tannenbaum for reading the manuscript and for his judicious remarks.

Lecture I. The commutant lifting theorem paradigm

In this lecture we present several basic interpolation problems occuring in H^∞-control theory and the operator theoretic method, based on the commutant lifting theorem, for solving those problems.

1. Interpolation problems

Throughout we shall denote by L^2 the Lebesgue space $L^2(\partial\mathbb{D})$, where $\partial\mathbb{D} = \{z \in \mathbb{C} : |z| = 1\}$, normed by

$$\|f\| = \left[\int_0^{2\pi} |f(e^{it})|^2 \frac{dt}{2\pi} \right]^{1/2},$$

and by $H^2 = H^2(\mathbb{D})$, where $\mathbb{D} = \{z \in \mathbb{C} : |z| < 1\}$, the Hardy spaces of the analytic functions

$$F(z) = a_0 + za_1 + \cdots + z^n a_n + \cdots \quad (z \in \mathbb{D})$$

on $\mathbb{D}$ such that

$$\|F\|(= \text{norm in } H^2) = [|a_0|^2 + |a_1|^2 + \cdots]^{1/2} < \infty .$$

We recall that

$$F(e^{it}) = \Sigma\, a_n e^{int} \qquad (0 \leq t < 2\pi)$$

makes sense in L^2 as the boundary values of F and that in this way H^2 is also identified with a subspace of L^2. Also we shall denote by U the canonical bilateral shift on L^2, that is the multiplication operator

$$(Uf)(z) := zf(z) \qquad (z \in T,\ f \in L^2)$$

and by S the canonical unilateral shift on H^2, i.e. $S = U|H^2$:= the restriction of U to H^2. The Lebesgue space $L^\infty(\partial\mathbb{D})$ of essentially bounded (classes of) measurable functions on $\partial\mathbb{D}$ will be denoted by L^∞, while that of bounded analytic functions on $\mathbb{D}$ by H^∞. Obviously $H^\infty \subset H^2$, $L^\infty \subset L^2$ and by the inclusion $H^2 \subset L^2$ we also have $H^\infty \subset L^\infty$. Moreover, if

$$\|f\|_\infty := \text{ess. sup}\{|f(\zeta)| : \zeta \in \partial\mathbb{D}\} \qquad (f \in L^\infty)$$

denotes the norm on L^∞ then for $F \in H^\infty$ we have

$$\|F\|_\infty = \sup\{|F(z)| : z \in \mathbb{D}\} .$$

If $\varphi \in L^\infty$ (resp. $\phi \in H^\infty$) then the multiplication operator by φ on L^∞ (resp. by ϕ on H^∞) will be denoted by $\phi(U)$ (resp. $\phi(S)$). We recall that any bounded operator on L^2 (resp. on H^2) commuting with U (resp. S) is of the form $\varphi(U)$ with some $\varphi \in L^\infty$ (resp. $\phi(S)$ with some $\phi \in H^\infty$). In both cases the norm of the operators equals that of the respective functions in L^∞. (For all these classical facts see [Sz.-NF2], [B1] or [RR1].)

We will recall now several classical interpolation problems and their underlying operator theoretical framework.

(1) The Carathéodory problem. Given $a_o, a_1, \ldots, a_n \in \mathbb{C}$, define

$$\phi = \{f \in H^\infty : f(z) = a_o + a_1 z + \cdots + a_n z^n + O(z^{n+1})\} .$$

Find

$$\mu = \inf\{\|f\|_\infty : f \in \mathcal{I}\} .$$

<u>Answer.</u> μ is the norm of the operator

$$A = \begin{bmatrix} a_0 & & & 0 \\ a_1 & \ddots & & \\ \vdots & \ddots & \ddots & \\ a_n & \cdots & a_1 & a_0 \end{bmatrix}$$

on the Euclidian space $\mathbb{C}^{n+1}$, that is $\mu = \|A\|$.

(2) <u>The Nevanlinna-Pick problem.</u> Given $z_1, \ldots, z_n \in \mathbb{D}$ and $a_1, a_2, \ldots, a_n \in \mathbb{C}$, define

$$\mathcal{I} = \{f \in H^\infty : f(z_k) = a_k \quad (1 \leq k \leq m)\}.$$

Find

$$\mu = \inf\{\|f\|_\infty : f \in \mathcal{I}\} .$$

<u>Answer</u>. Consider the operator A_* on the space

$$H = \sum_{k=1}^{m} \mathbb{C}y_k \subset H^2, \text{ where } y_k = \frac{1}{1-z\bar{z}_k} \quad (z \in \mathbb{D},\ 1 \leq k \leq m),$$

by

$$A_* \sum_{k=1}^{m} \beta_k y_k = \sum_{k=1}^{m} \beta_k \bar{a}_k y_k \quad (\beta_1, \ldots, \beta_n \in \mathbb{C}) .$$

Then

$$\mu = \|A_*\| .$$

(3) <u>The Hermite-Fejér problem.</u> Given $z_k \in \mathbb{D}$, $a_{k0}, a_{k1}, \ldots, a_{kn_k} \in \mathbb{C}$ $(1 \leq k \leq m)$ define

$$\mathcal{I} = \{f \in H^\infty : f(z) = \sum_{h=0}^{n_k} a_{kh}(z - z_k)^h + O((z-z_k)^{n_k+1}) \ (1 \leq k \leq m)\} .$$

Find

$$\mu = \inf\{\|f\|_\infty : f \in \mathcal{I}\} .$$

<u>Answer.</u> Define

$$y_{j,k}(z) = \frac{z^j}{(1-\bar{z}_k z)^{j+1}} \quad (z \in \mathbb{D},\ j = 0,1,\ldots, 1 \leq k \leq m)$$

(so $y_{0,k} = y_k$) and

$$H = \sum_{k=1}^{m} \sum_{j=0}^{n_k} \mathbb{C} y_{j,k} \quad (\subset H^2) .$$

Define the linear operator A_* on H by

$$A_* y_{j,k} = \sum_{j \le h \le N_k} \bar{a}_{kh} y_{j-h,k} \quad \text{for} \quad 0 \le j \le n_k, \ 1 \le k \le m .$$

Then $\mu = \|A_*\|$.

Obviously, the Problems (1) and (2) are special cases of the Problem (3). In particular if we view (1) as a special case of (3), then the matrix of the operator A_* with respect to the orthonormal basis $y_{0,1} \equiv 1, \ y_{1,1} \equiv z, \ \ldots \ y_{n,1} \equiv z^n$ (with $z_1=0$, $m=1$, $n_1=n$) in

$$H = \sum_{j=0}^{n} \mathbb{C} y_{j,1}$$

is given by

$$\begin{bmatrix} \bar{a}_0 & \bar{a}_1 & \cdot & \cdot & \bar{a}_n \\ & \ddots & \ddots & & \vdots \\ & & \bar{a}_0 & \ddots & \vdots \\ & & & \ddots & \bar{a}_1 \\ 0 & & & & \bar{a}_0 \end{bmatrix}$$

so that the norm $\|A_*\|$ of A_* will equal that of the adjoint of the operator A considered in (1), hence $\|A_*\| = \|A\|$.

D. Sarason [Sa1] was the first to show that Problems (1) and (2) are very particular cases of a new intriguing and inspiring result in operator theory (stated below as Corollary (8)) an easy consequence of which is the following optimization problem:

(4) Given an inner function $m \in H^\infty$ (i.e. $|m(\zeta)| = 1$ almost everywhere on $\partial\mathbb{D}$) and $w \in H^\infty$, find

$$\mu = \inf\{\|w-mq\|_\infty : q \in H^\infty\} .$$

<u>Answer.</u> (Sarason [Sa1]): Let P denote the orthogonal projection of H^2 onto $H(m) = H^2 \ominus mH^2$ (= the space of all functions $f \in H^2$ orthogonal on the subspace $mH^2 = \{mg : g \in H^2\}$) and let $A_* = w(S)^* | H(m)$. Then $\mu = \|A_*\|$.

We notice that Problems (1), (2) and (3) become particular cases of Problem (4) by choosing

$$m(z) = z^{n+1}, \ m(z) := \prod_{k=1}^{m} \frac{z-z_k}{1-\bar{z}_k z} \ \text{and} \ m(z) := \prod_{k=1}^{m} \left(\frac{z-z_k}{1-\bar{z}_k z}\right)^{n_k+1} \quad (z \in \mathbb{D}),$$

respectively.

A similar (albeit more complicated) optimization problem was solved by Feintuch-Francis [FF1].

(5) Given an inner function $m \in H^\infty$ and $w, f, g, h \in H^\infty$, find

$$\mu = \inf\left\{ \left\| \begin{bmatrix} w-mq & f \\ g & h \end{bmatrix} \right\|_\infty : q \in H^\infty \right\} .$$

(Here for a matrix-valued function

$$\varphi(z) = \begin{bmatrix} a(z) & b(z) \\ c(z) & d(z) \end{bmatrix}$$

with the entries $a, b, c, d \in H^\infty$,

$$\|\varphi\|_\infty = \sup_{z \in \mathbb{D}} \left\| \begin{bmatrix} a(z) & b(z) \\ c(z) & d(z) \end{bmatrix} \right\|$$

where the norm of an 2×2 matrix is the operator norm on the Euclidian space $\mathbb{C}^2$.)

<u>Answer.</u> (Feintuch-Francis [FF1]): Consider

$$H' = (L^2 \ominus mH^2) \oplus L^2, \quad H = H^2 \oplus L^2$$

and A the operator from H into H' defined by

$$A \begin{bmatrix} h_1 \\ h_2 \end{bmatrix} = P' \begin{bmatrix} w & f \\ g & h \end{bmatrix} \begin{bmatrix} h_1 \\ h_2 \end{bmatrix} \quad \left(\begin{bmatrix} h_1 \\ h_2 \end{bmatrix} \in H \right)$$

where P' denotes the orthogonal projection of $L^2 \oplus L^2$ onto H'. Then $\mu = \|A\|$.

The Problem (4) (and (5) as well) can be also treated with techniques developed in the theory of the Nehari interpolation problem. To see this we recall first:

(6) <u>Nehari interpolation problem</u> [Ne1]. Given $c_{-1}, \dots, c_{-n}, \dots \in \mathbb{C}$ define

$$\mathcal{I} = \{ f \in L^\infty : \frac{1}{2\pi} \int_0^{2\pi} e^{int} f(e^{it}) dt = c_{-n} \quad (n=1,2,\dots) \}.$$

Find

$$\mu = \inf\{ \|f\|_\infty : f \in \mathcal{I} \}.$$

<u>Answer.</u> (Nehari [N1]; see also [AAK1]) Consider the linear operator A given by the matrix

$$\begin{bmatrix} c_{-1} & c_{-2} & c_{-3} & \cdot & \cdot \\ c_{-2} & c_{-3} & \cdot & \cdot & \cdot \\ c_{-3} & \cdot & \cdot & \cdot & \cdot \\ \cdot & \cdot & \cdot & \cdot & \cdot \\ \cdot & \cdot & \cdot & \cdot & \cdot \end{bmatrix}$$

on ℓ^2 (= Hilbert space of square summable sequences).

Then $\mathcal{I} \neq \emptyset$ if $\|A\| < \infty$ and in this case $\mu = \|A\|$.

Proof. By (7) there exists an operator B on H^2 commuting with S such that

$$B^*|H = A_*, \quad \|B\| = \|B^*\| = \|A_*\| .$$

But because of the commutativity property, B is of the form $b(S)$ with some $b \in H^\infty$ satisfying $\|b\|_\infty = \|B\|$.

(9) Corollary (Adamjan-Arov-Krein [AAK1]). Let P denote the orthogonal projection of L^2 onto H^2 and let $S' = (I-P)U|L^2 \ominus H^2$. For any operator A from H^2 into $L^2 \ominus H^2$ intertwining S and S' (i.e. $AS = S'A$) there exists $b \in L^\infty$ such that

$$A = (I-P)b(U)|H^2, \quad \|b\|_\infty = \|A\| .$$

Proof. Apply (7) to $K_1 = L^2$, $K_2 = H^2$, $H_1 = L^2 \ominus H^2$, $H_2 = K_2$, $U_1 = U$, $U_2 = S$, $T_{1*} = U^*|H_1$, $T_{2*} = S^*$ and $A_* = A^*$ (= the adjoint of A), to obtain an operator B_* from L^2 into H^2 such that

$$B_*|L^2 \ominus H^2 = A^*, \quad B_*U^* = S^*B_* \quad \text{and} \quad \|B_*\| = \|A^*\| = \|A\| .$$

Apply (7) once again to $K_1 = L^2$, $K_2 = L^2$, $H_1 = L^2$, $H_2 = H^2$, $U_1 = U$, $U_2 = U$, $T_{1*} = U^*$, $T_{2*} = S^*$ and $A_* = B_*$ to obtain an operator B on L^2 commuting with U such that

$$B^* = B_*, \quad \|B^*\| = \|B_*\| = \|A\| .$$

By its commutativity with U, B is of the form $b(U)$ with some $b \in L^\infty$, $\|b\|_\infty = \|B\|$.

The answer to Problem (4) follows at once by applying (8) with $A_* = w(S)^*|H$, while the answer to Problem (6) by applying (8) with $A : H^2 \mapsto L^2 \ominus H^2$ defined by

$$A\zeta^n = \sum_{m=n+1}^{\infty} c_{-n}\zeta^{-n} \qquad (n=0,1,2,\ldots)$$

where we denoted by ζ also the identical function $\zeta \mapsto \zeta$ on $\partial\mathbb{D}$. Finally the Feintuch-Francis answer to Problem (5) was obtained by applying (7) with

$$K_1 = H^2 \oplus L^2 = H_1, \; K_2 = L^2 \oplus L^2, \; H_2 = (L^2 \ominus mH^2) \oplus L^2$$

$$U_1 = S \oplus U, \; U_2 = U \oplus U$$

and

$$A_* = \begin{bmatrix} Pw(U)^*|L^2\ominus mH^2 & Pg(U)^* \\ f(U)^*|L^2\ominus mH^2 & h(U)^* \end{bmatrix}$$

where P denotes the orthogonal projection of L^2 onto H^2 .

Notice (together with Arov-Adamjan-Krein [AAK2]) that if in Problem (4) we set $f = \bar{m}w$ on $\partial\mathbb{D}$ and then define

$$c_{-n} = \frac{1}{2\pi}\int_0^{2\pi} e^{int} f(e^{it})dt \quad (n=1,2,\ldots),$$

we reduced Problem (4) to Problem (6). Similarly one can reduce Problem (5) to an interpolation problem considered by Dym-Gohberg in [DG1].

2. The commutant lifting theorem

Let K_1, K_2 be two Hilbert spaces and let U_1, U_2 be isometric operators on K_1, K_2, respectively. Assume that $H_1 \subset K_1$, $H_2 \subset K_2$ are invariant subspaces for U_1^*, U_2^* (i.e. $U_1^* H_1 \subset H_1$, $U_2^* H_2 \subset H_2$) and denote by

$$T_{1*} = U_1^*|H_1, \; T_{2*} = U_2^*|\, H_2$$

the restrictions of U_1^*, U_2^* to H_1, H_2, respectively.

Finally let A_* be an operator from H_1 to H_2 intertwining T_{1*} and T_{2*}, i.e.

$$A_* T_{1*} = T_{2*} A_* .$$

(7) Commutant lifting theorem (Sarason [Sa1], Sz.-Nagy-Foias [Sz.-NF1]; see also [Sz.-NF2], [DMP1], [Pa1], [Sz.NF3]). Given A_* as above, there exists an operator B from K_1 to K_2, intertwining U_1^* and U_2^* (i.e. $BU_1^* = U_2^* B$) such that

$$A_* = B|H_1 \quad \text{and} \quad \|B\| = \|A_*\| .$$

For a rather complete discussion of this theorem, we refer to [FoF1]. Here we will only illustrate how it provides the answers to Problems (4) (hence to (1), (2), (3) too), (5) and (6). We start with the following two consequences of (7)

(8) Corollary (Sarason [Sa1]). Given an inner function $m \in H^\infty$, let $T_* = S^*|H$ where $H = H^2 \ominus mH^2$. Then if A_* is any operator commuting with T_* (i.e. $A_* T_* = T_* A_*$) there exists a function $b \in H^\infty$ such that

$$A_* = b(S)^*|H, \quad \|b\|_\infty = \|A_*\| .$$

For these lectures, there is no need to go deeper into the theory of the commutant lifting theorem (7). Indeed our exploitation of this theorem for devising algorithms solving Problems (4) and (5) will be based only on the geometrical properties of the operators A_*, T_{1*} and T_{2*} occurring in the commutant lifting theorem approaches (described above) to these problems.

3. <u>Connection to the sensitivity problem in control theory</u>

A time-invariant causal stable linear system can be modelled by its transfer function $P \in H^\infty$. It is graphically represented by the diagram

The corresponding operator $u \mapsto y = Pu$ is either $P(U)$ or $P(S)$, i.e. $u, y \in L^2$ or $u \in H^2$, $y \in L^2$, etc. The energy of an input, resp. output, will be $\|u\|^2$, resp. $\|y\|^2$ (where $\|\cdot\|$ denotes the norm in L^2). In the linear system

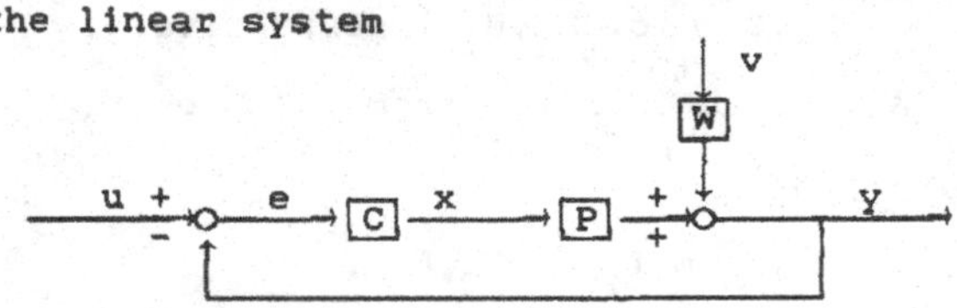

$P \in H^\infty$ represents a given plant, v a disturbance, $W \in H^\infty$ a given filter and u, resp. y, the input, resp. output. These are connected by

$$y = PC(u-y) + Wv \ (= P(U)C(U)(u-y) + W(U)v) \ .$$

The system is required to be <u>internally stable</u>, i.e. that the linear map from the u,v's to e,x,y's be bounded and causal, or equivalently that

$$\begin{gathered}(1+PC)^{-1}, \quad (1+PC)^{-1}W , \\ C(1+PC)^{-1}, \quad C(1+PC)^{-1}W , \\ PC(1+PC)^{-1}, \ PC(1+PC)^{-1}W\end{gathered}$$

be in H^∞ . Since $P, W \in H^\infty$ these conditions reduce to

$$(1+PC)^{-1} \in H^\infty \text{ and } Z = C(1+PC)^{-1} \in H^\infty \ .$$

The relative contribution in energy to y due to a particular disturbance v is

$$\|(1+PC)^{-1}Wv\|^2/\|v\|^2$$

and for any disturbance

$$\sup_{v\neq 0} \frac{\|(1+PC)^{-1}Wv\|^2}{\|v\|^2} = \|((I+PC)^{-1}W)(U)\|^2 = \|(I+PC)^{-1}W\|_\infty^2 .$$

So $\|(I+PC)^{-1}W\|_\infty$ represents the <u>sensitivity</u> of the output to the disturbance and the <u>sensitivity problem</u> consists in minimizing the sensitivity by an appropriate "minimization controller" C, that is:

(10) Find C_0 such that

$$\|(1+PC_0)^{-1}W\|_\infty = \mu := \inf_C \|(1+PC)^{-1}W\|_\infty$$

Using the fact that if $Z = C(1+PC)^{-1}$ then $(I+PC)^{-1} = I-PZ$ and assuming that $P = m\theta$ with m inner $\in H^\infty$, θ, θ^{-1} and $W^{-1} \in H^\infty$, Problem (10) reduces to a bit more elaborate version of Problem (4) namely:

(11) Find $q_0 \in H^\infty$ such that

$$\|W-mq_0\|_\infty = \mu := \inf\{\|W-mq\|_\infty = q \in H^\infty\} .$$

In H^∞-control theory, this problem is usually called the (scalar) <u>one block sensitivity problem</u>.

A similar but more complicated synthesis problem in Control theory leads to the following version of problem (5) (see [Fr1]), usually referred to as the (scalar) <u>four block optimization problem</u>:

(12) Find $q_0 \in H^\infty$ such that

$$\left\|\begin{bmatrix} W-mq_0 & f \\ g & h \end{bmatrix}\right\|_\infty \left[= \sup_{z\in\mathbb{D}} \left\|\begin{bmatrix} (W-mq_0)(z) & f(z) \\ g(z) & h(z) \end{bmatrix}\right\|\right] =$$

$$= \mu := \inf\left\{\left\|\begin{bmatrix} W-mq_0 & f \\ g & h \end{bmatrix}\right\| : q \in H^\infty\right\} .$$

Algorithms to solve Problems (11) and (12) will be given in the following Lectures II and III, respectively.

Lecture II. <u>Skew Toeplitz operators</u>

This lecture is devoted to the concrete algorithm for solving the scalar one block sensitivity problem, introduced in [FT2]. The method used is based on (scalar) skew Toeplitz operators defined in Section 4 below.

4. Optimal controllers and skew Toeplitz operators

The aim of this section is to present the computation of an optimal controller in the scalar one block sensitivity problem, i.e. Problem (11). So let m be any inner function $\in H^\infty$ and let $W \in H^\infty$. By Sarason's theorem (8) as well by his solution to Problem (4), we know that the infimum in Problem (11) is given by

$$\mu = \|W(S)^* | H\| \quad (\text{where } H = H^2 \ominus mH^2)$$

and that there exists $b \in H^\infty$ such that

$$(b(S)^* - W(S)^*) | H = 0 \quad \text{and} \quad \|b\|_\infty = \mu .$$

The first relation implies that $W - b = (W(S) - b(S))1$ is orthogonal on H hence $W - b \in mH^2$. Thus there exists $q_0 \in H^2$ such that $mq_0 = W-b$ in H^2 hence $q_0 = \bar{m}(W-b)$ is in L^∞ and therefore $q_0 \in H^\infty$. Since

$$\|W-mq_0\|_\infty = \|b\|_\infty = \|W(S)^* | H\| = \mu ,$$

q_0 is a solution to the sensitivity problem (11). In this general setting an explicit computation of such a solution seems to be presently out of reach. However, there is a very ubiquitous particular case in which this is possible just because of the functional geometry already set in place around Problem (11). Indeed we have the following simple fact

(13) **Lemma.** If the norm of $A = (w(S)^* | H)^*$ is attained, i.e. if there is a $h_0 \in H$, $h_0 \neq 0$ such that $\|Ah_0\| = \|A\|\|h_0\|$, then a solution q_0 of the sensitivity problem (11) is provided by

(13.a) $$q_0 = \bar{m}(Wh_0 - h_1)/h_0 ,$$

where $h_1 = Ah_0$ and

(13.b) $$(\|A\|^2 I - AA^*)h_1 = 0 , \; A^*h_1 = \|A\|^2 h_0 .$$

Proof. In this case for the above function $b \in H^\infty$ given by Sarason's theorem (8), we have (by using the projection P of H^2 onto H)

$$\|A\|^2\|h_0\|^2 = \|Ah_0\|^2 = \|Pb(S)h_0\|^2 \leq \|b(S)h_0\|^2 =$$

$$= \frac{1}{2\pi}\int_0^{2\pi} |b(e^{it})|^2 \; |h_0(e^{it})|^2 dt \leq \|b\|_\infty^2\|h_0\|^2 = \|A\|^2\|h_0\|^2$$

hence

$$h_1 = Ah_0 = Pb(S)h_0 = b(S)h_0 = bh_0$$

and $|b| = 1$ almost everywhere on the unit circle $\partial\mathbb{D}$. Moreover

(13.c) $\|Ah_0\| = \|A\| \, \|h_0\|$ if and only if $(\|A\|^2 I - A^* A)h_0 = 0$.

Indeed if the last equality above holds, then taking the scalar product with h_0 we obtain

$$0 = \|A\|^2\|h_0\|^2 - (A^*Ah_0, h_0) = \|A\|^2\|h_0\|^2 - \|Ah_0\|^2 .$$

Conversely, if the norm of A is attained by h_0, then since $R = \|A\|^2 I - A^*A \geq 0$ (i.e. $(Rh,h) \geq 0 \ \forall h \in H$) we have

$$\|Rh\|^2 \leq (Rh,h)^{1/2}(R^2h,Rh)^{1/2} = (\|A\|^2\|h\|^2 - \|Ah\|^2)^{1/2}(R^2h,Rh)^{1/2}$$

which is zero for $h = h_0$. Now (13.b) immediately follows from (13.c).

Note that the optimal $q_0 \in H^\infty$ always exists regardless of the existence of a maximal vector $h_0 \in H$. In order to give an easy way to check conditions for the existence of a maximal vector, let us introduce the operator $T = (T_*|H)^*$ (see (8)), and remark that the spectrum of T,

$$\sigma(T) := \{\lambda \in \mathbb{C} : (T-\lambda I)^{-1} \text{ does not exist on } H\} ,$$

is formed by the zeros of m in $\mathbb{D}$ as well as by those points on $\partial\mathbb{D}$ which are singular for m.

(14) <u>Lemma</u> [FTZ3]. Let $w \in H^\infty$ be continuous on $\partial\mathbb{D} \cap \sigma(T)$. Then

$$\|A\|_{ess} := \max\{|w(\zeta)| : \zeta \in \partial\mathbb{D} \cap \sigma(T)\} \leq \|A\|$$

and the norm of A is attained if $\|A\|_{ess} < \|A\|$.

We shall not give the proof of (14) since it will take us deeper into operator theory. However, we shall shortly exploit the framework provided by this theory in order to compute $\|A\|$. To this end we shall assume from now on that W <u>is a rational function in</u> H^∞, that is

$$W = \frac{p}{q} \quad \text{where} \quad p(z) = \sum_0^n a_j z^j, \ q(z) = \sum_0^n b_j z^j$$

are coprime polynomials of degree $\leq n$, and the zeros of q are off $\bar{\mathbb{D}} = \mathbb{D} \cup \partial\mathbb{D}$. (Thus W is continuous on $\bar{\mathbb{D}}$.) Using a bit of operator theory [Sz.-NF2], Ch.III, it is easy to check that $q(T)^{-1}$ exists and that

$$A = q(T)^{-1}p(T) = p(T)q(T)^{-1} .$$

On the other hand if $\|A\|_{ess} < \|A\|$ then the norm of A is attained, say by h_0, and thus the norm of A^* is also attained, for instance by $h_1 = A^*h_0$.

Finally we also remark that for any $\rho \geq 0$ we have

$$\rho^2 I - AA^* = q(T)^{-1}(\rho^2 q(T)q(T)^* - p(T)p(T)^*q(T)^{*-1}.$$

From the above discussion we readily obtain the following:

(15) <u>Lemma</u>. The norm $\|A\|$ of A is either $\|A\|_{ess}$ or the largest $\rho > \|A\|_{ess}$ such that the operator

$$Q = \rho^2 q(T)q(T)^* - p(T)p(T)^*$$

is singular (i.e. it has a non-zero kernel).

The operator Q has a remarkable form, namely

$$Q = \sum_{j,k=0}^{n} c_{jk}T^jT^{*k}$$

where

$$c_{jk} := \rho^2 a_j\bar{a}_k - a_j\bar{a}_k = \bar{c}_{kj} \quad (0 \leq j,k \leq n).$$

(16) <u>Definition</u> [BFT1]. Given an absolutely convergent series $\omega = \sum_{j,k=0}^{\infty} \gamma_{jk}z^j\bar{\zeta}^k$ in the complex variables z, $\bar{\zeta}$ (= the conjugate of ζ) in $\partial\mathbb{D} \times \partial\mathbb{D}$, the <u>skew Toeplitz operator</u> associated with W (as well as with T) is defined by

$$Q_\omega = \sum_{j,k=0}^{\infty} \gamma_{jk}T^jT^{*k}.$$

Obviously the definition depends also on the operator T; thus if $T = S$, and if $\gamma_{jk} = 0$ for $j \cdot k \neq 0$ or if $T = S^*$ one obtains the usual Toeplitz operators with absolutely convergent symbol. Also we can conclude that the computation of the optimum $\mu = \|A\|$ as well as of a maximal vector h_0 and hence of the optimal q_0 is reduced to the computation of the kernel of skew Toeplitz operators with polynomial symbol ω satisfying

$$\overline{\omega(z,\bar{\zeta})} = \omega(\zeta,\bar{z}) \quad (\forall z, \zeta \in \partial\mathbb{D}).$$

For convenience a symbol ω satisfying this condition will be called <u>hermitian</u>.

5. The reduction of the spectral study of a skew Toeplitz operator to linear algebra

In this section we shall consider a hermitian polynomial symbol $\psi = \sum_{j,k=0}^{n} c_{jk} z^j \bar{\zeta}^k$. Then $Q = Q_\psi$ is self-adjoint so that the spectrum of Q is necessarily contained in the real line $\mathbb{R}$. For $r \in \mathbb{R}$, $rI - Q$ is again a skew Toeplitz operator associated to the hermitian polynomial symbol $r - \psi(z,\bar{\zeta})$. Thus the spectral study of the skew Toeplitz operators with hermitian polynomial symbol is reduced to study of their invertibility.

(17)<u>Lemma</u>. If $\psi(\zeta,\bar{\zeta}) \neq 0$ for $\zeta \in \sigma(T) \cap \partial\mathbb{D}$ then $Q = Q_\psi$ is not invertible on H if and only if $\ker Q \neq \{0\}$.

As for the case of (14), we shall skip the proof of (16). Nevertheless, we remark that if the inner function m is rational or equivalently if H is of finite dimension, then the conclusion of Lemma (16) is obvious (and by the way $\sigma(T) \cap \partial\mathbb{D} = \phi$).

For the study of $\ker Q$, it is useful to associate to the symbol ψ the following system of polynomials (in one variable z)

(17.a) $$C_\ell(z) := \sum_{0\leq j\leq n} \sum_{\ell\leq k\leq n} c_{jk} z^{n+j-k+\ell} \quad (0\leq\ell\leq n-1)$$

and

(17.b) $$C(z) := \sum_{i=0}^{2n} c_i z^i := \sum_{j,k=0}^{n} c_{jk} z^{n+j-j}$$

which are also connected by the relations

$$C_\ell(z) = c_0 z^\ell + c_1 z^{\ell+1} + \cdots + c_{n-\ell-1} z^{n-1} + \cdots \quad (0\leq\ell\leq n-1),$$

that is

(17.c) $$C_\ell(z) - z^\ell C(z) = O(z^n) \quad (0\leq\ell\leq n-1).$$

We note the useful fact

$$c_{2n-i} = \bar{c}_i \quad (0\leq i\leq 2n).$$

Also, by some simple computations (left as an exercise for the reader) one proves the following:

(18)<u>Lemma</u>. For $f \in H$ the following conditions are equivalent:

(i) $Qf = 0$;

(ii) $\sum c_{jk} S^j S_*^k = mg$ for some $g \in H^2$;

(iii) $C(z)f(z) - \sum_{\ell=0}^{n-1} C_\ell(z)f_\ell = z^n m(z)g(z)$ almost everywhere on $\partial\mathbb{D}$ (and hence everywhere in $\mathbb{D}$) for the same $g \in H^2$ as in (ii), where $f_0, f_1, \dots, f_{n-1}$ are the first n Taylor coefficients of f at 0, that is

$$f(z) = f_0 + zf_1 + \cdots + z^{n-1}f_{n-1} + O(z^n) \quad (\text{in } \mathbb{D}) .$$

We can now state and give the full proof the main result in this section (and in [FT2] where many details of the proof were not explicated).

(19) **Theorem** [FT2]. Let $C(z) \neq 0$ for all $z \in \sigma(T)$ and for $z = 0$. Then there exists unique $Y_\ell \in H$ and $X_\ell = \sum_{j=0}^{2n-1} x_{\ell,j} z^j$ polynomials of degree $\leq 2n-1$ such that

$$CY_\ell - C_\ell = mX_\ell \quad (0 \leq \ell \leq n-1) .$$

Moreover:

(i) Q is not invertible if and only if $\ker Q \neq \{0\}$.

(ii) $f \in \ker Q$ if and only if

$$f = \sum_{j=0}^{n-1} \xi_j Y_j$$

where $(\xi_0, \dots, \xi_{n-1}) \in \mathbb{C}^n$ satisfies

$$\text{(19.a)} \qquad \begin{bmatrix} x_{0,0} & x_{1,0} & \cdots & x_{n-1,0} \\ x_{0,1} & x_{1,1} & \cdots & x_{n-1,1} \\ \cdots & \cdots & \cdots & \cdots \\ x_{0,n-1} & x_{1,n-1} & \cdots & x_{n-1,n-1} \end{bmatrix} \begin{bmatrix} \xi_0 \\ \cdot \\ \cdot \\ \xi_{n-1} \end{bmatrix} = 0 .$$

Also the Taylor expansion of f at 0 is of the form

$$f = \xi_0 + z\xi_1 + \cdots + z^{n-1}\xi_{n-1} + O(z^n) .$$

(Thus $f = 0$ if and only if $(\xi_0, \dots, \xi_{n-1}) = 0$.)

Proof. Let $C(z) = c_{2n}(z-\alpha_1)\dots(z-\alpha_{2n})$. Then $\alpha_j \notin \sigma(T)$ $(1 \leq j \leq 2n)$ so $C(T) = c_{2n}(T-\alpha_1 T)\dots(T-\alpha_{2n} I)$ is invertible. For $\ell = 0, \dots, n-1$, let $Y_\ell = C(T)^{-1}PC_\ell$ where P denotes the orthogonal projection of H^2 onto H. Then $Y_\ell \in H$ and $C(T)Y_\ell = PC_\ell$, that is $P(C(S)Y_\ell - C_\ell) = 0$ or equivalently $CY_\ell - C_\ell = mX_\ell$ for some $X_\ell \in H^2$. Conversely if this holds for some $Y_\ell \in H$ and $X_\ell \in H^2$

then taking the projection on H we see that $Y_\ell = C(T)^{-1}PC_\ell$ and thus Y_ℓ, and hence X_ℓ too, are uniquely determined. Since $Y_\ell \in H$ we have $\bar{m}Y_\ell \in L^2 \ominus H^2$ so that $C\bar{m}Y_\ell \in S^{2n-1}(L^2\ominus H^2)$. Obviously $C_\ell \in S^{2n-1}(L^2\ominus H^2)$. Therefore $X_\ell \in H^2 \cap S^{2n-1}(L^2\ominus H^2)$ what means that X_ℓ is a polynomial of degree $\leq 2n-1$. This establishes the first statement in the theorem.

The statement (i) is a direct consequence of Lemma (17). For the statement (ii) we will use Lemma (18)(iii). Indeed if $f \in \ker Q$ then from (18)(iii) we obtain

$$C(T)f = \sum_{j=0}^{n-1} f_j PC_\ell, \text{ hence } f = \sum_{\ell=0}^{n-1} f_\ell Y_\ell .$$

Therefore

$$m(z) \sum_{\ell=0}^{n-1} f_\ell X_\ell(z) = C(z) \sum_{\ell=0}^{n-1} f_\ell Y_\ell(z) - \sum_{\ell=0}^{n-1} C_\ell(z) f_\ell = m(z)z^n g(z) \quad (z\in\mathbb{D})$$

that is

$$\sum_{\ell=0}^{n-1} f_\ell X_\ell(z) = z^n g(z) \quad (z \in \mathbb{D})$$

which obviously implies that $(\xi_0,\dots,\xi_{n-1}) := (f_0,\dots,f_{n-1})$ is a solution of (19.a).

Conversely let $(\xi_0,\dots,\xi_{n-1})$ be a solution of (18.a) and define $f := \xi_0 Y_0 + \cdots + \xi_{n-1}Y_{n-1}$. Then $f \in H$ and

$$C(z)f(z) - \sum_{\ell=0}^{n-1} C_\ell \xi_\ell = m(z) \sum_{\ell=0}^{n-1} X_\ell \xi_\ell = mz^n g(z) \quad (z \in \mathbb{D})$$

for some $g \in H^2$. Moreover we also have

$$C(z)f(z) = \sum_{\ell=0}^{n-1} (z^\ell C(z) + O(z^n))\xi_\ell = O(z^n)$$

whence

$$C(z)(f(z)- \sum_{\ell=0}^{n-1} \xi_\ell z^\ell) = O(z^n)$$

and therefore

$$f(z) = \xi_0 + \xi_1 z + \cdots + \xi_{n-1}z^{n-1} + O(z^n) ,$$

that is $f_j = \xi_g$ $(0\leq j\leq n-1)$ and $f \in \ker Q$ by virtue of (18)(iii).

This concludes the proof.

(20) Remarks. (i) It is natural to call (18.a) a singular system of the skew Toeplitz operator $Q = Q_\omega$. Another singular system for $Q = Q_\omega$ was obtained previously in [FTZ3].

(ii) If $f \neq 0$, $Q_\psi f = 0$ and the ψ symbol is that of the operator considered in Lemma (15) with $\rho = \|A\|$ then $h_0 = A^* f$ is a maximal vector for A (i.e. $\|Ah_0\| = \|A\| \, \|h_0\| \neq 0$). In this way, Theorem (17) also provides, via (13), (16), an optimal solution q_0 to the sensitivity problem (10). We emphasize that, as we will show in the next section, the singular system can be explicitely computed.

6. Explicit computation of the singular system

In this section we shall give explicit formulae for the entries of the singular system; actually we shall give the explicit form of the polynomials X_l $(0 \leq l \leq n-1)$. We shall assume in this section that the polynomial $C(z) = \sum_{j=0}^{2n} c_j z^j$ satisfies the hypothesis in Theorem (18), that is

$$C(z) \neq 0 \quad \text{for} \quad z \in \sigma(T) \cup \{0\} .$$

Therefore using the property $c_{2n-i} = \bar{c}_i$ $(0 \leq i \leq 2n)$ one can easily check that the following fact is true.

(21) <u>Lemma</u>. If $\alpha_1, \ldots, \alpha_k$ are the zeros of C in $\mathbb{D}$ then $1/\bar{\alpha}_1, \ldots, 1/\bar{\alpha}_k$ are the zeros of C outside $\bar{\mathbb{D}}$ (with multiplicity accounted).

In order to simplify the presentation we shall make now a supplementary assumption on the polynomial $C(z)$, namely <u>that all zeros of</u> $C(z)$ <u>are simple</u>. (It is easy to drop this assumption but the exposition would become quite cumbersome; see Remark (23) below.) In this case we can denote by $\alpha_{k+1}, \ldots, \alpha_{n-k}$ the zeros of $C(z)$ on $\partial\mathbb{D}$ and we notice that $m(z)$ is analytic and non-zero at $\alpha_1, \ldots, \alpha_{2n-k} \in \bar{\mathbb{D}}$. Therefore making $z = z_j$ $(0 \leq j \leq 2n-k)$ in $CY_l - C_l = mX_l$ (see (18)) we obtain

$$X_l(z_j) = -C_l(z_j)/m(z_j) \quad (0 \leq l \leq n-1, \; 0 \leq j \leq 2n-k) .$$

Besides for $z = \zeta \in \partial\mathbb{D}$ we have

$$(\zeta^{2n-1} X_l(\zeta)) = (\zeta^{2n} C(\zeta)) \cdot (\overline{\zeta m(\zeta)} Y_l(\zeta)) - (\zeta^{2n-1} C_l(\zeta)(\overline{m(\zeta)}))$$

where the functions in the parentheses are all analytic in ζ (and in

H^2); in particular for $z \neq 0$ the first two functions and the last two are given by

$$z^{2n-1}X_\ell(\tfrac{1}{\bar{z}}),\ z^{2n}C(\tfrac{1}{\bar{z}}) \quad \text{and} \quad z^{2n-1}C_\ell(\tfrac{1}{\bar{z}}),\ \overline{m(\bar{z})}$$

respectively. Therefore taking $z = \bar{\alpha}_j$ and using $C(1/\bar{\alpha}_j) = 0$ we obtain

$$\bar{\alpha}_j^{\,2n-1}X_\ell(\frac{1}{\bar{\alpha}_j}) = -\bar{\alpha}_j^{\,2n-1}C_\ell(\frac{1}{\bar{\alpha}_j})\overline{m(\alpha_j)}, \text{ i.e. } X_\ell(\frac{1}{\bar{\alpha}_j}) = -\overline{m(\alpha_j)}C_\ell(\frac{1}{\bar{\alpha}_j})$$

$$(0 \leq j \leq n)$$

But the degree of $X_\ell(z)$ is $\leq$ 2n-1, so that we can conclude with

(22) <u>Proposition</u>. In case the zeros of $C(z)$ are simple, the polynomials X_ℓ $(0 \leq \ell \leq n-1)$ are the Langrange interpolating polynomials of degree 2n-1 satisfying the interpolation conditions

$$X_\ell(\alpha_j) = -\frac{1}{m(\alpha_j)}C_\ell(\alpha_j) \quad (1 \leq j \leq 2n-k)$$

$$X_\ell(\frac{1}{\bar{\alpha}_j}) = \overline{m(\alpha_j)}\,C_\ell(\frac{1}{\bar{\alpha}_j}) \quad (1 \leq j \leq k)$$

where $\alpha_1,\ldots,\alpha_{2n-k}$ and $1/\bar{\alpha}_1,\ldots,1/\bar{\alpha}_k$ are the zeros of $C(z)$.

We recall that if we denote those zeros by $z_1,\ldots,z_{2n}$ then

$$X_\ell(z) = \sum_{j=1}^{2n} \frac{(z-z_1)\cdots(z-z_{j-1})(z-z_{j+1})\cdots(z-z_{2n})}{(z_j-z_1)\cdots(z_j-z_{j-1})(z_j-z_{j+1})\cdots(z_j-z_{2n})} X_\ell(z_j)$$

(for $0 \leq \ell \leq n-1$) .

(23) <u>Remark</u>. (i) As already noticed, the assumption of simplicity for the zeros of $C(z)$ is not essential. Indeed if for instance α_1 is a zero of multiplicity 2 , then from $CY_\ell - C_\ell = mX_\ell$ we deduce also $C'Y_\ell + CY'_\ell - C'_\ell = m'X_\ell + mX'_\ell$ and thus we obtain the two interpolation conditions at $z = \alpha_1$ namely

$$X_\ell(\alpha_1) = -\frac{C_\ell(\alpha_1)}{m(\alpha_1)} \text{ and } X'_\ell(\alpha_1) = [-C'_\ell(\alpha_1) - m'(\alpha_1)X_\ell(\alpha_1)]\frac{1}{m(\alpha_1)} =$$

$$= -[C'_\ell(\alpha_1) - C_\ell(\alpha_1)\frac{m'(\alpha_1)}{m(\alpha_1)}]\frac{1}{m(\alpha_1)},$$

and similarly other two interpolating conditions at $1/\bar{\alpha}_1$. It should be now clear how one handles the general case.

(ii) We emphasize once again that the discussion in this section was made under the assumption that $C(z) \neq 0$ on $\sigma(T) \cup \{0\}$. It is not difficult but quite space consuming to drop this condition. We leave this task to our students, since this is not an easy matter (see [Gu1]).

Lecture III. Skew Toeplitz operators and the 4 block problem

The algorithm introduced in [FT3],[FT4] for solving the (scalar) 4 block problem is quite involved. Therefore in this lecture we will present in Section 7 the mathematical facts and computations on which this algorithm is based, while in Section 8, we will only outline the way the algorithm is obtained.

7. Some general facts on skew Toeplitz operators

We start by considering again a skew Toeplitz operator with a hermitian polynomial symbol

$$Q = \sum_{j,k=0}^{n} c_{jk} T^j T^{*k} \ , \ c_{jk} = \bar{c}_{kj} \quad (0 \leq j,k \leq n)$$

where $T = (S^*|H)^*$, $H = H^2 \ominus mH^2$ and m is an inner function in H^2. Let $C(z)$ and $C_\ell(z)$ $(0 \leq \ell \leq n-1)$ be the polynomials introduced in (17.a), (17.b). As in the sections 5 and 6 we assume that $C(z) \neq 0$ for $z \in \sigma(T) \cup \{0\}$. Then, as we already noted in section 5, the operator $C(T)^{-1}$ exists on H. Our first aim is to solve explicitly the equation $C(T)y = h$ in H. In order to avoid purely technical complications we assume again that the zeros of the polynomial $C(z)$ are simple.

(24) Lemma. For $h \in H$ let $y = C(T)^{-1}h$. Then $Cy = h + mg$ where g is a polynomial of degree $\leq 2n-1$ uniquely determined by the interpolating conditions

$$g(\alpha_j) = -\frac{h(\alpha_j)}{m(\alpha_j)} \quad (1 \leq j \leq 2n-k) \quad \text{(see (21))} \tag{24.a}$$

and

$$g\left(\frac{1}{\bar{\alpha}_j}\right) = -\overline{m(\alpha_j)}\, h\left(\frac{1}{\bar{\alpha}_j}\right) \quad (1 \leq j \leq k) \tag{24.b}$$

where h is defined outside $\mathbb{D}^-$ by

(24.c) $$h(z) := \frac{1}{\overline{m(\frac{1}{\bar z})}} \frac{1}{z} h_*(\frac{1}{z}) \quad (|z| > 1,\ m(\frac{1}{\bar z}) \neq 0)$$

<u>Proof</u>. Let P denote the orthogonal projection of H^2 onto H ; then since $P(Cy-h) = C(T)y - h = 0$ we have $Cy = h + mg$ with some $g \in H^2$. Also, since h and y are in H ,

$$\overline{m(\zeta)}\, h(\zeta) = \bar\zeta h_*(\bar\zeta) := h_{-1}\bar\zeta + h_{-2}\bar\zeta^2 + \cdots$$

and similarly $(\bar m y)(\zeta) = \bar\zeta y_*(\bar\zeta)$, with $h_*, y_* \in H^2$. Thus $g = C\bar m y - \bar m h \in C(U)(L^2 \ominus H^2) - (L^2 \ominus H^2) \subset U^{2n-1}(L^2 \ominus H^2)$, hence $g \in S^{2n-1}H^2$ that is g is a polynomial of degree $\leq 2n-1$. The conditions (24.a) follow at once by setting $z = \alpha_j$ $(1 \leq j \leq 2n-k)$ in $Cy = h + mg$. As for (24.b) we use

$$C(\zeta)\, \frac{1}{\zeta} y_*(\frac{1}{\zeta}) = \frac{1}{\bar\zeta} h_*(\frac{1}{\bar\zeta}) + g(\zeta)$$
$$= \frac{1}{\zeta} h_*(\frac{1}{\zeta}) + g(\zeta) \quad (\text{for } \zeta \in \partial\mathbb{D})$$

where all functions are analytic in $\zeta \in \mathbb{C}\backslash\bar{\mathbb{D}}$. So we can take $\zeta = \frac{1}{\bar\alpha_j}$ $(1 \leq j \leq k)$ obtaining

$$g(\frac{1}{\bar\alpha_j}) = -\bar\alpha_j h_*(\bar\alpha_j) \quad (1 \leq j \leq k)$$

whence (24.b), by virtue of (24.c).

We want now to discuss the equation

(25.a) $$Qf = h \quad \text{with} \quad h \in H \quad \text{given} .$$

Since $Q = Q^*$ this is possible if and only if $h \perp \ker Q$ (in H) that is

$$\sum_{j=0}^{n-1} (h, Y_j)\, \bar\xi_j = 0 \quad (\text{where} \quad (h, Y_j) = \frac{1}{2\pi}\int_0^{2\pi} h(e^{it})\overline{Y_j(e^{it})}\,dt \ (0 \leq j \leq n-1))$$

for all solutions $(\xi_0, \ldots, \xi_{n-1})$ of the singular system (19.a). However we will need later a more explicit form for the above condition as well as an explicit form for the solution f . We start by remarking that the relation (25.a) is equivalent to

(25.b) $$Cf - \sum_{\ell=0}^{n-1} c_\ell \xi_\ell = S^n h + mS^n g_1$$

for some $\xi_0, \ldots, \xi_{n-1} \in \mathbb{C}$ and $g_1 \in H^2$ (where the polynomials c_ℓ are given by (17.a)). Here, as in section 5, (25.b) implies (see (17.c))

(25.c) $$f(z) = \xi_0 + z\xi_1 + \cdots + z^{n-1}\xi_{n-1} + O(z^n) .$$

Introducing the functions $Y_j \in H$ $(0 \le j \le n-1)$ and the polynomials X_j $(0 \le j \le n-1)$ defined in Theorem (19) and using Lemma (24), we obtain that (25.b) is equivalent to

$$C(f - \sum_{j=0}^{n-1} Y_j \xi_j - S^n y) = m(\sum_{j=0}^{n-1} X_j \xi_j + S^n g_2) \tag{25.d}$$

for some polynomial g_2 ; here $y = C(T)^{-1} h$ (see (24)). By virtue of our assumptions on the polynomial $C(z)$, (25.d) is equivalent to the relations

$$f - \sum_{j=0}^{n-1} Y_j \xi_j - S^n y = m g_3 \tag{25.d.1}$$

$$C g_3 = \sum_{j=0}^{n-1} X_j \xi_j + S^n g_2 \tag{25.d.2}$$

with g_3 a polynomial of degree $\le n-1$; to see this last fact multiply (25.d.1) with $\overline{m(\zeta)}$ for $\zeta \in \partial \mathbb{D}$ and use the fact that $\bar{m}f$, $\bar{m}Y_j$'s and $\bar{m}y$ are in $L^2 \ominus H^2$. Denoting

$$(\bar{m}y)(\zeta) = \bar{\zeta} y_*(\bar{\zeta}) = y_{-1}\bar{\zeta} + y_{-2}\bar{\zeta}^2 + \cdots \quad (\zeta \in \partial\mathbb{D}) \tag{25.e.1}$$

we see that

$$g_3(z) = -y_{-n} - z y_{-n+1} - \cdots - z^{n-1} y_{-1} \tag{25.e.2}$$

and thus (25.d.2) becomes

$$\begin{bmatrix} c_0 & & & 0 \\ c_1 & \ddots & & \\ \vdots & \ddots & \ddots & \\ c_{n-1} & \cdots & c_1 & c_0 \end{bmatrix} \begin{bmatrix} y_{-n} \\ y_{-n+1} \\ \vdots \\ y_{-1} \end{bmatrix} + \begin{bmatrix} X_{0,0} & X_{1,0} & & X_{n-1,0} \\ X_{0,1} & X_{1,1} & & X_{n-1,1} \\ \cdots & & & \\ X_{0,n-1} & X_{1,n-1} & \cdots & X_{n-1,n-1} \end{bmatrix} \begin{bmatrix} \xi_0 \\ \xi_1 \\ \vdots \\ \xi_{n-1} \end{bmatrix} = 0 \,, \tag{25.f}$$

where the c_j's are the coefficients of the polynomial $C(z)$ (see (17.b)).

We can now conclude with the following

(26) <u>Proposition</u>. The equation (25.a) is solvable if and only if the equation (25.f) has a solution $(\xi_0, \xi_1, \ldots, \xi_{n-1})$. In this case the solutions f of (25.a) are given by (25.d.1), (25.e.1), (25.e.2)

where y is introduced and computed in (24) and $(\xi_0,\dots,\xi_{n-1})$ is a solution of (25.f).

It is noteworthy that the vector $(y_{-1},y_{-2},\dots,y_{-n})$ can be computed without the use of formula (25.e.1) which yields the y_{-j}'s as series sums. Indeed using $C(\zeta)y_*(\bar\zeta) = h_*(\bar\zeta) + \zeta g(\zeta)$ we easily obtain

$$\begin{bmatrix} c_0 & & & 0 \\ \bar c_1 & \ddots & & \\ \vdots & & \ddots & \\ \bar c_{2n-1} & \cdots & \bar c_1 & \bar c_0 \end{bmatrix} \begin{bmatrix} y_{-1} \\ y_{-2} \\ \vdots \\ y_{-2n} \end{bmatrix} =$$

$$\begin{bmatrix} c_{2n} & & & 0 \\ c_{2n-1} & \ddots & & \\ \vdots & & \ddots & \\ c_1 & \cdots & c_{2n-1} & c_{2n} \end{bmatrix} \begin{bmatrix} y_{-1} \\ \vdots \\ \\ y_{-2n} \end{bmatrix} = \begin{bmatrix} g_{2n-1} \\ g_{2n-2} \\ \vdots \\ g_0 \end{bmatrix}$$

where $g(z) = g_0 + g_1 z + \cdots + g_{2n-1}z^{2n-1}$ is explicitely computed in (24).

We shall now show that the method developed here for computing inverses of skew Toeplitz operators can be also applied to Toeplitz operators as an alternative to the classical spectral factorization method. So we take a symbol $\omega = \Sigma\, c_{jk} z^j \bar\zeta^k$ satisfying $\omega(\zeta,\bar\zeta) > 0$ on $\partial\mathbb{D}$ and $c_{jk} = 0$ whenever $j\cdot k \neq 0$. The Toeplitz operator associated to ω is

$$Q = \sum_{k\geq 0} c_{k0} S^k + \sum_{k<0} c_{0k} S^{*k} .$$

Again we will assume that the polynomial C is of degree $2n$ and has simple zeros. Thus by virtue of (21), C has precisely n zeros $\alpha_1,\dots,\alpha_n$ in $\mathbb{D}$.

(27) Proposition. The equation $Qf = h$ with h given in H^2 has a unique solution $f = (z^n h + g)/C$ where g is a polynomial of degree $\leq n-1$ uniquely determined by the interpolating conditions

(27.a) $$g(\alpha_j) = -\alpha_j^n h(\alpha_j) \quad (1 \le j \le n) .$$

Proof. $Qf = h$ is equivalent to $Cf = S^n h + U^n k$ with some k in $L^2 \ominus H^2$. It follows $k \in U^{-n}H^2$, thus $k(\zeta) = \bar\zeta g_{n-1} + \cdots + \bar\zeta^n g_0$ (for $\zeta \in \partial\mathbb{D}$) and therefore setting $g(z) = g_0 + g_1 z + \cdots + g_{n-1}z^{n-1}$ (for $z \in \mathbb{C}$) we obtain

(27.b) $$C(z)f(z) = z^n h(z) + g(z) \quad (z \in \mathbb{D}) .$$

In order that (27.b) should define an element $f \in H^2$ it is obviously sufficient and necessary that the conditions (27.a) hold.

8. The computation of the optimal solution of the 4 block problem

In this section we shall consider the 4 block optimization problem (12) under the assumption that W, f, g, h are all rational functions in H^∞. According to (5), the optimum μ is the norm of the operator

(28.a) $$A = \begin{bmatrix} W(T)P & f(T_1)P_1 \\ g(S) & h(U) \end{bmatrix}$$

where P and P_1 denote the orthogonal projections of H^2 and L^2 onto $H = H^2 \ominus mH^2$ and $H_1 = L^2 \ominus mH^2$ respectively; moreover $T = PS|H = T_1|H$, $T_1 = P_1 U|H_1$. Recall that S and U denote the canonical shifts on H^2 and L^2, respectively.

We define

(28.b) $$\alpha = \max\left\{\left\|\begin{bmatrix} W(\zeta) & f(\zeta) \\ g(\zeta) & h(\zeta) \end{bmatrix}\right\| : \zeta \in \partial\mathbb{D} \cap \sigma(T)\right\}$$

(28.c) $$\beta = \max\{\|[g(\zeta)\ h(\zeta)]\| : \zeta \in \partial\mathbb{D}\}$$

(28.d) $$\gamma = \max\left\{\left\|\begin{bmatrix} f(\zeta) \\ h(\zeta) \end{bmatrix}\right\| : \zeta \in \partial\mathbb{D}\right\}$$

Then we have the following analog of Lemma (14).

(29) Lemma [FT4]. Let

$$\|A\|_{ess} = \max\{\alpha, \beta, \gamma\} .$$

Then $\|A\| \ge \|A\|_{ess}$ and the norm of A is attained if $\|A\| > \|A\|_{ess}$.

As in Section 4, Lemma (29) tells us that either $\mu = \|A\|_{ess}$ or if $\mu > \|A\|_{ess}$ the operator $\mu^2 I - AA^*$ has a nontrivial kernel. In this last case μ is the largest $\rho > \|A\|_{ess}$ for which

(29.a) $$\ker(\rho^2 I - AA^*) \neq \{0\} .$$

Representing $W = a/q$, $f = b/q$, $g = c/q$, $h = d/q$ with polynomials a,b,c,d,q (of degree $\leq n$ and $q \neq 0$ on $\bar{D}$) we easily obtain (as in Section 4) that (27.a) is equivalent to

(29.b) $$\ker\begin{bmatrix} Q_{11} & Q_{12} \\ Q_{21} & Q_{22} \end{bmatrix} \neq \{0\}$$

where (with P_+ now denoting the orthogonal projection of L^2 onto H ; note $P_+H_1 = H$)

(29.b.1) $$Q_{11} = \rho^2 q(T_1)q(T_1)^* - b(T_1)b(T_1)^* - a(T)a(T)^*P$$

(29.b.2) $$Q_{22} = \rho^2 q(U)q(U)^* - d(U)d(U)^* - c(S)c(S)^*P_+$$

(29.b.3) $$Q_{21} = -d(U)b(T_1)^* - c(S)a(T)^*P_+ \quad \text{and} \quad Q_{12} = Q_{21}^* .$$

The operator (29.b) acts on $H_1 \oplus L^2$. By decomposing this space into the orthogonal sum of three spaces namely

$$H_1 \oplus L^2 = [(L^2 \ominus H^2) \oplus (L^2 \ominus H^2)] \oplus (H \oplus H(\oplus (\{0\} \oplus mH^2) ,$$

that operator takes the matrical form

(29.c) $$\begin{bmatrix} D_- & E & E_+ \\ F_- & D & F_+ \\ G_- & G & D_+ + G_+ \end{bmatrix}$$

where D_- , D_+ and respectively D are invertible Toeplitz operators (because of the condition $\rho > \|A\|_{ess} \geq \beta$ and γ) and respectively an invertible skew Toeplitz operator (because $\rho > \|A\|_{ess} \geq \alpha$) . Actually D_- and F_+ have 2×2 - matrix valued symbols but the methods developed in Section 7 can be easily extended to this case too. Moreover E, E_+, G_-, F_+, G_-, G are very simple explicit finite rank operators. For instance

(29.c.1) $$G_+ = c(U)(I-P_+)c(U)^*|mH^2$$

In order to determine if (29.b) holds, one has to study the solvability of the equation

(29.c.2) $$\begin{bmatrix} D_- & E & E_+ \\ F_- & D & F_+ \\ G_- & G & D_+ + G_+ \end{bmatrix}\begin{bmatrix} h_- \\ h_0 \\ h_+ \end{bmatrix} = 0$$

with

(29.c.3) $h_- \in (L^2 \ominus H^2) \oplus (L^2 \ominus H^2)$, $h_0 \in H \oplus H$, $h_+ \in mH^2$.

Through laborious computations using the techniques presented in the preceding 3 sections, we can show that the relation (29.c.1) can be given the form

$$(29.d)\qquad \begin{bmatrix} I - R_{11} & R_{12} \\ R_{21} & I - R_{22} \end{bmatrix} \begin{bmatrix} h_0 \\ h_+ \end{bmatrix} = 0 \ , \quad \begin{bmatrix} h_0 \\ h_+ \end{bmatrix} \neq 0$$

where the R_{ij}'s are <u>explicit</u> finite rank operators of rank $\leq n$. To illustrate this we shall present here only part of the computation of

$$(29.d.1)\qquad R_{22} = D_+^{-1}(G_{\neq} + G_- D_-^{-1} E_+) = D_+^{-1}(G_+ + E_+^* D_-^{-1} E_+)$$

Namely, for simplicity we shall only show how to compute D_+^{-1} and E_+ . (For a detailed but messy presentation of all the body of computations we refer to [FT4].)

(30) <u>Lemma.</u> Let Q denote the Toeplitz operator with symbol

$$(30.a)\qquad \omega = \rho^2|q|^2 - |c|^2 - |d|^2 > 0 \quad \text{on} \quad \partial\mathbb{D} \ .$$

Then

$$(30.b)\qquad D_+h_+ = mQ(\bar{m}h_+),\ D_+^{-1}h_+ = mQ^{-1}(\bar{m}h_+) \quad (h_+ \in mH^2)$$

where for the computation of Q^{-1} see (27).

<u>Proof.</u> We have (identifying as in (29.c.1), $\{0\}\oplus mH^2$ with mH^2)

$$D_+h_+ = mP_+((\rho^2|q|^2-|d|^2)\bar{m}h_+) - (I-P_1)[c(S)c(S)^*h_+ + c(U)(I-P_+)c(U)^*]h_+$$
$$= mP_+[(\rho^2|q|^2-|d|^2)\bar{m}h_+] - (I-P_1)c(U)c(U)^*h_+ =$$
$$= mP_+[(\rho^2|q|^2-|d|^2-|c|^2)\bar{m}h_+] = mQ(\bar{m}h_+) \quad (h_+ \in mH^2) \ .$$

This establishes the first relation (30.b). Since $\rho > \beta$ in $\partial\mathbb{D}$ and $q \neq 0$ in $\bar{\mathbb{D}}$, we have

$$\omega = |q|^2(\rho^2-|g|^2-|h|^2) > |q|^2(\rho^2-\beta^2) > 0 \ ,$$

so that Q^{-1} exists by (27). □

To any polynomial of degree $\leq n$

$$r(z) = r_0 + r_1 z + \cdots + r_n z^n$$

we associate the following two matrices

$$(31)\qquad \Gamma_r = \begin{bmatrix} r_1 & r_2 & \cdots & r_n & 0 \\ r_2 & & \cdot & 0 & \\ \vdots & \cdot & & & \\ r_n & & & & \\ 0 & & & & 0 \end{bmatrix}, \theta_r = \begin{bmatrix} r_0 & & & 0 \\ r_1 & \ddots & & \\ \vdots & & \ddots & \\ r_n & \cdots & r_1 & r_0 \end{bmatrix}$$

(32) **Lemma.** For $h_+ = h_{+0} + h_{+1}z + h_{+2}z^2 + \cdots$ define

$$(32.a.1)\qquad \xi = [\xi_1\ \xi_2 \cdots \xi_{n+1}]^{tr} = -\Gamma_d^*\theta_b[h_{+0}\ h_{+1} \cdots h_{+n}]^{tr}$$

$$(32.a.2)\qquad \eta = [\eta_1 \cdots \eta_{n+1}]^{tr} = (\rho^2\Gamma_q^*\theta_q - \Gamma_d^*\theta_d)[h_{+0} \cdots h_{+n}]^{tr} .$$

Then

$$(32.b)\qquad (E_+h_+)(\zeta) = \begin{bmatrix} \xi_1\bar{\zeta} + \cdots + \zeta_n\bar{\zeta}^n \\ \eta_1\bar{\zeta} + \cdots + \eta_n\bar{\zeta}^n \end{bmatrix} \qquad (\zeta \in \partial\mathbb{D})$$

Proof. We have

$$E_+h_+ = \begin{bmatrix} (I-P_+)[-b(T_1)P_1d(U)^* - a(T)Pc(S)^*]h_+ \\ (I-P_+)[\rho^2q(U)q(U)^* - d(U)d(U)^* - c(S)c(S)^*]h_+ \end{bmatrix} =$$

$$= \begin{bmatrix} -(I-P_+)b(U)P_1d(U)^*h_+ \\ (I-P_+)[\rho^2q(U)q(U)^* - d(U)d(U)^*]h_+ \end{bmatrix} =$$

$$= \begin{bmatrix} -(I-P_+)b(U)(I-P_+)d(U)^*h_+ \\ \eta \end{bmatrix} = \begin{bmatrix} -(I-P_+)b(U)d(U)^*h_+ \\ \eta \end{bmatrix} =$$

$$\begin{bmatrix} -(I-P_+)d(U)^*b(U)^*h_+ \\ \eta \end{bmatrix} = \begin{bmatrix} \xi \\ \eta \end{bmatrix},$$

where we used the fact that for any two polynomials r_1, r_2 of degree $\leq n$ and $k = k_0 + k_1z + \cdots \in H^2$ we have

$$(32.c)\qquad (I-P_+)r_1(U)^*r_2(U_2)k = [\zeta\ \zeta^2 \cdots \zeta^{n+1}[\Gamma_{r_1}^*\theta_{r_2}[k_0 \cdots k_n]^{tr} . \quad \square$$

Lecture IV. <u>Nonlinear plants and nonlinear controllers.</u>

In this last lecture we shall give a short presentation of a natural extension of the H^∞-sensitivity problem to nonlinear analytic plants considered in [FT5],[FT6]. We conclude with a recent development [FT7] which enhances the practical value of our extension.

9. The formulation of optimality for nonlinear analytic plants

In order to carry out the extension of H^∞ synthesis theory to nonlinear systems, we will need first to discuss a few standard results about analytic mappings on Hilbert spaces. We are essentially following the treatments of [BFHT1],[BFHT2] to which the reader may refer for all of the details. In particular, input/output operators which admit Volterra expansions are special cases of the operators which we study here; see [BCh1].

Let G and H denote complex Hilbert spaces. Set

$$B_{r_o}(G) := \{g \in G : \|g\| < r_o\}$$

(the open ball of radius r_o in G with origin as center).

(33) <u>Definition.</u> A mapping $\Phi : B_{r_o}(G) \to H$ is <u>analytic</u> if the complex function $(z_1,\ldots,z_n) \mapsto (\Phi(z_1g_1 + \ldots + z_ng_n),h)$ is analytic in a neighborhood of $(1,1,\ldots,1) \in C^n$ as a function of the complex variables $z_1,\ldots,z_n$ for all $g_1,\ldots,g_n \in G$ such that $\|g_1 + \ldots + g_n\| < r_o$, for all $h \in H$, and for all $n > 0$. (Recall that (k,h) denotes the scalar product of h with k in the Hilbert space in which h and k live.)

Without loss of generality we will assume that $\Phi(0) = 0$. It is easy to see that if $\Phi : B_{r_o}(G) \to H$ is analytic, then Φ admits a convergent Taylor series expansion, i.e.

$$\Phi(g) = \Phi_1(g) + \Phi_2(g,g) + \cdots + \Phi_n(g,\cdots,g) + \cdots$$

where $\Phi_n : G \times \cdots \times G \to H$ is an n-linear map. Clearly, without loss of generality we take the n-linear map $(g_1,\cdots,g_n) \to \Phi(g_1,\cdots,g_n)$ to be symmetric in the arguments $(g_1,\cdots,g_n)$. This assumption will be made throughout this lecture. For Φ a Volterra series, Φ_n is basically the nth-Volterra kernel. The map

$$\Phi_n(g_1\otimes \cdots \otimes g_n) \mapsto \Phi_n(g_1,\cdots,g_n)$$

extends in a unique manner to a H-valued linear map on the algebraic tensor product $G \otimes G \otimes \ldots \otimes G$. By abuse of notation we shall denote this map also by Φ_n.

Let $G^{\otimes n}$ denote the Hilbert space completion of the algebraic tensor product. If Φ_n has finite norm on $G \otimes \ldots \otimes G$, then Φ_n extends by continuity to a bounded linear operator, still denoted Φ_n, on $G^{\otimes n}$ into H. We denote by $\|\Phi_n\|$ the norm of this operator. By a <u>majorizing sequence</u> for the holomorphic map Φ, we mean a positive sequence of numbers $a_n (n = 1,2,\ldots)$ such that $\|\Phi_n\| \leq a_n$ for $n \geq 1$. If $\rho := \limsup a_n^{1/n} < \infty$, then the Taylor series expansion of Φ converges at least on the ball $B_r(G)$ of radius $r = 1/\rho$. If Φ admits a majorizing sequence as above, then we will say that Φ is <u>majorizable</u>. An analytic map $\Phi : B_{r_o}(G) \to H$, $\Phi(0) = 0$ has <u>fading memory</u> if its nonlinear part $\Phi - \Phi'(0)$ admits a factorization

$$\Phi - \Phi'(0) = \Psi \circ \Sigma$$

where Ψ is an analytic map defined in some neighborhood of $0 \in G$, and Σ is a linear Hilbert-Schmidt operator. It is easy to show that a fading memory operator is majorizable. All Volterra expansions are with fading memory and hence majorizable.

(34) For the majorizable analytic maps $W : B_r(G_1) \mapsto H$ (the weight), $\pi : B_r(G_1) \mapsto H$ (the plant), and $q : B_r(G) \mapsto G_1$ (the compensating parameter), satisfying $W(0) = 0$, $\pi(0) = 0$ and $q(0) = 0$, we define the <u>sensitivity function</u> $S(q)$,

(34.a) $$S(q)(r) := \sum_{n=1} \rho^n \|(W - \pi \circ q)_n\|$$

for all $\rho > 0$ such that the sum converges. We write $S(q) \preceq S(\tilde{q})$, if there exists a $\rho_o > 0$ such that $S(q)(\rho) \leq S(\tilde{q})(\rho)$ for all $\rho \in [0,\rho_o]$. If $S(q) \preceq S(\tilde{q})$ and $S(\tilde{q}) \preceq S(q)$, we write $S(q) \cong S(\tilde{q})$. This means that $S(q)(\rho) = S(\tilde{q})(\rho)$ for all $\rho > 0$ sufficiently small, i.e. $S(q)$ and $S(\tilde{q})$ are equal as germs of functions. If $S(q) \preceq S(\tilde{q})$, but $S(\tilde{q}) \preceq S(q)$, we will say that q <u>ameliorates</u> $\tilde{q}$. Note that this means $S(q)(\rho) < S(\tilde{q})(\rho)$ for all $\rho > 0$ sufficiently small.

(35) We will denote by $\mathcal{TI}$ (which stands for $\mathcal{T}$ime $\mathcal{I}$nvariant) the class of all analytic maps $\Phi : B_r(H^2(\mathbb{C}^K)) \mapsto H^2(\mathbb{C}^M)$ (where the integers K, M, and the real $r > 0$ dependend of Φ but otherwise are arbitrary), $\Phi(0) = 0$ which are time invariant, i.e.

$$\Phi(S_K f) = S_M \Phi(f) \quad \forall f \in B_r(H^2(\mathbb{C}^K))$$

where S_K and S_M denote the canonical shifts

$$(S_\square f)(z) = zf(z) \quad (z \in \mathbb{D},\ f \in H^2(\mathbb{C}^\square))$$

on the respective H^2-spaces. We will consider only weights W, plants π and compensators q which belong to $\mathcal{T}\mathcal{s}$. Here $H^2(\mathbb{C}^N) = H^2 \oplus \ldots \oplus H^2$ (n-copies) or equivalently the space $H^2(\mathbb{D};\mathbb{C}^N)$ of H^2-function valued in $\mathbb{C}^N$ (see [Sz.-NF2],Ch.V).

(36) An element $q_o \in \mathcal{T}\mathcal{s}$ is called <u>optimal</u> if $S(q_o) \preceq S(q)$ for all $q \in \mathcal{T}\mathcal{s}$. We say $q \in \mathcal{T}\mathcal{s}$ is <u>optimal with respect to its n-th term</u> q_n, if for every n-linear $\tilde{q}_n \in \mathcal{T}\mathcal{s}$, we have

$$S(q_1+\ldots+q_{n-1}+q_n+q_{n+1}) \preceq S(q_1+\ldots+q_{n-1}+\tilde{q}_n+q_{n+1}+\ldots) .$$

If $q \in \mathcal{T}\mathcal{s}$ is optimal with respect to all of its terms, then we say that it is <u>partially optimal</u>.

10. <u>The iterative commutant lifting method</u>

In this section, we discuss the construction by which we will derive both partially optimal and optimal compensators relative to the sensitivity function given above. As before, π will denote the plant, and W the weighting operator, both of which are assumed in $\mathcal{T}\mathcal{s}$. As in the linear case, we always suppose that π_1 is an isometry, i.e. π_1 is <u>inner</u> and that $K = N$ (see (35)).

(37) <u>The iterative commutant lifting procedure</u>. Let $P : H^2(C^N) \to H^2(C^N) \ominus \Pi_1 H^2(C^N)$ denote the orthogonal projection of $H^2(\mathbb{C}^N)$ onto $H' = H^2(\mathbb{C}^N) \ominus \pi_1 H^2(\mathbb{C}^N)$. Using the linear commutant lifting theorem (7) for $T' = PS_N|H'$ and $T = S_M$ (see [FT5] for the details), we can choose q_1 such that

$$\|W_1 - \pi_1 q_1\| = \|PW_1\| .$$

Now given this q_1, we choose q_2 (by using again the commutant lifting theorem (7) for T' as before but $T = S_M^{\otimes 2} = S_M \otimes S_M$ on $H^2(\mathbb{C}^M)^{\otimes 2}$) such that

$$\|W_2 - \pi_2(q_1 \otimes q_1) - \pi_1 q_2\| = \|P(W_2 - \pi_2(q_1 \otimes q_1))\| .$$

Inductively, given $q_1, \ldots, q_{n-1}$, set

$$A_n := W_n - \sum_{1 \le j \le n} \sum_{i_1+\ldots+i_j=n} \pi_j(q_{i_1} \otimes \ldots \otimes q_{i_j})$$

for $n \ge 2$. Then applying the commutant lifting theorem (7) for the same T' but $T = S_M^{\otimes n} = S_M \otimes \ldots \otimes S_M$ (n-times) on $H^2(\mathbb{C}^M)^{\otimes n}$, we choose q_n such that

$$\|A_n - \pi_1 q_n\| = \|PA_n\| ,$$

and so on. The first key point of the iterative commutant lifting method is the following convergence fact.

(38) Proposition. With the above notation, let $q_0 := q_1+q_2+\ldots$. Then $q_0 \in \mathcal{T}\mathcal{S}$.

Note that given any $q \in \mathcal{T}\mathcal{S}$, we can apply the iterative commutant lifting procedure to $W - \pi \circ q$. Now set

(38.a) $$S_P(q)(\rho) := \sum_{n=1} \rho^n \|P(W - \pi \circ q)_n\| .$$

Clearly, $S_P(q) \le S(q)$ (as germs of functions). The following result is immediate from the above discussion:

(39) Proposition. Given $q \in \mathcal{T}\mathcal{S}$, there exists $\tilde{q} \in \mathcal{T}\mathcal{S}$, such that $S(\tilde{q}) \equiv S_P(q)$. Moreover $\tilde{q}$ may be constructed from the iterated commutant lifting procedure. Also q is partially optimal if and only if $S(q) \cong S_P(q)$.

We can now summarize the above discussion with the following.

(40) Theorem. For given W and π as above, any $q \in \mathcal{T}\mathcal{S}$ is either partially optimal or can be ameliorated by a partially optimal compensating parameter.

Proof. Immediate from Propositions (38) and (39). □

It is important to emphasize that a partially optimal compensating parameter need not be optimal (see (33)). The question of optimality will be considered in the next section.

11. Explicit computation in the rational case

In this section we will derive our main results about optimal compensators given in [FT5],[FT6]. Basically, we will show that in the single input/single output setting (SISO), the iterated commutant lifting procedure leads to an optimal design.

For the construction of an optimal compensator below, we will need one more technical result. First recall that $H^2 := H^2(\mathbb{C})$, and that $S_1 = S$ is the canonical unilateral shift on H^2 (see Section 1). Let $m \in H^\infty$ be a nonconstant inner function, let $P : H^2 \mapsto H = H^2 \ominus mH^2$ denote orthogonal projection, and let $T := PS|H$. (In our iterative construction: $m = \pi_1$.) For G a complex separable Hilbert space, let $S_\infty : G \to G$ denote a unilateral shift, i.e. an isometric operator with no unitary part. (This means that $S_\infty^{*n} g \to 0$ for all $g \in G$ as $n \to \infty$; see [Sz.-NF2]Ch.I.) The proof of Lemma (13) above can be adapted for proving the following useful generalization.

(41) Lemma [FT5]. Let $A : G \to H$ be a bounded linear operator intertwining S_∞ and T and attaining its norm. Then there exists a unique operator $B : G \mapsto H^2$ such that $BS_\infty = SB$, $\|A\| = \|B\|$, and $PB = A$.

This operator B can be explicitely computed in terms of a maximal vector h_0, by a formula analogous to (13.a).

The main result of this section is:

(42) Theorem. Let the weight W and the plant π be in $\mathcal{R}\mathcal{H}$ and SISO (in our setting this means that $N = M = K = 1$). Suppose that PW_j and π_k are compact for $j \geq 1$ and for $k \geq 2$, respectively. Let q_{opt} be a partially optimal compensating parameter constructed by the iterated commutant lifting procedure described in Section 10. Then q_{opt} is optimal.

Proof. First of all, since PW_1 attains its norm, from (38) we have that the optimal q_1 constructed relative to W_1 and π_1 is unique. Now from our above hypotheses, each PA_n is compact for

$n \geq 2$, and hence each PA_n attains its norm. Therefore by (42) each optimal q_n constructed by the iterated commutant lifting procedure is unique. (It is very important to notice that although we treat only SISO systems, the handling of the higher degree terms PA_n involves the infinite multiplicity unilateral shift $S^{\otimes n}$. This is the reason for which in Lemma (41) no assumption is made about the multiplicity of the shift S_∞ .) The theorem now follows immediately from (38). □

(43) Corollary. Let W and π be as in (42) and besides let the linear part π_1 of the plant be rational. Then the partially optimal compensating parameter q_{opt} constructed by the iterated commutant lifting procedure is optimal.

Proof. Indeed, since π_1 is SISO and rational (recall that we also always assume that π_1 is inner), $H = H^2 \ominus \pi_1 H^2$ is finite dimensional, and all the conditions in Theorem (42) are met. □

(44) Remark. Corollary (43) gives a constructive procedure for finding the optimal compensator under the given hypotheses. Indeed, when π_1 is SISO rational, the iterative commutant lifting procedure can be reduced to finite dimensional matrix calculations. In the paper [FT6], we have shown that when the hypotheses of Corollary (43) are satisfied, the skew Toeplitz theory (elements of which we sketched in Lectures 2 and 3) provides an algorithmic design procedure for nonlinear systems as well.

12. The problem of causality

A major difference between the linear and the nonlinear case is the separation of the concepts of time invariance and causality in the nonlinear case. Indeed if Φ is a SISO in $\mathcal{TI}$, i.e.

$$\Phi : B_r(H^2) \mapsto H^2 \text{ is analytic (where } r > 0) ,$$

$$\Phi(0) = 0$$

$$\Phi(Sf) = S\Phi(f) \quad (f \in B_r(H^2)) ,$$

then the following definition of causality is the natural one.

(45) <u>Definition</u>. Φ is called <u>causal</u> if

(45.a) $$P_k\Phi(f) = P_k\Phi(P_k f) \quad \forall f \in B_r(H^2)$$

where $P_k = I - S^k S^{*k} \quad (k \geq 1)$.

We shall denote by $\mathcal{C}$ the class of all causal SISO's in $\mathcal{T}\!\mathcal{s}$. The difference referred above is exhibited by the following two simple facts.

(46) Let Φ be a linear SISO in $\mathcal{T}\!\mathcal{s}$. Then Φ is causal.

Indeed $\Phi S = S\Phi$ implies $S^kS^{*k}\Phi S^kS^{*k} = S^kS^{*k}S^k\Phi S^{*k} = S^k\Phi S^{*k} = \Phi S^kS^{*k}$ which is equivalent to (46.a), because Φ is linear. □

(47) Let $\Phi : (H^2)^{\otimes 2} \mapsto H^2$ be the linear operator intertwining $S^{\otimes 2} = S\otimes S$ and S defined by

(47.a) $$(\Phi(f\otimes g))(z) = \sum_{k=0}^{\infty} (f_{k+1}g_k + f_k g_k + f_k g_{k+1})z^k$$

where

(47.b) $$f(z) = \sum_{k=0}^{\infty} f_k z^k \ , \ g(z) = \sum_{k=0}^{\infty} g_k z^n \ .$$

Then

$$\Phi(f) = \Phi(f\otimes f)$$

is an analytic map (actually a homogeneous polynomial of degree 2) in $\mathcal{T}\!\mathcal{s}$ which is not causal. Indeed

$$(P_1\Phi(f))(z) = 2f_1f_0 + f_0^2 \quad (z\in\mathbb{D})$$

$$(P_1\Phi(P_1 f))(z) = f_0^2 \quad (z\in\mathbb{D}) \ ,$$

so that $P_1\Phi(f) \neq P_1\Phi(P_1 f)$ for instance for

$$f(z) = 1 + z \quad (z\in\mathbb{D}) \ .$$

The iterative commutant lifting procedure based on the sensitivity function $S(q)$ defined in (34.a) may lead to a noncausal (although time invariant) optimal compensator even if W and π are in $\mathcal{C}$. To avoid this mishap a new, more involved variant of the commutant lifting theorem is needed. Very recently (together with A. Tannenbaum) we were able to find and to prove this variant [FT7]. We will conclude by presenting a particular but illuminating case of this new theorem.

(48) For n fixed, let $\Phi : (H^2)^{\otimes n} \mapsto H^2$ be any linear operator intertwining $S^{\otimes n} = S \otimes S \otimes \ldots \otimes S$ (n-times) with S . Let m (equal

to π_1 when applied to the optimization problem) be an inner function such that $m(0) \neq 0$ (this is a simplifying technical assumption) and let $H = H^2 \ominus mH^2$, $T = PS|H$. Finally let $\mu = \min\|\Phi - mQ\|$ where Q runs over all operators from $(H^2)^{\otimes n}$ to H^2 satisfying the time invariance condition

$$QS^{(n)} = SQ$$

and the causality condition

$$P_k^{(n)} Q P_k^{(n)} = P_k^{(n)} Q \quad (\forall k \geq 1)$$

where

$$P_k^{(n)} = P_k \otimes \ldots \otimes P_k \quad \text{(n-times)} .$$

Then

(48.a) $$\mu = \mu(P\Phi) := \min(\gamma \geq 0 : \|P\Phi\| \leq \gamma , \|(I-P_k^{(n)})\Phi^* h\| \leq \leq \gamma \|T^{*k} h\| \quad (\forall h \in H, \forall k=1,2,\ldots)) .$$

A little reflection immediately will suggest that in order to obtain compensators in $\mathcal{C}$ when W and π are in $\mathcal{C}$ we only must substitute in the results in Sections 10 and 11 (starting with Proposition (39)) the role of the basic function $s_P(a)$ with the following

(48.b) $$s_{P,\text{causal}} = \sum_{n=1}^{\infty} \mu(P(\Phi - \pi \text{O} q)_n) .$$

We hope that the approach to nonlinear H^∞-control theory described in this lecture may develop into a rich and useful subject for mathematicians and engineers.

References:

[AAK1] V.M. Adamjam, D.Z. Arov and M.G. Krein Infinite Hankel matrices and generalized problems of Caratheodory-Fejer and F. Riesz, Functional Anal. Appl. 2(1968), 1-18.

[AAK2] V.M. Adamjam, D.Z. Arov and M.G. Krein, Analytic properties of Schmidt pairs for Hankel operator and the generalized Schur-Takagi problem, Math. USSR-Sb. 15(1972), 31-78.

[AAK3] V.M. Adamjam, D.Z. Arov aned M.G. Krein, Infinite block Hankel matrices and related extension problems, Amer. Math. Soc. Transl. 111(1978), 133-156.

[AD1] V. Anantharam and C. Desoer, On the stabilization of nonlinear systems, IEEE Trans. Automat. Control, AC-29(1984), 569-573.

[BFHT1] J. Ball, C. Foias, J.W. Helton and A. Tannenbaum, On a local nonlinear commutant lifting theorem, I.U. Math. Journal, 36(1987), 693-709.

[BFHT2] J. Ball, C. Foias, J.W. Helton and A. Tannenbaum, A Poincaré-Dulac approach to a nonlinear Beurling-Lax-Halmos theorem, J. Math. Anal. Appl., Vol. 139, 2(1989), 496-514.

[BFT1] H. Bercovici, C. Foias and A. Tannenbaum, On skew Toeplitz operators, I, Operator Theory: Adv. Appl., 29(1988), 21-44.

[B1] H. Bercovici, Operator theory and arithmetic in H^∞, AMS, Providence, RI, (1988).

[BCh1] S. Boyd and L. Chua, Fading memory and the problem of approximating nonlinear operators with Volterra series, IEEE Trans. Circuits and Systems, CAS-32(1985), 1150-1161.

[DMP1] R.G. Douglas P.S. Muhly and C. Pearcy, Lifting commuting operators, Michigan Math. J., 15(1968), 385-395.

[DG1] H. Dym and I. Gohberg, A new class of contractive interpolants and maximum entropy principles, Operator Theory: Adv. and Appl., 29(1988), 117-150.

[FF1] A. Feintuch and B.A. Francis, Uniformally optimal control of linear feedback systems, Automatica, 21(1985), 563-574.

[FF2] A. Feintuch and B.A. Francis, Distance formulas for operator alegebras arising in optimal control problems, Topics in Operator Theory and Interpolation; Operator Theory: Adv. and Appl., 29(1988), 151-179.

[F1] C. Foias, Contractive intertwining dilations and waves in layers media, Proceedings of International Congress of Mathematicians, Helsinki, Vol. 2(1978), 605-613.

[FT1] C. Foias and A. Tannenbaum, On the Nehari problem for a certain class of L^∞ functions appearing in control theory, J. Funct. Anal., 74(1987), 146-159.

[FT2] C. Foias and A. Tannenbaum, Some remarks on optimal interpolation, Systems and Control Letters, 11(1988), 259-264.

[FT3] C. Foias and A. Tannenbaum, On the four block problem I, Operator Theory: Adv. and Appl., 32(1988), 93-112.

[FT4] C. Foias and A. Tannenbaum, On the four block problem, II: the singular system, Integral Equ. Oper. Theory, 11(1988), 726-767.

[FT5] C. Foias and A. Tannenbaum, Weighted optimization theory for nonlinear systems, SIAM J. Control and Optimiz., Vol.27, 4(1989), 842-860.

[FT6] C. Foias and A. Tannenbaum, Iterative commutant lifting for systems with rational symbol, Operatory Theory: Adv. and Appl., 41(1989), 255-277.

[FT7] C. Foias and A. Tannenbaum, A causal commutant lifting theorem and H^{∞}-optimization for nonlinear systems (in preparation).

[FTZ1] C. Foias, A. Tannenbaum and G. Zames, On the H^{∞}-optimal sensitivity problem for systems with delays, SIAM J. Control and Optimiz., Vol.25, 3(1987), 686-705.

[FTZ2] C. Foias, A. Tannenbaum and G. Zames, Sensitivity minimization for arbitrary SISO distributed plants, Systems Control Lett., 8(1987), 189-195.

[FTZ3] C. Foias, A. Tannenbaum and G. Zames, Some explicit formulae for the singular values of certain Hankel operators with factorizable symbol, SIAM J. Math. Anal., 19(1988), 1081-1089.

[Fr1] B.A. Francis, A course in H^{∞} Control Theory, Lecture Notes in Control and Informaion Science, Springer, New York, 1987.

[Gu1] C. Gu, Eliminating the genericity conditions in a skew Toeplitz algorithm for the H^{∞}-optimal sensitivity problem (in preparation).

[N1] Z. Nehari, On bounded bilinear forms, Ann. of Math. 65(1957), 153-162.

[Pa1] S. Parrott, Unitary dilations for commuting contractions, Pacific J. Math., 34(1979), 481-490.

[RR1] M. Rosenblum and J. Rovnyak, Hardy classes and operator theory, Oxford Univ. Press, New York, 1985.

[Sa1] D. Sarason, Generalized interpolation in H^{∞}, Trans. Amer. Math. Soc., 127(1967), 179-203.

[Sz.-NF1] B. Sz.-Nagy and C. Foias, Dilation des commutants d'opérateurs, C.R. Acad. Sci. Paris, Ser. A 265(1968), 493-495.

[Sz.-NF2] B. Sz.-Nagy and C. Foias, Harmonic analysis of operators on Hilbert space, North-Holland, Amsterdam, 1970.

[Sz.-NF3] B. Sz.-Nagy and C. Foias, The "lifting theorem" for intertwining operators and some new applications, I.U. Math. Journal, 20(1971), 901-904.

[Z1] G. Zames, Feedback and optimal sensitivity: model reference transformations, multiplicative seminorms, and approximate inverses, IEEE Trans. Automat. Control, AC-26(1981), 301-320.

Lectures on $\mathcal{H}_\infty$ Control and Sampled-Data Systems

Bruce Francis
Department of Electrical Engineering
University of Toronto
Toronto
Canada M5S 1A4

Contents

Preface

These lectures are divided into three groups:

- Lecture 1 is intended as an introduction to the main issues in feedback control theory. It starts with norms for signals and systems, proceeds through the basic issues in feedback design, and ends up with a big picture and a summary of some standard control problems.
- Lecture 2 presents an elementary treatment of the simplest $\mathcal{H}_\infty$ control problem, the weighted sensitivity problem for single-input, single-output systems. Lecture 2 concludes with a design example.
- Lectures 3-6 present recent work on multivariable sampled-data control systems from the operator-theoretic viewpoint. The main problems studied are $\mathcal{H}_2$ and $\mathcal{H}_\infty$ optimization.

Lectures 1 and 2 are extracted from the book "Feedback Control Theory," by J.C. Doyle, B.A. Francis, and A.R. Tannenbaum, to be published by Macmillan.

It is my pleasure to acknowledge that Lectures 3-6 are based on the work of my students: Chen Tongwen, Gary Leung, and Tony Perry.

I am grateful to Professors Roberto Conti, Edoardo Mosca, and Luciano Pandolfi for inviting me to participate in this CIME Session on $\mathcal{H}_\infty$ control theory.

Finally, I would like to dedicate these lectures to Boyd Pearson on the occasion of his sixtieth birthday.

This work was partially supported by the Natural Sciences and Engineering Research Council of Canada.

Lecture 1

Introduction

1 Norms for signals and systems

We consider systems which are linear, time-invariant, causal, and finite-dimensional. In the time domain an input-output model for such a system has the form of a convolution equation,

$$y(t) = \int_{-\infty}^{\infty} g(t-\tau)u(\tau)\, d\tau$$

Causality means that $g(t) = 0$ for $t < 0$. Such a system has an equivalent state-space model

$$\begin{aligned} \dot{x}(t) &= Ax(t) + Bu(t) \\ y(t) &= Cx(t) + Du(t) \end{aligned}$$

where A, B, C, D are real matrices of appropriate sizes.

Let $G(s)$ denote the transfer matrix, the Laplace transform of $g(t)$. Then G is rational (by finite-dimensionality) with real coefficients. We'll say that G is ***stable*** if it is analytic in the closed right half-plane (Re $s \geq 0$), ***proper*** if $G(j\infty)$ is finite, and ***strictly proper*** if $G(j\infty) = 0$. The transfer matrix for the state-space model is

$$G(s) = D + C(sI - A)^{-1}B$$

One way to describe the performance of a control system is in terms of the size of certain signals of interest. For example, the performance of a tracking system could be measured by the size of the error signal. This section looks at several ways of defining a signal's size, i.e., at several norms for signals. Which norm is appropriate depends on the situation at hand. Also introduced are norms for a system's transfer function. For simplicity, all signals are assumed to be scalar-valued.

Norms for signals

We consider signals mapping $(-\infty, \infty)$ to $\mathbf{R}$. For simplicity, they are assumed to be piecewise continuous.

$\mathcal{L}_p$-norm, $1 \leq p < \infty$ The $\mathcal{L}_p$-norm of a signal $u(t)$ is

$$\|u\|_p := \left(\int_{-\infty}^{\infty} |u(t)|^p\, dt \right)^{1/p}$$

From a physical viewpoint, probably the most important of these is the $\mathcal{L}_2$-norm. For example, suppose u is the current through a 1-ohm resistor. Then the instantaneous power equals $u(t)^2$ and the total energy equals the integral of this, namely, $\|u\|_2^2$. We shall generalize this interpretation: the *instantaneous power* of a signal $u(t)$ is defined to be $u(t)^2$ and its *energy* is defined to be the square of its $\mathcal{L}_2$-norm.

$\mathcal{L}_\infty$-norm

$$\|u\|_\infty := \sup_t |u(t)|$$

power signals

The *average power* of u is the average over time of its instantaneous power:

$$\lim_{T\to\infty} \frac{1}{2T} \int_{-T}^{T} u(t)^2 \, dt$$

The signal u will be called a *power signal* if this limit exists, and then the square-root of the average power will be denoted $pow(u)$:

$$pow(u) := \left(\lim_{T\to\infty} \frac{1}{2T} \int_{-T}^{T} u(t)^2 \, dt \right)^{1/2}$$

Note that a nonzero signal can have zero average power, so *pow* isn't a norm.

Norms for systems

We introduce norms for a stable transfer function G:

$\mathcal{H}_p$-norm, $1 \leq p < \infty$

$$\|G\|_p := \left(\frac{1}{2\pi} \int_{-\infty}^{\infty} |G(j\omega)|^p \, d\omega \right)^{1/p}$$

$\mathcal{H}_\infty$-norm

$$\|G\|_\infty := \sup_\omega |G(j\omega)|$$

More generally, $\mathcal{H}_\infty$ denotes the space of bounded analytic functions in the open right half-plane; prefix $\mathcal{R}$ denotes real-rational.

An important property of the $\mathcal{H}_\infty$-norm is that it is submultiplicative:

$$\|GH\|_\infty \leq \|G\|_\infty \|H\|_\infty$$

(The other $\mathcal{H}_p$-norms are not.)

It is easy to tell when these two norms are finite: *The $\mathcal{H}_2$-norm of G is finite iff G is strictly proper; the $\mathcal{H}_\infty$-norm is finite iff G is proper.*

Input-output relationships

Consider a linear system with input u, output y, and transfer function G, assumed stable and strictly proper. Suppose u is not a fixed signal, but can be any signal of $\mathcal{L}_2$-norm ≤ 1. It turns out that the $\mathcal{L}_2$ induced norm, i.e.,

$$\sup\{\|y\|_2 : \|u\|_2 \leq 1\}$$

equals the $\mathcal{H}_\infty$-norm of G; this provides entry (1,1) in Table 1. The other entries are the other induced norms. (The ∞ in the the various entries is true as long as $G \not\equiv 0$, that is, as long as there is some ω for which $G(j\omega) \neq 0$.)

Notice that the $\mathcal{H}_\infty$-norm of the transfer function appears in several entries in the tables. This norm is therefore an important measure for system performance.

	$\|u\|_2$	$\|u\|_\infty$	$pow(u)$
$\|y\|_2$	$\|G\|_\infty$	∞	∞
$\|y\|_\infty$	$\|G\|_2$	$\|g\|_1$	∞
$pow(y)$	0	$\leq \|G\|_\infty$	$\|G\|_\infty$

Table 1: System gains

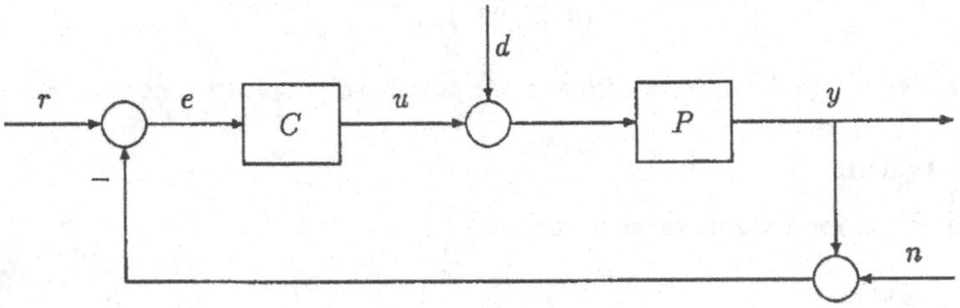

Figure 1: Basic feedback loop

2 Performance

Now we turn to the basic feedback loop in Figure 1. The signals are as follows:

r	reference input
u	control signal
y	plant output
d	disturbance input
n	sensor noise
e	tracking error

The symbols P and C denote the plant and controller transfer functions. All signals are assumed to be scalar-valued.

Internal stability of the system in Figure 1 means that the transfer functions from the exogenous inputs r, d, n to the internal signals e, u are stable.

In this section we look at tracking a *set* of reference signals with a view to obtaining a bound on the steady-state error. This performance objective will be quantified in terms of a weighted norm bound.

Let L denote the loop transfer function, $L := PC$. The transfer function from reference input r to tracking error e is $S := 1/(1+L)$, called the *sensitivity function.* In the analysis to follow, it will always be assumed that the feedback system is internally stable, so S is a stable, proper transfer function. Observe that since L is strictly proper (since P is), $S(j\infty) = 1$.

Now we have to decide on a performance specification, a measure of goodness of tracking. This decision depends on two things: what we know about r; what measure we choose to assign to the tracking error. Usually, r is not known in advance—few control systems are designed for one and only one input. Rather, a set of possible rs will be known, or at least postulated for the purpose of design.

Let's first consider sinusoidal inputs. Suppose r can be any sinusoid of amplitude ≤ 1 and we want e to have amplitude $< \epsilon$. Then the performance specification can be expressed succinctly as

$$\|S\|_\infty < \epsilon$$

Here we used the fact that the maximum amplitude of e equals the $\mathcal{H}_\infty$-norm of the transfer function. Or if we define the (trivial, in this case) weighting function $W_1(s) = 1/\epsilon$, then the performance specification is $\|W_1 S\|_\infty < 1$.

The situation becomes more realistic and more interesting with a frequency-dependent weighting function. Assume $W_1(s)$ is real-rational and stable. The following is a typical scenario giving rise to an $\mathcal{H}_\infty$-norm bound on $W_1 S$.

Recall that

$$\|r\|_2^2 = \frac{1}{2\pi}\int_{-\infty}^{\infty} |R(j\omega)|^2 \, d\omega$$

and that $\|r\|_2^2$ is a measure of the energy of r. Thus, we may think of $|R(j\omega)|^2$ as *energy spectral density*, or energy spectrum. Suppose the set of all rs is

$$\{r : R = W_1 R_{pf}, \|R_{pf}\|_2 \leq 1\}$$

i.e.,

$$\left\{r : \frac{1}{2\pi}\int_{-\infty}^{\infty} |R(j\omega)/W_1(j\omega)|^2 \, d\omega \leq 1\right\}$$

Thus, r has an energy constraint and its energy spectrum is weighted by $1/|W_1(j\omega)|^2$. For example, if W_1 were a bandpass filter, then the energy spectrum of r would be confined to the passband. More generally, W_1 could be used to shape the energy spectrum of the expected class of reference inputs.

Now suppose the tracking error measure is the $\mathcal{L}_2$-norm of e. Then from Table 1,

$$\sup_r \|e\|_2 = \sup\{\|SW_1 R_{pf}\|_2 : \|R_{pf}\|_2 \leq 1\} = \|W_1 S\|_\infty$$

so $\|W_1 S\|_\infty < 1$ means $\|e\|_2 < 1$ for all rs in the above set.

There is a nice graphical interpretation of the norm bound $\|W_1 S\|_\infty < 1$. Note that

$$\|W_1 S\|_\infty < 1 \;\Leftrightarrow\; \left|\frac{W_1(j\omega)}{1 + L(j\omega)}\right| < 1 \quad \forall\omega$$
$$\Leftrightarrow\; |W_1(j\omega)| < |1 + L(j\omega)| \quad \forall\omega$$

The last inequality says that, at every frequency, the point $L(j\omega)$ on the Nyquist plot lies outside the disk of center -1, radius $|W_1(j\omega)|$:

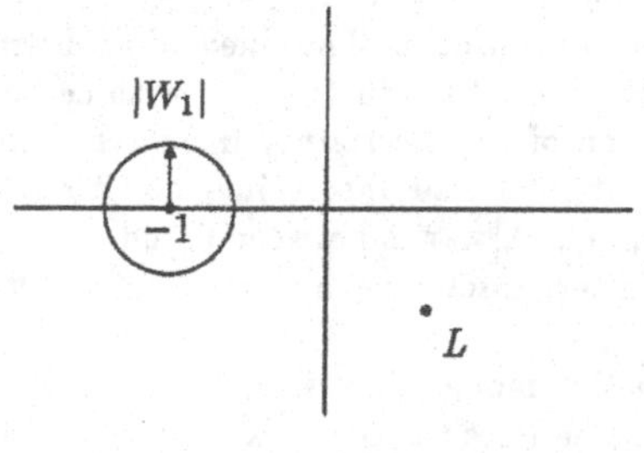

Other performance problems could be posed by focusing on the response to the other two exogenous inputs, d and n.

3 Uncertainty and robustness

No mathematical system can exactly model a physical system. For this reason we must be aware of how modelling errors might adversely affect the performance of a control system.

Plant uncertainty

The basic technique is to model the plant as belonging to a set $\mathcal{P}$. Such a set can be either *structured* or *unstructured*.

For an example of a structured set consider the plant model

$$\frac{1}{s^2 + as + 1}$$

This is a standard second-order transfer function with natural frequency 1 rad/s and damping ratio $a/2$—it could represent, for example, a mass-spring-damper setup or an R-L-C circuit. Suppose the constant a is known only to the extent that it lies in some interval $[a_{min}, a_{max}]$. Then the plant belongs to the structured set

$$\mathcal{P} = \left\{ \frac{1}{s^2 + as + 1} : a_{min} \leq a \leq a_{max} \right\}$$

Thus a structured set is parametrized by a finite number of scalar parameters (one parameter, a, in this example). So structured sets are finite-dimensional.

For us, unstructured sets are more important. They are infinite-dimensional. The basic starting point for an unstructured set is that of disk-like uncertainty.

In what follows, multiplicative perturbation is chosen for detailed study. This is only one type of unstructured perturbation. The important point is that we use disk uncertainty instead of more complicated uncertainty.

Suppose the nominal plant transfer function is P and consider perturbed plant transfer functions of the form $\tilde{P} = (1 + \Delta W_2)P$. Here W_2 is a fixed stable transfer function, the weight, and Δ is a variable stable transfer function satisfying $\|\Delta\|_\infty \leq 1$. Such a perturbation Δ is said to be *allowable*.

The idea behind this uncertainty model is that ΔW_2 is the normalized plant perturbation away from 1:

$$\frac{\tilde{P}}{P} - 1 = \Delta W_2$$

Hence if $\|\Delta\|_\infty \leq 1$, then

$$\left| \frac{\tilde{P}(j\omega)}{P(j\omega)} - 1 \right| \leq |W_2(j\omega)| \quad \forall \omega$$

so $|W_2(j\omega)|$ provides the uncertainty profile. This inequality describes a disk in the complex plane: at each frequency the point $\tilde{P}/P$ lies in the disk with center 1, radius $|W_2|$. Typically, $|W_2(j\omega)|$ is an increasing function of ω: uncertainty increases with increasing frequency. The main purpose of Δ is to account for phase uncertainty and to act as a scaling factor on the magnitude of the perturbation, i.e., $|\Delta|$ varies between 0 and 1.

Thus, this uncertainty model is characterized by a nominal plant P together with a weighting function W_2.

(The multiplicative perturbation model is not suitable for every application because the disk covering the uncertainty set is sometimes too coarse an approximation. In this case a controller designed for the multiplicative uncertainty model would probably be too conservative for the original uncertainty model.)

Robust stability

The notion of robustness can be described as follows. Suppose the plant transfer function P belongs to a set $\mathcal{P}$, as in the previous section. Consider some characteristic of the feedback system, for example, that it is internally stable. A controller C is *robust* with respect to this characteristic if this characteristic holds for every plant in $\mathcal{P}$. The notion of robustness therefore requires a controller, a set of plants, and some characteristic of the system. For us, the two most important variations of this notion are robust stability and robust performance.

A controller C provides *robust stability* if it provides internal stability for every plant in $\mathcal{P}$. We might like to have a test for robust stability, a test involving C and $\mathcal{P}$. Or if $\mathcal{P}$ has an associated size, then the maximum size such that C stabilizes all of $\mathcal{P}$ might be a useful notion of stability margin.

The Nyquist plot gives information about stability margin. Note that the distance from the critical point -1 to the nearest point on the Nyquist plot of L equals $1/\|S\|_\infty$:

$$\begin{aligned}\text{distance from -1 to Nyquist plot} &= \inf_\omega |-1-L(j\omega)| \\ &= \inf_\omega |1+L(j\omega)| \\ &= \left[\sup_\omega \frac{1}{|1+L(j\omega)|}\right]^{-1} \\ &= \|S\|_\infty^{-1}\end{aligned}$$

Thus if $\|S\|_\infty \gg 1$, then the Nyquist plot comes close to the critical point, and the feedback system is nearly unstable. However, as a measure of stability margin this distance is not entirely adequate because it contains no frequency information.

Better stability margins are obtained by taking explicit frequency-dependent perturbation models. For example, the multiplicative perturbation model, $\tilde{P} = (1+\Delta W_2)P$. Fix a positive number β and consider the family of plants

$$\{\tilde{P} : \Delta \text{ is stable and } \|\Delta\|_\infty \le \beta\}$$

Now a controller C which achieves internal stability for the nominal plant P will stabilize this entire family if β is small enough. Denote by β_{sup} the least upper bound on β such that C achieves internal stability for the entire family. Then β_{sup} is a stability margin (with respect to this uncertainty model).

Now we look at a typical robust stability test, one for the multiplicative perturbation model. Assume the nominal feedback system, i.e., with $\Delta = 0$, is internally stable for controller C. Bring in the *complementary sensitivity function*

$$T = 1 - S = \frac{L}{1+L} = \frac{PC}{1+PC}$$

Theorem 1 *(Multiplicative uncertainty model) C provides robust stability iff $\|W_2T\|_\infty < 1$.*

The theorem can be effectively used to find the stability margin β_{sup} defined previously. The simple scaling technique

$$\begin{aligned}\{\tilde{P} = (1+\Delta W_2)P : \|\Delta\|_\infty \le \beta\} &= \{\tilde{P} = (1+\beta^{-1}\Delta\beta W_2)P : \|\beta^{-1}\Delta\|_\infty \le 1\} \\ &= \{\tilde{P} = (1+\Delta_1\beta W_2)P : \|\Delta_1\|_\infty \le 1\}\end{aligned}$$

together with the theorem shows that

$$\beta_{sup} = \sup\{\beta : \|\beta W_2 T\|_\infty < 1\} = 1/\|W_2T\|_\infty$$

The condition $\|W_2T\|_\infty < 1$ also has a nice graphical interpretation. Note that

$$\begin{aligned}\|W_2T\|_\infty < 1 \quad &\Leftrightarrow \quad \left|\frac{W_2(j\omega)L(j\omega)}{1+L(j\omega)}\right| < 1 \quad \forall\omega \\ &\Leftrightarrow \quad |W_2(j\omega)L(j\omega)| < |1+L(j\omega)| \quad \forall\omega\end{aligned}$$

The last inequality says that, at every frequency, the critical point, -1, lies outside the disk of center $L(j\omega)$, radius $|W_2(j\omega)L(j\omega)|$:

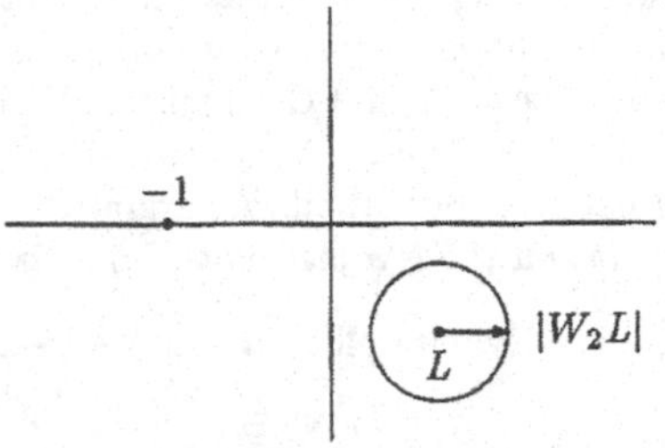

There's a simple way to see the relevance of the condition $\|W_2T\|_\infty < 1$. First, draw the block diagram of the perturbed feedback system, but ignoring inputs:

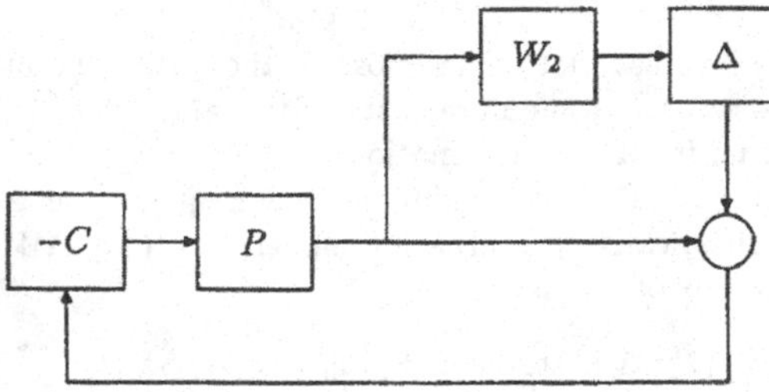

The transfer function from the output of Δ around to the input of Δ equals $-W_2T$, so the block diagram collapses to

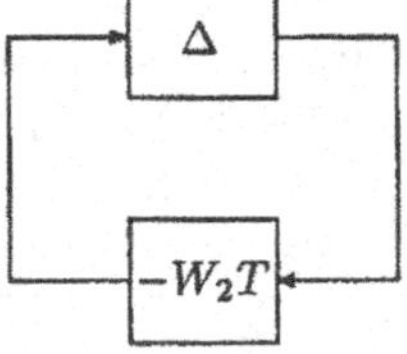

The maximum loop gain in the latter diagram equals $\|-\Delta W_2T\|_\infty$, which is < 1 for all allowable Δs iff the small-gain condition $\|W_2T\|_\infty < 1$ holds.

The foregoing discussion is related to the *small-gain theorem*, a special case of which is this: if L is stable and $\|L\|_\infty < 1$, then $(1+L)^{-1}$ is stable too.

4 Robust performance

Now we look into performance of the perturbed plant. Suppose the plant transfer function belongs to a set $\mathcal{P}$. The general notion of *robust performance* is that internal stability and performance, of a specified type, should hold for all plants in $\mathcal{P}$. Again we focus on multiplicative perturbations.

Recall that, when the nominal feedback system is internally stable, the *nominal performance* condition is $\|W_1S\|_\infty < 1$ and the *robust stability* condition is $\|W_2T\|_\infty < 1$. If P is perturbed to $(1+\Delta W_2)P$, S is perturbed to

$$\frac{1}{1+(1+\Delta W_2)L} = \frac{S}{1+\Delta W_2 T}$$

Clearly the *robust performance* condition should therefore be

$$\|W_2T\|_\infty < 1 \text{ and } \left\|\frac{W_1S}{1+\Delta W_2T}\right\|_\infty < 1 \quad \forall\Delta$$

Here Δ must be allowable. The next theorem gives a test for robust performance in terms of the function

$$s \mapsto |W_1(s)S(s)| + |W_2(s)T(s)|$$

which is denoted $|W_1S| + |W_2T|$.

Theorem 2 *A necessary and sufficient condition for robust performance is*

$$\||W_1S| + |W_2T|\|_\infty < 1 \tag{1}$$

Test (1) also has a nice graphical interpretation. For each frequency ω, construct two closed disks: one with center -1, radius $|W_1(j\omega)|$; the other with center $L(j\omega)$, radius $|W_2(j\omega)L(j\omega)|$. Then (1) holds iff for each ω these two disks are disjoint:

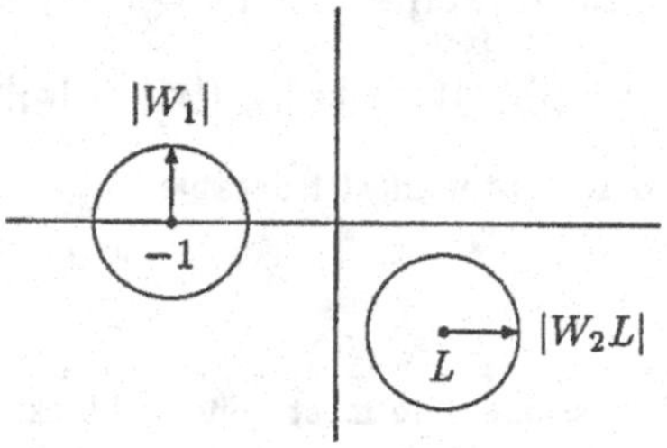

5 The mixed sensitivity problem

The nominal feedback system is assumed to be internally stable. The condition for simultaneously achieving nominal performance and robust stability is

$$\|\max(|W_1S|, |W_2T|)\|_\infty < 1 \tag{2}$$

The *robust performance* condition is

$$\|W_2T\|_\infty < 1 \text{ and } \left\|\frac{W_1S}{1+\Delta W_2T}\right\|_\infty < 1 \quad \forall\Delta$$

and the test for this is

$$\||W_1S| + |W_2T|\|_\infty < 1 \tag{3}$$

Since

$$\max(|W_1S|, |W_2T|) \leq |W_1S| + |W_2T| \leq 2\max(|W_1S|, |W_2T|) \tag{4}$$

conditions (2) and (3) aren't too far apart. For instance, if nominal performance and robust stability are obtained with a safety factor of 2, i.e.,

$$\|W_1S\|_\infty < 1/2, \quad \|W_2T\|_\infty < 1/2$$

then robust performance is automatically obtained.

A compromise condition is

$$\|(|W_1S|^2 + |W_2T|^2)^{1/2}\|_\infty < 1 \tag{5}$$

Designing a controller to achieve (5) (or minimization of the left-hand side) is called the *mixed sensitivity problem.* Simple plane geometry shows that

$$\max(|W_1S|, |W_2T|) \leq (|W_1S|^2 + |W_2T|^2)^{1/2} \leq |W_1S| + |W_2T| \tag{6}$$

and

$$\frac{1}{\sqrt{2}}(|W_1S| + |W_2T|) \leq (|W_1S|^2 + |W_2T|^2)^{1/2} \leq \sqrt{2}\max(|W_1S|, |W_2T|) \tag{7}$$

Thus (5) is a reasonable approximation to both (2) and (3).

To elaborate on this point, let's consider

$$x = \begin{pmatrix} x_1 \\ x_2 \end{pmatrix} = \begin{pmatrix} |W_1S| \\ |W_2T| \end{pmatrix}$$

as a vector in $\mathbf{R}^2$. Then (2), (3), and (5) correspond, respectively, to the three different norms

$$\max(|x_1|, |x_2|), \quad |x_1| + |x_2|, \quad (|x_1|^2 + |x_2|^2)^{1/2}$$

The third is the Euclidean norm and is the most tractable.

6 The big picture

Let's generalize the previous discussion. The most general block diagram of a control system is shown in Figure 2. The generalized plant consists of everything that is fixed at the start of the control design exercise: the plant, actuators which generate inputs to the plant, sensors measuring certain signals, analog-to-digital and digital-to-analog converters, etc. The controller consists of the designable part: it may be an electric circuit, a programmable logic controller, a general purpose computer, or some other such device. The signals w, z, y, and u are, in general, vector-valued functions of time. The components of w are all the exogenous inputs: references, disturbances, sensor noises, etc. The components of z are all the signals we wish to control: tracking errors between reference signals and plant outputs, actuator signals whose values must be kept between certain limits, etc. The vector y contains the outputs of all sensors. Finally, u contains all controlled inputs to the generalized plant.

Assume that the generalized plant and controller are modelled by linear systems G and K which are time-invariant and finite-dimensional. Presented next is a summary of some design problems.

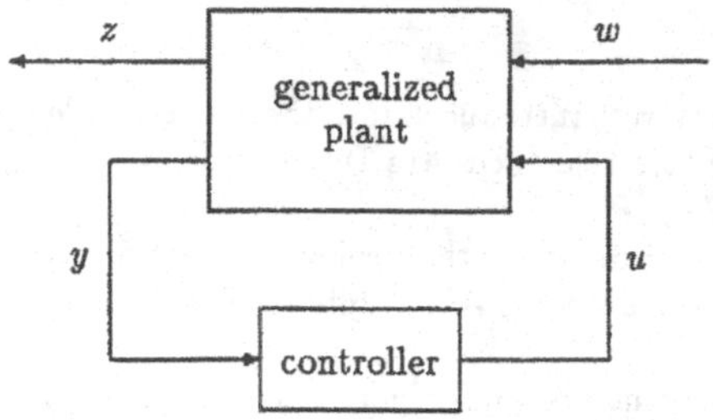

Figure 2: Most general control system

Standard design problems

1. **Nominal design**

   Given G, design K so that the system is internally stable and

   (1a) the closed-loop poles lie in some desired region, or

   (1b) $\|z\|_2$ is minimum for some fixed w (the $\mathcal{H}_2$ criterion), or

   (1c) $\sup\{\|z\|_2 : \|w\|_2 \leq 1\}$ is minimum (the $\mathcal{H}_\infty$ criterion), or

   (1d) $\sup\{\|z\|_\infty : \|w\|_\infty \leq 1\}$ is minimum (the $\mathcal{L}_1$ criterion), or

   (1e) the rms value of z is minimum when w is standard white noise (the LQG criterion), etc.

   Modifications are possible, for example, (1c) could be replaced by

   $$\sup\{\|z\|_2 : \|w\|_2 \leq 1\} \leq \gamma$$

   where γ is some prespecified performance level. Or we could have different spatial norms besides Euclidean, for example,

   $$\|z(t)\| = \max_i |z_i(t)|$$

2. **Robust design**

   Given a family $\mathcal{G}$, design K so that internal stability holds for all G in $\mathcal{G}$ and, for example,

   $$\|z\|_2 \leq \gamma, \quad \forall \|w\|_2 \leq 1, G \in \mathcal{G}$$

   Thus a performance level γ is achieved for all plants in a prespecified class. The family $\mathcal{G}$ could be described by either structured or unstructured perturbations.

There are reasonably complete theories for the nominal design problems, but not for the robust design problems.

7 Notes

Tools from functional analysis were introduced into the subject of feedback control around 1960 by G. Zames and others. Some references are Desoer and Vidyasagar 1975, Holtzman 1970, Mees 1981, and Willems 1971.

It was Zames (1981) who formulated the problem of optimizing $W_1 S$ with respect to the ∞-norm, stressing the role of the weight W_1. Additional motivation for this problem is offered in Zames and Francis 1983.

Doyle and Stein (1981) emphasized the importance of explicit uncertainty models, like multiplicative. Theorem 1 is stated in their paper, but a complete proof is due to Chen and Desoer (1982).

Lecture 2

The Weighted Sensitivity Problem

This lecture looks at the simplest $\mathcal{H}_\infty$ design problem, designing a controller to achieve $\|W_1 S\|_\infty \leq 1$.

1 Design constraints

Before we see how to design control systems, it is useful to see what are the basic limitations on achievable performance. In this section we study design constraints arising from the fact that closed-loop transfer functions must be stable, i.e., analytic in the right half-plane. It is assumed throughout this section that the feedback system is internally stable.

Bounds on the weight W_1

Suppose the loop transfer function L has a zero z in Re $s \geq 0$. Then

$$\|W_1 S\|_\infty \geq |W_1(z)| \tag{1}$$

This is a direct consequence of the maximum modulus theorem and the fact that $S(z) = 1$:

$$|W_1(z)| = |W_1(z)S(z)| \leq \sup_{\text{Re } s \geq 0} |W_1(s)S(s)| = \|W_1 S\|_\infty$$

So a necessary condition that the performance criterion $\|W_1 S\|_\infty < 1$ be achievable is that the weight satisfy $|W_1(z)| < 1$. In words, the magnitude of the weight at a right half-plane zero of P or C must be less than 1.

Inner and outer transfer functions

Two types of transfer functions play a critical role: inner and outer. A function in $\mathcal{RH}_\infty$ is *inner* if its magnitude equals 1 at all points on the imaginary axis. It is not difficult to show that such a function has pole-zero symmetry about the imaginary axis in the sense that a point s_0 is a zero iff its reflection, $-\bar{s}_0$, is a pole. Consequently, the function being stable, all its zeros lie in the right half-plane. Thus an inner function is, up to sign, the product of factors of the form

$$\frac{s - s_0}{s + \bar{s}_0}, \qquad \text{Re } s_0 > 0$$

Examples of inner functions are

$$1, \quad \frac{s-1}{s+1}, \quad \frac{s^2 - s + 2}{s^2 + s + 2}$$

A function in $\mathcal{RH}_\infty$ is *outer* if it has no zeros in Re $s > 0$. Examples of outer functions are

$$1, \quad \frac{1}{s+1}, \quad \frac{s}{s+1}, \quad \frac{s+2}{s^2+s+1}$$

It is a useful fact that every function in $\mathcal{RH}_\infty$ can be written as the product of two such factors. For example

$$\frac{4(s-2)}{s^2+s+1} = \left(\frac{s-2}{s+2}\right)\left(\frac{4(s+2)}{s^2+s+1}\right)$$

For each function G in $\mathcal{RH}_\infty$ there exist an inner function G_i and an outer function G_o such that $G = G_iG_o$. The factors are unique up to sign.

For technical reasons we assume for the remainder of this section that L has no poles on the imaginary axis. Factor the sensitivity function as

$$S = S_iS_o$$

Then S_o has no zeros on the imaginary axis (such zeros would be poles of L) and S_o is not strictly proper (since S is not). Thus $S_o^{-1} \in \mathcal{RH}_\infty$.

As a simple example of the use of inner functions, suppose P has a zero at z with $\text{Re } z > 0$, a pole at p with $\text{Re } p > 0$, and no other poles or zeros in the closed right half-plane, and suppose C has no right half-plane poles or zeros. Then

$$S_i(s) = \frac{s-p}{s+p}$$

Since $S(z) = 1$, we get

$$S_o(z) = S_i(z)^{-1} = \frac{z+p}{z-p}$$

and hence

$$\|W_1S\|_\infty = \|W_1S_o\|_\infty \geq |W_1(z)S_o(z)| = \left|W_1(z)\frac{z+p}{z-p}\right|$$

So a necessary condition that the performance criterion $\|W_1S\|_\infty < 1$ be achievable is that the weight satisfy

$$|W_1(z)| < \left|\frac{z-p}{z+p}\right|$$

So $|W_1(z)|$ must be even smaller than in (1).

The waterbed effect

Consider a tracking problem where the reference signals have their energy spectra concentrated in a known frequency range, say $[\omega_1, \omega_2]$. This is the idealized situation where W_1 is a bandpass filter. Let M_1 denote the maximum magnitude of S on this frequency band,

$$M_1 := \max_{\omega_1 \leq \omega \leq \omega_2} |S(j\omega)|$$

and let M_2 denote the maximum magnitude over all frequencies, that is, $\|S\|_\infty$. Then good tracking capability is characterized by the inequality $M_1 \ll 1$. On the other hand, we cannot permit M_2 to be too large: remember that $1/M_2$ equals the distance from the critical point to the Nyquist plot of L, so large M_2 means small stability margin. Notice that M_2 must be at least 1 because this is the value of S at infinite frequency. So the question arises, can we have M_1 very small and M_2 not too large? Or does it happen that very small M_1 necessarily means very large M_2? The latter situation might be compared to a waterbed: as $|S|$ is pushed down on one frequency range, it pops up somewhere else. It turns out that non-outer plants exhibit the waterbed effect.

Theorem 1 *Suppose P has a zero at z with Re $z > 0$. Then there exist positive constants c_1 and c_2, depending only on ω_1, ω_2, and z, such that*

$$c_1 \log M_1 + c_2 \log M_2 \geq \log |S_i(z)^{-1}| \geq 0$$

The proof requires a lemma.

Lemma 1 *For every point $s_0 = \sigma_0 + j\omega_0$ with $\sigma_0 > 0$,*

$$\log|S_o(s_0)| = \frac{1}{\pi}\int_{-\infty}^{\infty} \log|S(j\omega)| \frac{\sigma_0}{\sigma_0^2 + (\omega - \omega_0)^2}\, d\omega$$

Proof Set $F(s) := \ln S_o(s)$. Then F is analytic and of bounded magnitude in Re $s \geq 0$. (This follows from the properties $S_o, S_o^{-1} \in \mathcal{RH}_\infty$; the idea is that since S_o has no poles or zeros in the right half-plane, $\ln S_o$ is well-behaved there.) The Poisson integral formula says that

$$F(s_0) = \frac{1}{\pi}\int_{-\infty}^{\infty} F(j\omega) \frac{\sigma_0}{\sigma_0^2 + (\omega - \omega_0)^2}\, d\omega$$

Now take real parts of both sides:

$$\text{Re } F(s_0) = \frac{1}{\pi}\int_{-\infty}^{\infty} \text{Re } F(j\omega) \frac{\sigma_0}{\sigma_0^2 + (\omega - \omega_0)^2}\, d\omega \tag{2}$$

But

$$S_o = e^F = e^{\text{Re } F} e^{j\text{Im } F}$$

so

$$|S_o| = e^{\text{Re } F}$$

that is,

$$\ln|S_o| = \text{Re } F$$

Thus from (2)

$$\ln|S_o(s_0)| = \frac{1}{\pi}\int_{-\infty}^{\infty} \ln|S_o(j\omega)| \frac{\sigma_0}{\sigma_0^2 + (\omega - \omega_0)^2}\, d\omega$$

or, since $|S| = |S_o|$ on the imaginary axis,

$$\ln|S_o(s_0)| = \frac{1}{\pi}\int_{-\infty}^{\infty} \ln|S(j\omega)| \frac{\sigma_0}{\sigma_0^2 + (\omega - \omega_0)^2}\, d\omega$$

Finally, since $\log x = \log e \ln x$, the result follows upon multiplying the last equation by $\log e$. ■

Proof of theorem Since z is a zero of P, it follows from the previous section that $S(z) = 1$, and hence $S_o(z) = S_i(z)^{-1}$. Apply Lemma 1 with

$$s_0 = z = \sigma_0 + j\omega_0$$

to get

$$\log|S_i(z)^{-1}| = \frac{1}{\pi}\int_{-\infty}^{\infty} \log|S(j\omega)| \frac{\sigma_0}{\sigma_0^2 + (\omega - \omega_0)^2}\, d\omega$$

Thus

$$\log|S_i(z)^{-1}| \leq c_1 \log M_1 + c_2 \log M_2$$

where c_1 is defined to be the integral of

$$\frac{1}{\pi}\frac{\sigma_0}{\sigma_0^2 + (\omega - \omega_0)^2}$$

over the set

$$[-\omega_2, -\omega_1] \cup [\omega_1, \omega_2]$$

and c_2 equals the same integral but over the complementary set.

It remains to observe that $|S_i(z)| \leq 1$ by the maximum modulus theorem, so

$$\log|S_i(z)^{-1}| \geq 0$$

■

Example As an illustration of the theorem consider the plant transfer function

$$P(s) = \frac{s-1}{(s+1)(s-p)}$$

where $p > 0$, $p \neq 1$. Now S must interpolate zero at the unstable poles of P, so $S(p) = 0$. Thus the inner factor of S must contain the factor

$$\frac{s-p}{s+p}$$

that is,

$$S_i(s) = \frac{s-p}{s+p}G(s)$$

for some inner function G. Since $|G(1)| \leq 1$ (maximum modulus theorem), there follows

$$|S_i(1)| \leq \left|\frac{1-p}{1+p}\right|$$

So the theorem gives

$$c_1 \log M_1 + c_2 \log M_2 \geq \log\left|\frac{1+p}{1-p}\right|$$

Note that the right-hand side is very large if p is close to 1. This example illustrates again a general fact: the waterbed effect is amplified if the plant has a pole and a zero close together in the right half-plane. We would expect such a plant to be very difficult to control.

It is emphasized that the waterbed effect applies to non-outer plants only. In fact, the following can be proved: if P has no zeros in Re $s > 0$ nor on the imaginary axis in the frequency range $[\omega_1, \omega_2]$, then for every $\epsilon > 0$ and $\delta > 1$ there exists a controller C so that the feedback system is internally stable, $M_1 < \epsilon$, and $M_2 < \delta$. As a very easy example, take

$$P(s) = \frac{1}{s+1}$$

The controller $C(s) = k$ is internally stabilizing for all $k > 0$, and then

$$S(s) = \frac{s+1}{s+1+k}$$

So $\|S\|_\infty = 1$ and, for every $\epsilon > 0$ and ω_2, if k is large enough, then

$$|S(j\omega)| < \epsilon, \quad \omega \leq \omega_2$$

The area formula

There is a formula for the area bounded by the graph of $|S(j\omega)|$ (log scale) plotted as a function of ω (linear scale). The formula is valid when the relative degree of L is large enough. *Relative degree* equals degree of denominator minus degree of numerator.

Let $\{p_i\}$ denote the set of poles of L in $\mathrm{Re}\ s > 0$.

Theorem 2 *Assume the relative degree of L is at least 2. Then*

$$\int_0^\infty \log |S(j\omega)|\, d\omega = \pi(\log e)(\sum \mathrm{Re}\ p_i)$$

The proof also is based on the Poisson integral formula, but is quite intricate.

Example Take the plant and controller

$$P(s) = \frac{1}{(s-1)(s+2)}, \quad C(s) = 10$$

The feedback system is internally stable and L has relative degree 2. The plot of $|S(j\omega)|$, log scale, versus ω, linear scale, is shown in Figure 1. The area below the line $|S| = 1$ is negative,

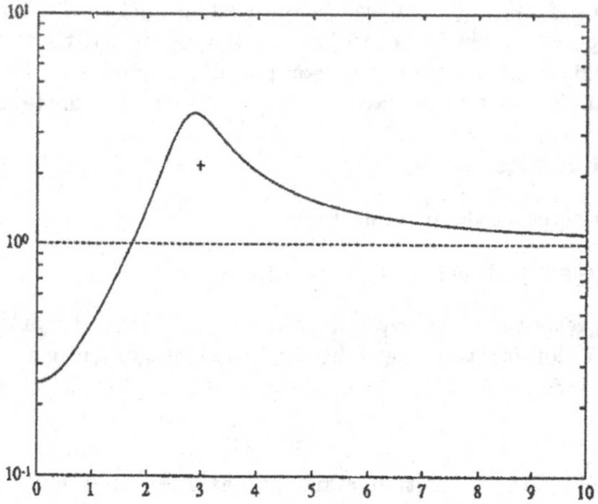

Figure 1: $|S(j\omega)|$, log scale, versus ω, linear scale

the area above, positive. The theorem says that the net area is positive, equaling

$$\pi(\log e)(\sum \mathrm{Re}\ p_i) = \pi(\log e)$$

So the negative area, required for good tracking over some frequency range, must unavoidably be accompanied by some positive area.

The waterbed effect applies to non-outer systems, whereas the area formula applies in general (except for the relative degree assumption). In particular, the area formula doesn't itself imply a peaking phenomenon, only an area conservation. However, one can infer a type of peaking phenomenon from the area formula when another constraint is imposed, namely, controller bandwidth, or more precisely, the bandwidth of the loop transfer function PC. For example, suppose the constraint is

$$|PC| < \frac{1}{\omega}, \quad \omega \geq \omega_1$$

where $\omega_1 > 1$. This is one way of saying the loop bandwidth is constrained to be $\leq \omega_1$. Then for $\omega \geq \omega_1$

$$|S| \leq \frac{1}{1-|PC|} < \frac{1}{1-\omega^{-1}} = \frac{\omega}{\omega-1}$$

Hence

$$\int_{\omega_1}^{\infty} \log|S(j\omega)|\, d\omega \leq \int_{\omega_1}^{\infty} \log\frac{\omega}{\omega-1}\, d\omega < \infty$$

and so the possible positive area over the interval $[\omega_1, \infty)$ is limited. Thus if $|S|$ is made smaller and smaller over some subinterval of $[0, \omega_1]$, incurring a larger and larger debt of negative area, then $|S|$ must necessarily become larger and larger somewhere else in $[0, \omega_1]$. Roughly speaking, with a loop bandwidth constraint the waterbed effect applies even to outer plants.

2 Design to achieve $\|W_1 S\|_\infty < 1$

The performance criterion $\|W_1 S\|_\infty < 1$ was introduced in Lecture 1. The associated design problem is to find a proper C for which the feedback system is internally stable and $\|W_1 S\|_\infty < 1$. When does such a C exist and how can it be computed? These questions are easy when the inverse of the plant transfer function is stable, so we assume P^{-1} is unstable. This will require us to use interpolation theory.

To simplify matters we'll assume that

- P has no poles or zeros on the imaginary axis
- W_1 is stable and strictly proper

It would be possible to relax these assumptions, but the development would be messier.

We'll need to use a rolloff function. Let k be a positive integer and τ a positive real number, and consider the transfer function

$$J(s) := \frac{1}{(\tau s+1)^k}$$

Sketch the Bode plot of J: the magnitude starts out at 1, is relatively flat out to the corner frequency $\omega = 1/\tau$, and then rolls off to $-\infty$ with slope $-k$; the phase starts out at 0, is relatively flat up to, say, $\omega = 0.1/\tau$, and then rolls off to $-k\pi/2$ radians. So for low frequency, $J(j\omega) \approx 1$. This function has the useful property that it approximates 1 beside a strictly proper function.

Lemma 2 *If G is stable and strictly proper, then*

$$\lim_{\tau \to 0} \|G(1-J)\|_\infty = 0$$

(The easy proof is omitted.)

To motivate the procedure to follow, let's see roughly how the design problem of finding an internally stabilizing C so that $\|W_1 S\|_\infty < 1$ can be translated into an NP problem. The definition of S is

$$S = \frac{1}{1+PC}$$

For C to be internally stabilizing it is necessary and sufficient that $S \in \mathcal{RH}_\infty$ and PC have no right half-plane pole-zero cancellations. Thus, S must interpolate the value 1 at the right half-plane zeros of P and the value 0 at the right half-plane poles, i.e., S must satisfy the conditions

$$S(z) = 1, \quad z \text{ a zero of } P \text{ in Re } s > 0$$

$$S(p) = 0, \quad p \text{ a pole of } P \text{ in Re } s > 0$$

The weighted sensitivity function $G := W_1 S$ must therefore satisfy

$$G(z) = W_1(z), \quad z \text{ a zero of } P \text{ in Re } s > 0$$

$$G(p) = 0, \quad p \text{ a pole of } P \text{ in Re } s > 0$$

So the requirement of internal stability imposes interpolation constraints on G. The performance spec $\|W_1 S\|_\infty < 1$ translates into $\|G\|_\infty < 1$. Finally, the condition $S \in \mathcal{RH}_\infty$ requires that G be analytic in the right half-plane.

One approach to the design problem might be to find a function G satisfying these conditions, then to get S, and finally to get C by back-substitution. This has a technical snag because of the requirement that C be proper. For this reason we proceed via controller parametrization.

Bring in a coprime factorization of P:

$$P = \frac{N}{M}, \quad NX + MY = 1, \quad N, M, X, Y \in \mathcal{RH}_\infty$$

The controller parametrization formula is

$$C = \frac{X + MQ}{Y - NQ}, \quad Q \in \mathcal{RH}_\infty$$

and for such C the weighted sensitivity function is

$$W_1 S = W_1 M(Y - NQ)$$

The parameter Q must be both stable and proper. Our approach is first to drop the properness requirement and find a suitable parameter, say, Q_{im}, which is improper but stable, and then to get a suitable Q by rolling Q_{im} off at high frequency. The reason this works is that W_1 is strictly proper, so there is no performance requirement at high frequency. The method is outlined as follows:

Procedure

Input: P, W_1

Step 1 Do a coprime factorization of P: find four functions in $\mathcal{RH}_\infty$ satisfying the equations

$$P = N/M, \quad NX + MY = 1$$

Step 2 Find a stable function Q_{im} such that

$$\|W_1 M(Y - NQ_{im})\|_\infty < 1$$

Step 3 Set

$$J(s) := \frac{1}{(\tau s+1)^k}$$

where k is just large enough that $Q_{im}J$ is proper and τ is just small enough that

$$\|W_1 M(Y - NQ_{im}J)\|_\infty < 1$$

Step 4 Set $Q = Q_{im}J$.

Step 5 Set $C = (X + MQ)/(Y - NQ)$.

That Step 3 is feasible follows from the equation

$$W_1 M(Y - NQ_{im}J) = W_1 M(Y - NQ_{im})J + W_1 MY(1 - J)$$

The first term on the right-hand side has $\mathcal{H}_\infty$-norm less than 1 from Step 2 and the fact that $\|J\|_\infty \leq 1$, while the $\mathcal{H}_\infty$-norm of the second term goes to 0 as τ goes to 0 by Lemma 2.

Step 2 is the model-matching problem, find a stable function Q_{im} to minimize

$$\|T_1 - T_2 Q_{im}\|_\infty$$

where $T_1 := W_1 MY$ and $T_2 := W_1 MN$. This problem is discussed in the next section.

3 The model-matching problem

Let $T_1(s)$ and $T_2(s)$ be functions in $\mathcal{RH}_\infty$. The *model-matching problem* is to find a stable transfer function $Q(s)$ to minimize the $\mathcal{H}_\infty$-norm of $T_1 - T_2 Q$. The interpretation is that T_1 is a model, T_2 is a plant, and Q is a cascade controller to be designed so that $T_2 Q$ approximates T_1. Thus, $T_1 - T_2 Q$ is the error transfer function. The transfer function Q is required to be stable but not necessarily proper (this makes the problem easier). We'll *assume* that T_2 has no zeros on the imaginary axis. Define the minimum model-matching error

$$\gamma_{opt} := \min \|T_1 - T_2 Q\|_\infty$$

where the minimum is taken over all stable Qs. It turns out that the minimum is achieved by virtue of the assumption on T_2; a Q achieving the minimum is said to be *optimal.*

The trivial case of the problem is when T_1/T_2 is stable, for then the unique optimal Q is $Q = T_1/T_2$ and $\gamma_{opt} = 0$.

The simplest nontrivial case is when T_2 has only one zero in the right half-plane, say at $s = s_0$. If Q is stable and $T_2 Q$ has finite $\mathcal{H}_\infty$-norm, i.e., $T_2 Q \in \mathcal{RH}_\infty$, then by the maximum modulus theorem

$$\|T_1 - T_2 Q\|_\infty \geq |T_1(s_0)|$$

and so $\gamma_{opt} \geq |T_1(s_0)|$. On the other hand, the function

$$Q = \frac{T_1 - T_1(s_0)}{T_2} \tag{3}$$

is stable and yields the value $|T_1(s_0)|$ for the model-matching error. The conclusion is that $\gamma_{opt} = |T_1(s_0)|$ and (3) is an optimal Q, in fact, the unique optimal Q.

There are many ways to solve the model-matching problem when T_2 has two or more zeros in the right half-plane. There isn't time to go into any here.

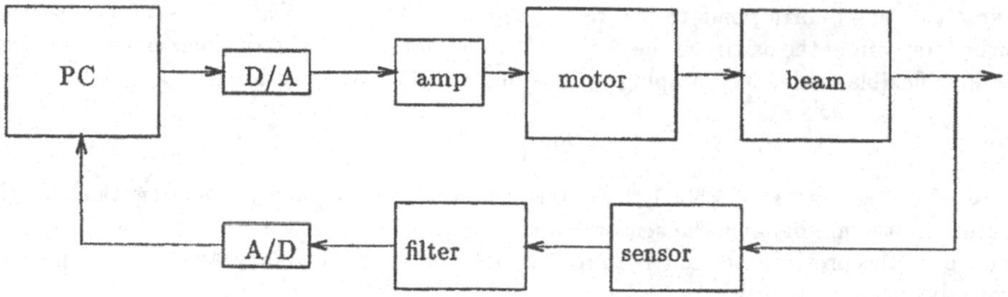

Figure 2: Flexible beam setup

4 Design example: flexible beam

This section presents an example to illustrate the procedure of the previous sections. The example is based on a real experimental setup at the University of Toronto.

The control system, depicted in Figure 2, has the following components: a flexible beam; a high-torque dc motor at one end of the beam; a sonar position sensor at the other end; a digital computer as the controller with analog-to-digital interface hardware; a power amplifier to drive the motor; an anti-aliasing filter. The objective is to control the position of the sensed end of the beam.

A plant model was obtained as follows. The beam is pinned to the motor shaft and is free at the sensed end. First the beam itself was modeled as an ideal Euler-Bernoulli beam with no damping; this yielded a partial differential equation model, reflecting the fact that the physical model of the beam has an infinite number of modes. The model is therefore linear but infinite-dimensional. The corresponding transfer function from torque input at the motor end to tip deflection at the sensed end has the form

$$\sum_{i=0}^{\infty} \frac{c_i}{s^2 + \omega_i^2}$$

Then damping was introduced, yielding the form

$$\sum_{i=0}^{\infty} \frac{c_i}{s^2 + 2\zeta_i\omega_i s + \omega_i^2}$$

The first term is c_0/s^2 and corresponds to the rigid body slewing motion about the pinned end. The second term,

$$\frac{c_1}{s^2 + 2\zeta_1\omega_1 s + \omega_1^2}$$

corresponds to the first flexible mode. And so on. The motion was found to be adequately modeled by the first four flexible modes. Then the damping ratios and natural frequencies were determined experimentally. Finally, the amplifier, motor, and sensor were introduced into the model. The anti-aliasing filter was ignored for the purpose of design.

For simplicity we shall take the plant transfer function to be

$$P(s) = \frac{-6.4750s^2 + 4.0302s + 175.7700}{s(5s^3 + 3.5682s^2 + 139.5021s + 0.0929)}$$

The poles are

$$0, \quad -0.0007, \quad -0.3565 \pm 5.2700j$$

The first two poles correspond to the rigid body motion; the one at $s = -0.0007$ has been perturbed away from the origin by the back emf in the motor. The two complex poles correspond to the first flexible mode, the damping ratio being 0.0675. The zeros are

$$-4.9081, \quad 5.5308$$

Because of the zero at $s = 5.5308$ the plant is non-minimum phase, reflecting the fact that the actuator (the motor) and the sensor are not located at the same point on the beam. The procedure of the previous section requires no poles on the imaginary axis, so the model is (harmlessly) perturbed to

$$P(s) = \frac{-6.4750s^2 + 4.0302s + 175.7700}{5s^4 + 3.5682s^3 + 139.5021s^2 + 0.0929s + 10^{-6}}$$

A common way to specify desired closed-loop performance is by a step response test. For this flexible beam the spec is that a step reference input (r) should produce a plant output (y) satisfying

$$\text{settling time} \approx 8 \text{ sec}$$
$$\text{overshoot} \leq 10\%$$

We'll accomplish this by shaping $T(s)$, the transfer function from r to y, so that it approximates a standard second-order system: the ideal $T(s)$ is

$$T_{id}(s) := \frac{\omega_n^2}{s^2 + 2\zeta\omega_n s + \omega_n^2}$$

A settling time of 8 sec requires

$$\frac{4.6}{\zeta\omega_n} \approx 8$$

and an overshoot of 10% requires

$$\exp\left(-\zeta\pi/\sqrt{1-\zeta^2}\right) = 0.1$$

The solutions are $\zeta = 0.5912$ and $\omega_n = 0.9583$. Let's round to $\zeta = 0.6$ and $\omega_n = 1$. So the ideal $T(s)$ is

$$T_{id}(s) = \frac{1}{s^2 + 1.2s + 1}$$

Then the ideal sensitivity function is

$$S_{id}(s) := 1 - T_{id}(s) = \frac{s(s+1.2)}{s^2 + 1.2s + 1}$$

Now take the weighting function $W_1(s)$ to be $S_{id}(s)^{-1}$, i.e.,

$$W_1(s) = \frac{s^2 + 1.2s + 1}{s(s+1.2)}$$

The rationale for this choice is a rough argument which goes as follows. Consider Step 2 of the procedure in the Section 2; from it the function

$$F := W_1 M(Y - NQ_{im})$$

equals a constant times an inner function. The procedure then rolls off Q_{im} to result in the weighted sensitivity function

$$W_1 S := W_1 M(Y - NQ_{im}J)$$

So $W_1 S \approx F$ except at high frequency, i.e.,

$$S \approx FS_{id}$$

Now F behaves approximately like a time-delay except at high frequency (this is a property of inner functions). So we arrive at the rough approximation

$$S \approx (\text{time-delay}) \times S_{id}$$

Hence our design should produce

$$\text{actual step response} \approx \text{delayed ideal step response}$$

One further adjustment is required in the problem setup: W_1 must be stable and strictly proper, so the above function is modified to

$$W_1(s) = \frac{s^2 + 1.2s + 1}{(s + 0.001)(s + 1.2)(0.001s + 1)}$$

The procedure can now be applied.

Step 1 Since $P \in \mathcal{RH}_\infty$ we take $N = P, M = 1, X = 0, Y = 1$.

Step 2 The model-matching problem is to minimize

$$\|W_1 M(Y - NQ_{im})\|_\infty = \|W_1(1 - PQ_{im})\|_\infty$$

Since P has only one right half-plane zero, at $s = 5.5308$, we have from Section 3 that

$$\min \|W_1(1 - PQ_{im})\|_\infty = |W_1(5.5308)| = 1.0210$$

Thus the spec $\|W_1 S\|_\infty < 1$ is not achievable for this P and W_1. Let's scale W_1 as

$$W_1 \leftarrow \frac{0.9}{1.0210} W_1$$

Then $|W_1(5.5308)| = 0.9$ and the optimal Q_{im} is

$$Q_{im} = \frac{W_1 - 0.9}{W_1 P} \quad \Rightarrow$$

$$Q_{im}(s) = \frac{s(0.0008s^5 + 0.0221s^4 + 0.1768s^3 + 0.7007s^2 + 3.8910s + 0.0026)}{s^3 + 6.1081s^2 + 6.8897s + 4.9801}$$

Step 3 Set

$$J(s) := \frac{1}{(\tau s + 1)^3}$$

Compute $\|W_1(1 - PQ_{im}J)\|_\infty$ for decreasing values of τ until the norm is < 1:

τ	∞-norm
0.1	1.12
0.05	1.01
0.04	0.988

Take $\tau = 0.04$.

Step 4 $Q = Q_{im}J$

Step 5 $C = Q/(1 - PQ)$

A Bode magnitude plot of the resulting sensitivity function is shown in Figure 3. Figure 4 shows

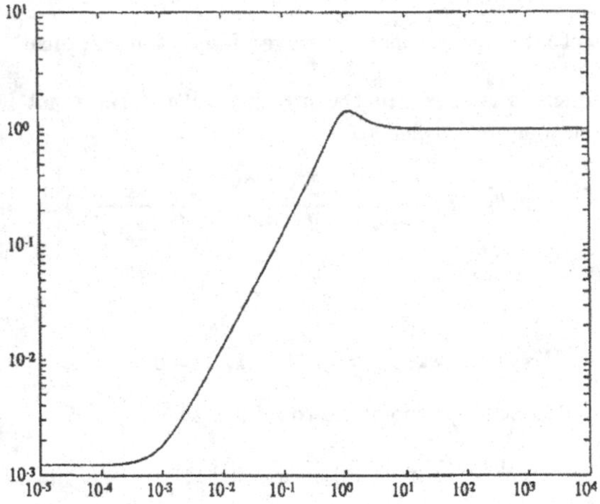

Figure 3: Bode plot of $|S|$

the step response of the plant output together with the ideal step response, i.e., that of T_{id}. The performance specs are met.

The above design, while achieving the step response goal, may not be satisfactory for other reasons; for example, internal signals may be too large (in fact for this design the input to the power amplifier would saturate during the step response test). The problem is that we have placed no limits on controller gain or bandwidth. To handle that we should go to a more complicated formulation, such as the mixed sensitivity problem.

Another disadvantage of the above design is that C is of high order. The common way to alleviate this is to approximate Q by a lower order transfer function (it's difficult to reduce C directly because of the internal stability constraint, whereas Q only has to be stable and proper).

5 Notes

The first section is in the spirit of Bode's book (Bode 1945) on feedback amplifiers. Bode showed that electronic amplifiers must have certain inherent properties by virtue of the fact that stable

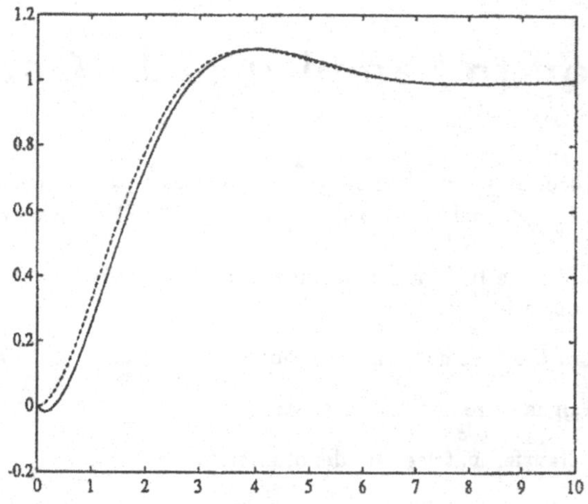

Figure 4: actual (solid) and ideal (dash) step responses

network functions are analytic. Bode's work was generalized to control systems by Bower and Schultheiss (1957) and Horowitz (1963).

The interpolation constraints on S were first obtained by Raggazini and Franklin (1958). Theorem 1 is due to Francis and Zames (1984), but the proof here is due to Freudenberg and Looze (1985). Theorem 2 was proved by Bode (1945) in case L is stable and by Freudenberg and Looze (1985) in the general case.

The material in Section 2 is drawn from Zames 1981, Zames and Francis 1983, Bensoussan 1984, Francis and Zames 1984, Khargonekar and Tannenbaum 1985, and Francis 1987. The extraordinarily useful controller parametrization is due to Youla, Jabr, and Bongiorno Jr. (1976).

Lecture 3

Introduction to Sampled-Data Control Systems

This lecture introduces a general framework for sampled-data control systems and studies some properties of the sample and hold operations. The context from now on is multivariable systems.

We shall work in both the time and frequency domains. So we'll have to be careful with notation. The conventions are as follows:

1. Linear operators in the time domain are denoted P, F, etc.
2. Continuous-time signals are denoted y, u, etc.
3. Continuous-time transfer matrices are denoted $\hat{g}(s)$, etc.
4. Discrete-time signals are denoted ψ, η, etc.
5. Discrete-time transfer matrices are denoted $\hat{k}(\lambda)$, etc. The λ-transform of a sequence $\psi(k)$, i.e., a discrete-time signal, is
$$\hat{\psi}(\lambda) = \sum_i \psi(k)\lambda^i$$
(engineers use $z = 1/\lambda$).

1 The big sampled-data picture

The standard digital setup

From the big picture in Lecture 1 we get the following configuration of a digital control system:

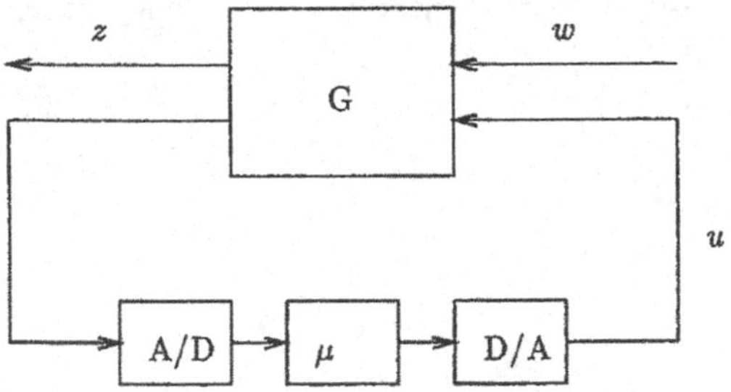

Let's think of a PC-type controller, in which case the blocks would be

A/D: analog-digital card

D/A: digital-analog card

μ: the CPU of the PC

This type of controller has the following properties:

- finite wordlength

- finite time for data acquisition and reconstruction
- overflow and underflow
- finite time for computing

The mathematical idealization of this controller is

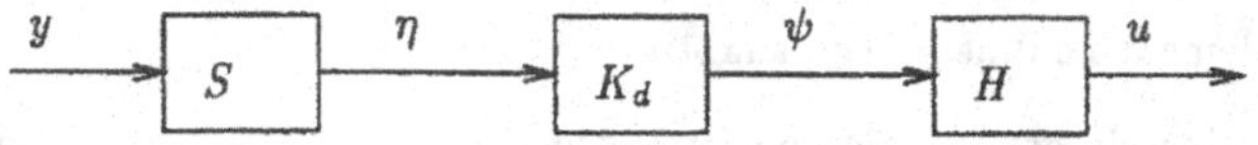

where

S: ideal sampler with fixed sampling period h. Thus

$$\eta(k) = y(kh)$$

K_d: linear, time-invariant, finite-dimensional, discrete-time

H: ideal zero-order hold, synchronized with S. Thus

$$u(t) = \psi(k), \quad kh \leq t < (k+1)h$$

Note that SH is the discrete-time identity operator, but HS isn't the continuous-time identity operator.

This mathematical idealization leads to the next figure.

The standard sampled-data setup

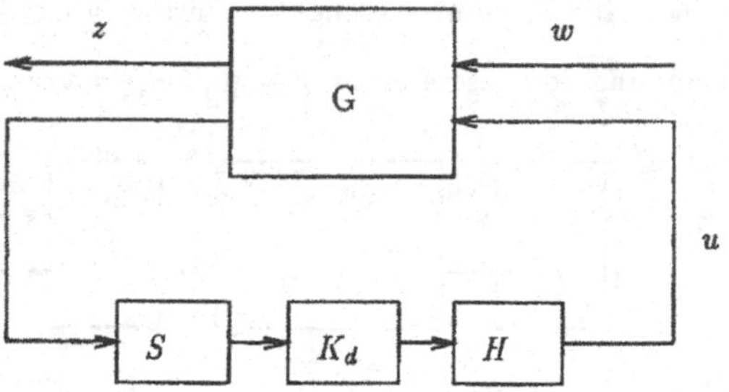

Here G is

finite-dimensional, linear, time-invariant (fdlti), continuous-time

and K_d is

fdlti, discrete-time

Note that the controller, $K := HK_dS$, is not time-invariant, so the standard sampled-data setup represents a linear time-varying system.

There are several reasons for using a digital controller:

1. (relatively) fixed complexity, that is, higher dimensional controllers don't require more hardware
2. reliability
3. flexibility: you can tune on-line by re-setting parameters
4. you can store various controller settings for gain scheduling
5. you can implement nonlinear, e.g., adaptive, controllers
6. you can in principle get superior performance since it's time-varying; for example, you can get deadbeat response with a linear digital controller, but not an analog (finite-dimensional) one

Approaches to digital design

There are essentially three approaches.

1. Do an analog design then a digital implementation. Many such techniques are disreputable (some give an unstable digital controller for a stable analog one). A common reputable implementation is as follows: suppose the analog controller is K; approximate it by $HSKHS$; then set $K_d := SKH$ so that $K = HK_dS$. Advantages of this method:

   - the design is for an lti system in the continuous-time domain, where most intuition resides
   - you can expect to recover the analog performance as $h \longrightarrow 0$

   Disadvantages:

   - the digital performance can't be expected to be better than the analog one
   - to recover the analog performance might require very fast sampling

2. Discretize the plant and do a discrete-time design, i.e., do a discrete-time design for

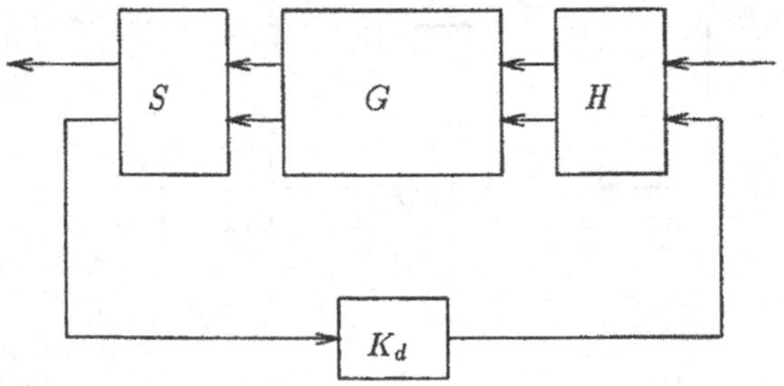

   Advantage:

   - the design is for an lti system (but in the discrete-time domain)

   Disadvantages:

   - intersample behaviour is ignored

- continuous-time performance specs don't always readily translate over to discrete-time specs

3. Direct sampled-data design, i.e., design K_d for the continuous-time hybrid system.

   Advantage:

   - it's the real problem

   Disadvantage:

   - this way is harder because the overall system is time-varying

Our main goal in these lectures is to develop some theory for the third method.

2 Properties of S and H

In this section we're interested in viewing the sampling and hold operations as linear operators.

2.1 Preliminary functional analysis

Definitions

1. For $1 \leq p \leq \infty$, $\mathcal{L}_p$ denotes the Lebesgue space of functions from $(-\infty, \infty)$ to $\mathbf{C}^n$ such that the following norm is finite:

$$\|u\|_p = \begin{cases} \left[\int_{-\infty}^{\infty} \|u(t)\|^p dt\right]^{1/p}, & 1 \leq p < \infty \\ \sup_t \|u(t)\|, & p = \infty \end{cases}$$

   Here the norm on $\mathbf{C}^n$ is the Euclidean norm.

2. $\mathcal{C}$ denotes the space of continuous functions from $(-\infty, \infty)$ to $\mathbf{C}^n$.

3. For $1 \leq p \leq \infty$, $\mathbf{l}_p$ denotes the space of sequences from the integers to $\mathbf{C}^n$ such that the following norm is finite:

$$\|\psi\|_p = \begin{cases} [\sum \|\psi(k)\|^p]^{1/p}, & 1 \leq p < \infty \\ \sup_k \|\psi(k)\|, & p = \infty \end{cases}$$

We're primarily interested in the cases $p = 2, \infty$, for then the p-norm has a physical significance—$\|u\|_2^2$ or $\|\psi\|_2^2$ is energy and $\|u\|_\infty$ or $\|\psi\|_\infty$ is maximum value.

Lti systems

Now we summarize some stability theory for lti systems. Let G be an lti, causal, continuous-time system, that is, it has a representation of the form

$$y(t) = g_0 u(t) + \int_{-\infty}^{t} g_1(t-\tau)u(\tau)d\tau$$

where g_0 is a constant matrix and $g_1(t) = 0$ for $t < 0$. The impulse response function is therefore $\delta(t)g_0 + g_1(t)$.

Theorem 1 *The following are equivalent:*

1. *$G : \mathcal{L}_1 \longrightarrow \mathcal{L}_1$ is bounded*
2. *$G : \mathcal{L}_\infty \longrightarrow \mathcal{L}_\infty$ is bounded*
3. *$G : \mathcal{L}_p \longrightarrow \mathcal{L}_p$ is bounded for every p*
4. *Each column of g_1 is in $\mathcal{L}_1$.*
5. *Assume G is finite-dimensional and let $\left[\begin{array}{c|c} A & B \\ \hline C & D \end{array}\right]$ be a minimal realization of the transfer matrix. Then A is stable (all eigs in the open left half-plane).*

So any one of these five conditions could qualify as a definition of stability of G.

Continuing with such G, define the induced norm

$$M_p := \sup_{\|u\|_p \leq 1} \|Gu\|_p$$

This is the gain of the system from $\mathcal{L}_p$ to $\mathcal{L}_p$. To get an upper bound on this gain, introduce a matrix N as follows: take the ij^{th} element of the impulse response matrix and write it as $g_0\delta(t) + g_1(t)$; then the ij^{th} element of N is defined as $|g_0| + \|g_1\|_1$.

Theorem 2 *Assume G is stable. Then $M_p \leq \sigma_{max}(N)$. Also, $M_2 = \|\hat{g}\|_\infty$.*

Now assume G is lti, strictly causal ($g_0 = 0$), and stable. If we want to sample its ouput, we are interested in knowing that its output is continuous. It is not hard to prove that $G\mathcal{L}_p \subset \mathcal{C}$.

M. Riesz convexity theorem

We shall want to be able to infer $\mathcal{L}_2$-stability from $\mathcal{L}_1$- and $\mathcal{L}_\infty$-stability. Theorem 1 provides this for lti systems. But sampled-data systems are not time-invariant so we need a more powerful result, the M. Riesz convexity theorem.

Let $\mathcal{L}_{1e}$ denote the extended $\mathcal{L}_1$ space, the space of all piecewise continuous functions from $(-\infty, \infty)$ to $\mathbf{C}^n$ such that $\|u(t)\|$ is integrable on every finite time interval. Let's say that u has *finite starting time* if $u(t) = 0$ for t less than some finite time.

Theorem 3 *Suppose G is a linear system with the property that $Gu \in \mathcal{L}_{1e}$ for every input u in $\mathcal{L}_{1e}$ having finite starting time. If*

$G : \mathcal{L}_1 \longrightarrow \mathcal{L}_1$ is bounded, with induced norm M_1, and

$G : \mathcal{L}_\infty \longrightarrow \mathcal{L}_\infty$ is bounded, with induced norm M_∞

then for every p

$G : \mathcal{L}_p \longrightarrow \mathcal{L}_p$ is bounded, with induced norm $M_p \leq M_1^{\frac{1}{p}} M_\infty^{1-\frac{1}{p}}$.

Why is this called a convexity theorem? Note that $M_p \leq M_1^{\frac{1}{p}} M_\infty^{1-\frac{1}{p}}$ iff

$$\ln M_p \leq \frac{1}{p} \ln M_1 + \left(1 - \frac{1}{p}\right) \ln M_\infty$$

or, with $x := 1/p$, $f(x) := \ln M_p$,

$$f(x) \leq x f(1) + (1 - x) f(0)$$

This says that f is a convex function on the interval $[0, 1]$.

2.2 Properties of S and H

Now we derive some properties of S and H.

Property 1 *$S : \mathcal{C} \cap \mathcal{L}_p \longrightarrow \mathbf{l}_p$ is bounded and of norm 1 for $p = \infty$, but it's not bounded for $p < \infty$.*

Proof The case $p = \infty$ is easy. Now let $p < \infty$ and consider the case of scalar-valued functions. Define u to be a series of ever-narrowing triangular pulses as follows:

$$u(t) = \begin{cases} 1 - \frac{2}{h}|t-h|, & h - \frac{h}{2} \le t \le h + \frac{h}{2} \\ 1 - \frac{2\times 2^2}{h}|t-2h|, & 2h - \frac{h}{2\times 2^2} \le t \le 2h + \frac{h}{2\times 2^2} \\ \text{etc.} \end{cases}$$

Then $(Su)(k) = 1$ for all $k \ge 1$, so $Su \notin \mathbf{l}_p$. Yet

$$\begin{aligned} \|u\|_p^p &= \sum_1^\infty 2\int_{kh}^{kh+\frac{h}{2k^2}} \left[1 - \frac{2k^2}{h}(t-kh)\right]^p dt \\ &= \sum_1^\infty 2\frac{h}{2k^2}\int_0^1 \tau^p d\tau \\ &= \frac{h}{p+1}\sum_1^\infty \frac{1}{k^2} \\ &< \infty \end{aligned}$$

So $u \in \mathcal{C} \cap \mathcal{L}_p$. ■

Property 2 *$H : \mathbf{l}_p \longrightarrow \mathcal{L}_p$ is bounded and of norm $h^{1/p}$.*

Proof Let $v \in \mathbf{l}_p$, $y = Hv$. Then

$$\begin{aligned} \|y\|_p^p &= \int_{-\infty}^{\infty} [y(t)^* y(t)]^{p/2} dt \\ &= \sum_k \int_{kh}^{(k+1)h} [v(k)^* v(k)]^{p/2} dt \\ &= \sum_k h[v(k)^* v(k)]^{p/2} \\ &= h\|v\|_p^p \end{aligned}$$

■

From the previous two properties we obtain immediately the third.

Property 3 *$HS : \mathcal{C} \cap \mathcal{L}_p \longrightarrow \mathcal{L}_p$ is bounded and of norm 1 for $p = \infty$, but it's not bounded for $p < \infty$. On the other hand, SH is the identity operator $\mathbf{l}_p \longrightarrow \mathbf{l}_p$.*

The next property relates to the setup

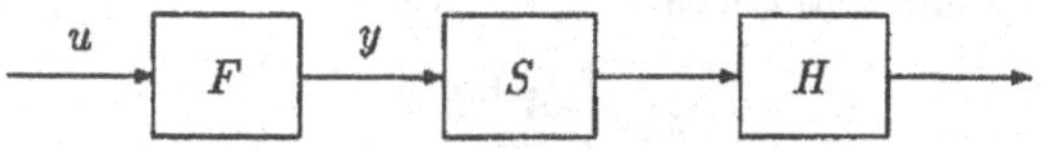

with a pre-filter F, assumed causal and lti with impulse response matrix $f \in \mathcal{L}_{1e}$. Then F maps $\mathcal{L}_{1e}$ to $\mathcal{C}$.

Define the sequence ϕ by

$$\phi(k) := \sup_{t \in [kh,(k+1)h)} \|f(t)\| \tag{1}$$

Property 4 *If the sequence ϕ is in l_1, then the operator HSF is bounded from $\mathcal{L}_p$ to $\mathcal{L}_p$ for every $1 \le p \le \infty$.*

Note that ϕ being in l_1 implies that f belongs to $\mathcal{L}_1$. So F itself is a bounded operator on $\mathcal{L}_p$ for every $1 \le p \le \infty$. The proof uses the standard fact that if g, u are in $\mathcal{L}_1$, so is their convolution $g * u$; moreover

$$\|g * u\|_1 \le \|g\|_1 \cdot \|u\|_1$$

Proof of Property 4 By Theorem 3, it suffices to show that HSF is bounded on both $\mathcal{L}_\infty$ and $\mathcal{L}_1$. It's easy to see that HSF is bounded on $\mathcal{L}_\infty$ with $\|HSF\| \le \|f\|_1$.

Now suppose $u \in \mathcal{L}_1$. For $k \ge 0$,

$$y(kh) = \int_0^{kh} f(kh-\tau)u(\tau)d\tau = \sum_{i=1}^{k} \int_{(i-1)h}^{ih} f(kh-\tau)u(\tau)d\tau$$

So

$$\begin{aligned} \|y(kh)\| &\le \sum_{i=1}^{k} \int_{(i-1)h}^{ih} \|f(kh-\tau)\| \cdot \|u(\tau)\| d\tau \\ &\le \sum_{i=1}^{k} \phi(k-i) \int_{(i-1)h}^{ih} \|u(\tau)\| d\tau \end{aligned}$$

Define the sequence v by $v(i) := \int_{(i-1)h}^{ih} \|u(\tau)\| d\tau$. Then the right-hand side of the above is the convolution of ϕ and v: $\phi \in l_1$ by the hypothesis and $v \in l_1$ since $\|v\|_1 = \|u\|_1 < \infty$. Thus

$$\sum_k \|y(kh)\| \le \|\phi\|_1 \cdot \|v\|_1 = \|\phi\|_1 \cdot \|u\|_1$$

Hence the operator SF is bounded from $\mathcal{L}_1$ to l_1 with $\|SF\| \le \|\phi\|_1$. Since for any $w \in l_1$ we have $\|Hw\|_1 = h\|w\|_1$, it follows that HSF is bounded on $\mathcal{L}_1$ with induced norm $\le h\|\phi\|_1$. ∎

It follows from the proof that an upper bound for the $\mathcal{L}_p$ induced norm of HSF is

$$(h\|\phi\|_1)^{1/p}(\|f\|_1)^{1/q}$$

where $\frac{1}{p} + \frac{1}{q} = 1$. Since $\|f\|_1 \le h\|\phi\|_1$, we obtain that $h\|\phi\|_1$ is an upper bound for $\|HSF\|$ on every $\mathcal{L}_p$ space.

Corollary 1 *If F is fdlti, strictly causal, and stable, then HSF is bounded from $\mathcal{L}_p$ to $\mathcal{L}_p$ for every $1 \le p \le \infty$.*

Proof Such a filter F admits a state-space representation

$$\hat{g} = \left[\begin{array}{c|c} A & B \\ \hline C & 0 \end{array}\right]$$

with A stable. Then $f(t) = Ce^{At}B$, $t \geq 0$. It can be verified that the corresponding ϕ belongs to l_1. ■

The next result establishes uniform convergence of HSF to F as the sampling period approaches zero. Define

$$f_h(t) := \sup_{a \in (0,h)} \|f(t) - f(t-a)\| \tag{2}$$

From equation (1),

$$\phi(0) = \sup_{t \in [0,h)} \|f(t)\|$$

Property 5 *If* $\lim_{h \to 0} \phi(0)$ *is finite and* $\lim_{h \to 0} \|f_h\|_1 = 0$, *then* HSF *converges to* F *as* h *tends to zero in the* $\mathcal{L}_p$ *induced norm for every* $1 \leq p \leq \infty$.

Proof It suffices to show that

$$\lim_{h \to 0} \|(I - HS)F\| = 0$$

if the domain and co-domain spaces are both $\mathcal{L}_p$ for $p = 1, \infty$ respectively.

First let $u \in \mathcal{L}_\infty$. For any $t > 0$, choose k such that $kh < t \leq (k+1)h$. Then since $y = Fu$,

$$\begin{aligned} [(I - HS)Fu](t) &= y(t) - y(kh) \\ &= \int_0^{kh} [f(t-\tau) - f(kh-\tau)]u(\tau)d\tau + \int_{kh}^{t} f(t-\tau)u(\tau)d\tau \end{aligned}$$

Hence

$$\begin{aligned} \|[(I - HS)Fu](t)\| &\leq \int_0^{kh} f_h(t-\tau)\|u(\tau)\|d\tau + \int_{kh}^{t} \|f(t-\tau)\| \cdot \|u(\tau)\|d\tau \qquad (3) \\ &\leq \{\|f_h\|_1 + h\phi(0)\}\|u\|_\infty \end{aligned}$$

The quantity in parentheses is independent of t. Thus it's an upper bound for the $\mathcal{L}_\infty$ induced norm of the operator $(I - HS)F$. By our hypothesis this upper bound tends to 0 as h tends to 0.

Next suppose $u \in \mathcal{L}_1$. Again, for any $t > 0$, choose k such that $kh < t \leq (k+1)h$. From inequality (3),

$$\begin{aligned} \|[(I - HS)Fu](t)\| &\leq \int_0^{t} f_h(t-\tau)\|u(\tau)\|d\tau + \int_{kh}^{t} \|f(t-\tau)\| \cdot \|u(\tau)\|d\tau \\ &\leq (f_h * \|u\|)(t) + \phi(0)\int_{kh}^{(k+1)h} \|u(\tau)\|d\tau \end{aligned}$$

Then

$$\|(I - HS)Fu\|_1 \leq \|f_h\|_1 \cdot \|u\|_1 + h\phi(0)\|u\|_1$$

Thus an upper bound for the $\mathcal{L}_1$ induced norm of the operator $(I - HS)F$ is $\|f_h\|_1 + h\phi(0)$, which tends to 0 as h tends to 0 by our hypothesis. ■

Corollary 2 *If* F *is fdlti, strictly causal, and stable, then* HSF *converges to* F *as* h *tends to zero in the* $\mathcal{L}_p$ *induced norm for every* $1 \leq p \leq \infty$.

Proof Bring in a realization for F as in the proof of Corollary 1 to get $f(t) = Ce^{At}B$, $t \geq 0$, where A is stable. It's clear that $\phi(0)$ is bounded for any $h > 0$. From (2)

$$\begin{aligned} f_h(t) &\leq \|B\| \cdot \|C\| \cdot \|e^{A}t\| \sup_{a \in (0,h)} \|I - e^{-Aa}\| \\ &\leq \|B\| \cdot \|C\|(e^{\|A\|h} - 1)\|e^{At}\| \end{aligned}$$

Since A is stable, the 1-norm of e^{At} is finite. It follows that $\|f_h\|_1 \longrightarrow 0$ at least as fast as $h \longrightarrow 0$. ■

2.3 Application: recovery of internal stability

This subsection looks at an application of the previous properties: we'll see that stability of the digital implementation of an analog system is recovered as the sampling period tends to zero. This is comforting to know, but it doesn't actually provide a very useful stability test.

Start with the setup

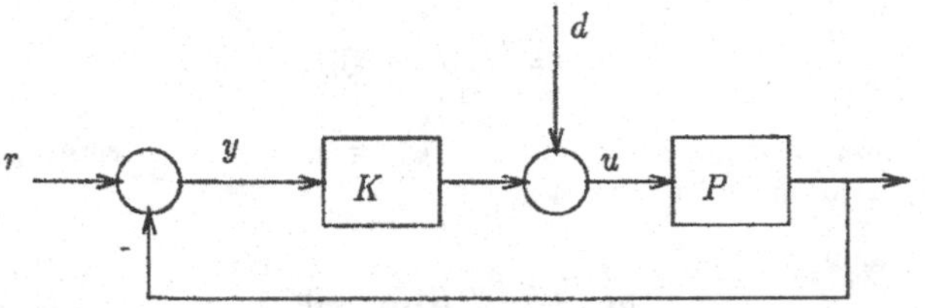

with P and K fdlti and strictly causal. Assume *internal stability* in the sense that the mapping

$$\begin{bmatrix} I & P \\ -K & I \end{bmatrix}^{-1} : \begin{bmatrix} r \\ d \end{bmatrix} \mapsto \begin{bmatrix} y \\ u \end{bmatrix} : \mathcal{L}_2 \longrightarrow \mathcal{L}_2$$

is bounded.

Now do a digital implementation as follows:

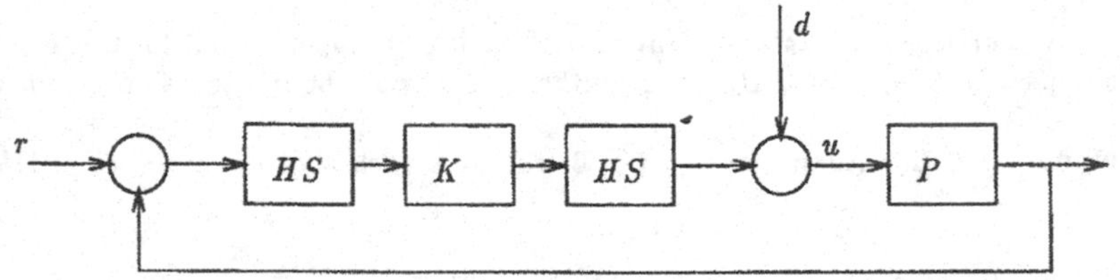

This isn't set up quite right for $\mathcal{L}_2$ stability because the input to the left-hand S isn't filtered. Instead, we'll study stability of

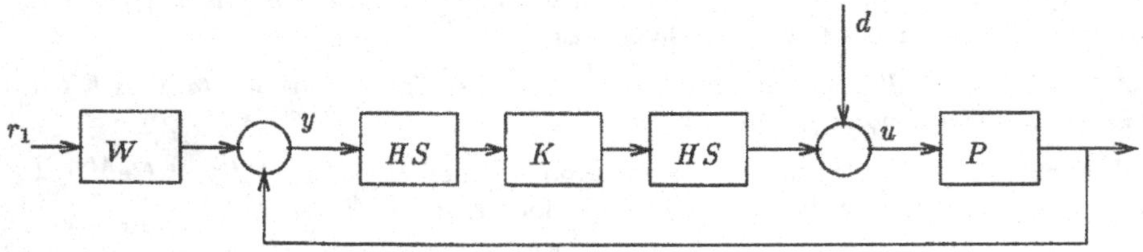

where W is fdlti, strictly causal, and stable.

Reconfigure the last figure to look like

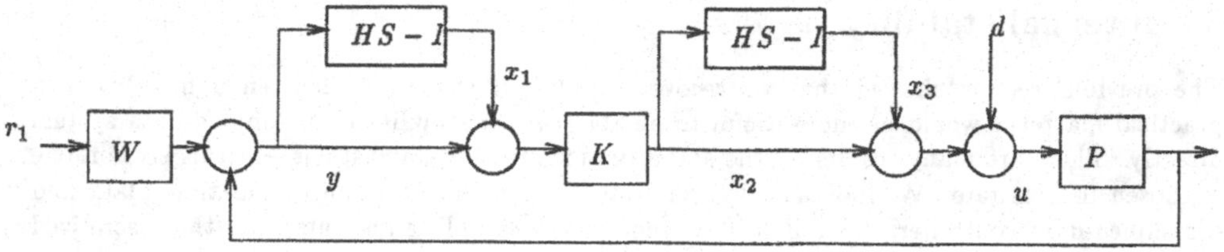

This in turn can be viewed as

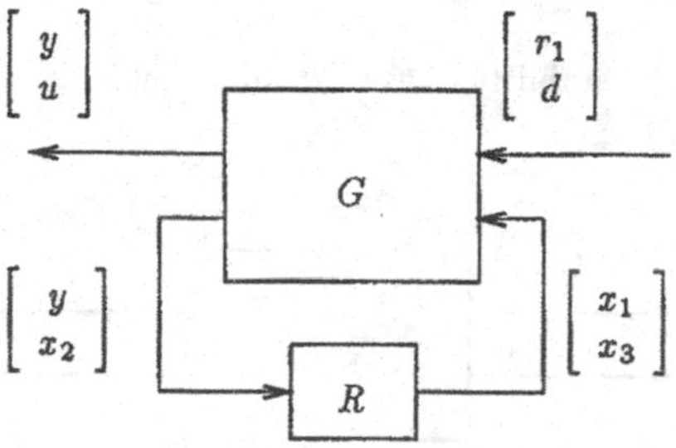

where

$$G = \begin{bmatrix} G_{11} & G_{12} \\ G_{21} & G_{22} \end{bmatrix}$$

$$G_{11} = \begin{bmatrix} (I+PK)^{-1}W & -P(I+KP)^{-1} \\ K(I+PK)^{-1}W & (I+KP)^{-1} \end{bmatrix}$$

$$G_{12} = \begin{bmatrix} -P(I+KP)^{-1}K & -P(I+KP)^{-1} \\ (I+KP)^{-1}K & (I+KP)^{-1} \end{bmatrix}$$

$$G_{21} = \begin{bmatrix} (I+PK)^{-1}W & -P(I+KP)^{-1} \\ K(I+PK)^{-1}W & -KP(I+KP)^{-1} \end{bmatrix}$$

$$G_{22} = \begin{bmatrix} -P(I+KP)^{-1}K & -P(I+KP)^{-1} \\ (I+KP)^{-1}K & -KP(I+KP)^{-1} \end{bmatrix}$$

and

$$R = \begin{bmatrix} HS-I & 0 \\ 0 & HS-I \end{bmatrix}$$

Thus G is fdlti and stable, and R is time-varying.

The mapping from $\begin{bmatrix} r_1 \\ d \end{bmatrix}$ to $\begin{bmatrix} y \\ u \end{bmatrix}$ is

$$G_{11} + G_{12}(I - RG_{22})^{-1}RG_{21}$$

Since G_{21} and G_{22} are strictly causal, we have from Property 4 that RG_{21} and RG_{22} are bounded on $\mathcal{L}_2$; then from Corollary 2 $\|RG_{22}\|$ tends to zero as $h \longrightarrow 0$. Thus by the small-gain theorem $(I - RG_{22})^{-1}$ is bounded for small enough h.

In conclusion, if h is small enough—namely, if $\|RG_{22}\| < 1$—then the sampled-data system is stable (bounded on $\mathcal{L}_2$).

3 Internal stability

The previous section showed that we recover stability if we sample fast enough. But as a practical matter it would be more useful to be able to test stability of a sampled-data system directly. There are many results on the stability of sampled-data systems—in this section one is chosen to illustrate. We look at a hybrid system composed of a continuous-time plant and a sampled-data controller. This hybrid system at the sampling instants, i.e., the discretized system, is time-invariant, hence easy to analyze. The main idea is

stability of the discretized system
+
non-pathological sampling
$\Longrightarrow$
stability of the hybrid system

The system to be studied is

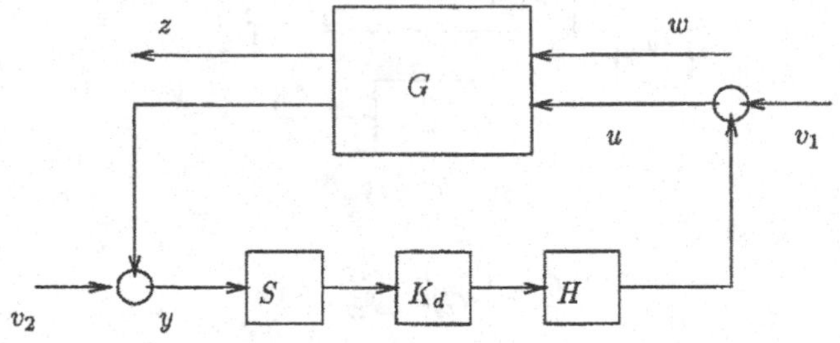

Figure 1

Here G is fdlti continuous-time and K_d is fdlti discrete-time; their transfer matrices, $\hat{g}(s)$ and $\hat{k}_d(\lambda)$, are proper, real-rational.

To define internal stability for this system bring in a low-pass filter F, either ideal or with transfer matrix $\hat{f}(s)$ strictly proper and real-rational. Then *internal stability* in Figure 1 means boundedness of the mapping

$$\begin{bmatrix} w \\ v_1 \\ v_2 \end{bmatrix} \mapsto \begin{bmatrix} z \\ u \\ y \end{bmatrix} : F\mathcal{L}_2 \longrightarrow \mathcal{L}_2$$

It turns out that this doesn't depend on the particular F; it's simply that the input to a sampler should be low-pass filtered.

Bring in minimal realizations

$$\hat{g}(s) = D + C(sI - A)^{-1}B =: \left[\begin{array}{c|c} A & B \\ \hline C & D \end{array}\right] = \left[\begin{array}{c|c} A & \begin{bmatrix} B_1 & B_2 \end{bmatrix} \\ \hline \begin{bmatrix} C_1 \\ C_2 \end{bmatrix} & \begin{bmatrix} D_{11} & D_{12} \\ D_{21} & D_{22} \end{bmatrix} \end{array}\right]$$

$$\hat{k}_d(\lambda) = D_K + \lambda C_K(I - \lambda A_K)^{-1}B_K =: \left[\begin{array}{c|c} A_K & B_K \\ \hline C_K & D_K \end{array}\right]$$

The system G is said to be *stabilizable* if (A, B_2) is stabilizable and (C_2, A) is detectable. The idea is that in Figure 1 the unstable modes of G should be controllable through u and observable from y.

In Figure 1 set the inputs to zero and look at the internal loop:

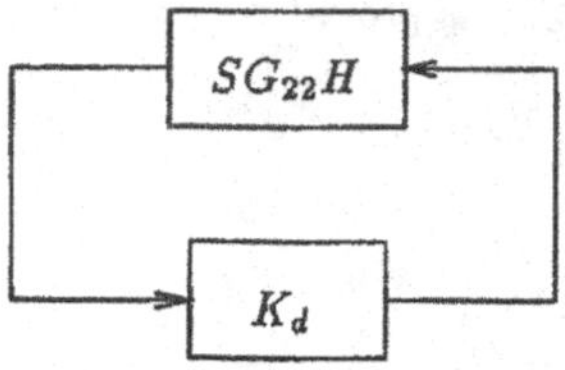

Figure 2

This system is lti discrete-time. The components have the realizations

$$\left[\begin{array}{c|c} e^{hA} & \int_0^h e^{\tau A} d\tau B_2 \\ \hline C_2 & D_{22} \end{array}\right] \text{ for } SG_{22}H$$

$$\left[\begin{array}{c|c} A_K & B_K \\ \hline C_K & D_K \end{array}\right] \text{ for } K$$

The system in Figure 2 is *well-posed* if the matrix

$$\begin{bmatrix} I & -D_K \\ -D_{22} & I \end{bmatrix}$$

is invertible. To find the A-matrix in Figure 2 it's useful to observe the following result.

If two systems with realizations

$$\left[\begin{array}{c|c} A_1 & B_1 \\ \hline C_1 & D_1 \end{array}\right], \quad \left[\begin{array}{c|c} A_2 & B_2 \\ \hline C_2 & D_2 \end{array}\right]$$

are connected—the output of the first connected to the input of the second, and vice versa—the A-matrix of the resulting system is

$$\begin{bmatrix} A_1 & 0 \\ 0 & A_2 \end{bmatrix} + \begin{bmatrix} B_1 & 0 \\ 0 & B_2 \end{bmatrix} \begin{bmatrix} I & -D_2 \\ -D_1 & I \end{bmatrix}^{-1} \begin{bmatrix} 0 & C_2 \\ C_1 & 0 \end{bmatrix}$$

This formula is valid for both continuous and discrete time.

Define the system in Figure 2 to be *internally stable* if all eigenvalues of its A-matrix are in the open unit disk.

Finally, we need the notion of *non-pathological sampling*: it means that the distance between any two eigenvalues of A on the same vertical line isn't a multiple of the sampling frequency, $2\pi/h$. This guarantees that there is no loss of controllability or observability through sampling.

Theorem 4 *Assume well-posedness. If G is stabilizable, the sampling is non-pathological, and the system in Figure 2 is internally stable, then the system in Figure 1 is internally stable.*

4 Notes

Section 2 is taken from Chen and Francis 1989. Theorems 1 and 2 can be found for example in Vidyasagar 1978. Theorem 3 is taken from Stein and Weiss 1971. The non-pathological sampling condition is due to Kalman, Ho and Narendra (1963). Theorem 4 is from Leung, Perry, and Francis 1989. A similar result for $\mathcal{L}_\infty$ was proved by Francis and Georgiou (1988).

Lecture 4

Norms of Sampled-Data Systems

As discussed in the previous lecture, the sampled-data system of interest is as follows:

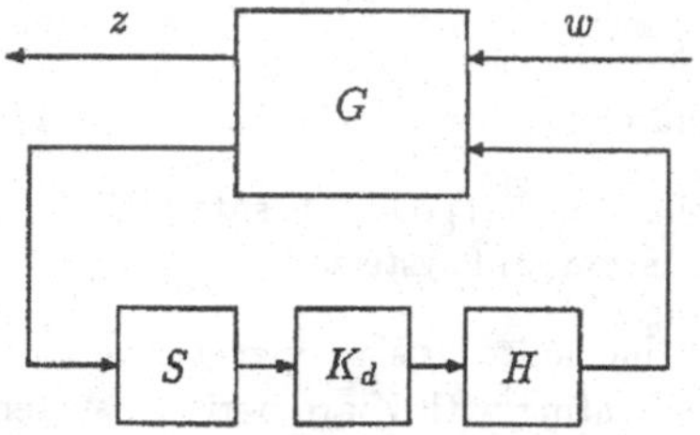

Figure 1

This system, being time-varying, doesn't have a transfer matrix whose $\mathcal{H}_\infty$-norm can be evaluated. Instead, it makes sense to look at the induced norm of the map $T_{zw} : w \mapsto z$ from $\mathcal{L}_2$ to $\mathcal{L}_2$. If the feedback system is internally stable (and under mild conditions on G), T_{zw} is a bounded operator from $\mathcal{L}_2$ to $\mathcal{L}_2$. How to minimize its norm, or even how to compute its norm for a given controller, is a subject of current research. The goal in this lecture is a first step: to derive formulas for the induced norms of the systems

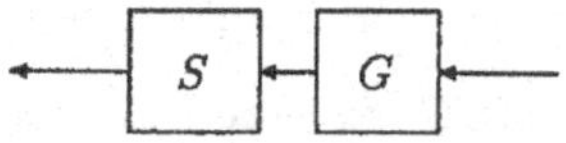

Figure 2

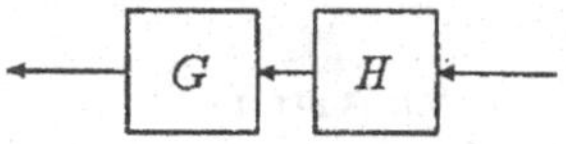

Figure 3

where G is a fdlti, stable, continuous-time system. In Figure 2 G must be strictly causal—the input to a sampler must be low-pass filtered for the resulting operator to be bounded—but in Figure 3 it need only be just causal. In Figure 2, SG maps $\mathcal{L}_2$ to l_2. Similarly, in Figure 3, GH maps l_2 to $\mathcal{L}_2$. Formulas for $\|SG\|$ and $\|GH\|$ lead in turn to a formula for the induced norm of the system

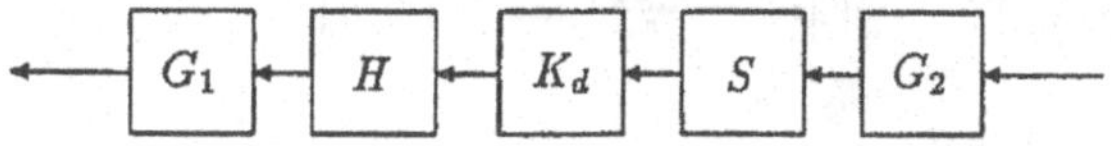

Figure 4

Here G_1 and G_2 are continuous-time and K_d is discrete-time: all three are linear time-invariant and stable. These formulas can be used to provide solutions to certain optimal sampled-data control problems, e.g., minimizing the induced norm from $\mathcal{L}_2$ to $\mathbf{l}_2$ (or from $\mathbf{l}_2$ to $\mathcal{L}_2$) of a pre-specified closed-loop operator over all stabilizing digital controllers K_d as in Figure 1.

1 Norm of SG

Introduce the following two operators:

1. U is the unit time delay (bilateral shift) on $\mathbf{l}_2$. It satisfies $U^*U = I$, $UU^* = I$. So operators commuting with U are time-invariant systems.

2. V denotes time delay by h time units. It's an operator on $\mathcal{L}_2$ and it too satisfies $V^*V = I$, $VV^* = I$. So operators commuting with V are periodic systems. Example: HH^*.

Suppose G is an operator on $\mathcal{L}_2$ which commutes with V and such that SG is bounded from $\mathcal{L}_2$ to $\mathbf{l}_2$. Then SG intertwines U and V, i.e., $(SG)V = U(SG)$. (**Proof** $SGV = SVG = USG$.) It follows that $(SG)(SG)^*$ is time-invariant (though non-causal).

Proof

$$
\begin{aligned}
(SG)V = U(SG) &\implies U^*(SG)V = (SG) \\
&\implies [U^*(SG)V][U^*(SG)V]^* = (SG)(SG)^* \\
&\implies U^*(SG)(SG)^*U = (SG)(SG)^*
\end{aligned}
$$

■

It follows that, for such G, the norm of SG can be computed by finding the transfer function of the lti system $(SG)(SG)^*$.

In detail, start with G having the transfer function

$$\hat{g}(s) = \left[\begin{array}{c|c} A & B \\ \hline C & 0 \end{array}\right]$$

with A stable (this G is of course lti, which is stronger than just commutativity with V). Let L denote the controllability grammian for the pair (A, B) and define the two discrete transfer functions

$$
\begin{aligned}
\hat{r}(\lambda) &= \left[\begin{array}{c|c} e^{hA} & LC' \\ \hline Ce^{hA} & 0 \end{array}\right] \\
\hat{q} &= CLC' + \hat{r} + \hat{r}^{\sim}
\end{aligned}
$$

where $\hat{r}^{\sim}(\lambda) := \hat{r}(1/\lambda)'$. Since A is stable, $\hat{r}$ is analytic on the closed unit disk; hence $\hat{q}$ is analytic on the unit circle and the norm

$$\|\hat{q}\|_\infty := \max_\theta \sigma_{\max}\left[\hat{q}\left(e^{j\theta}\right)\right]$$

is well-defined.

Theorem 1 *The transfer function of* $(SG)(SG)^*$ *is* $\hat{q}$.

Proof Letting $g(t)$ denote the impulse response function for G, we have

$$\psi = SGu \Longleftrightarrow \psi(k) = \int_{-\infty}^{\infty} g(kh-\tau)u(\tau)\, d\tau$$

Now we determine the adjoint operator

$$(SG)^* : \mathbf{l}_2 \longrightarrow \mathcal{L}_2$$

Let $v \in \mathbf{l}_2$, $u \in \mathcal{L}_2$, and set $y = (SG)^*v$. Then

$$\langle (SG)^*v, u\rangle = \int_{-\infty}^{\infty} y(\tau)' u(\tau)\, d\tau$$

and

$$\begin{aligned}\langle v, SGu\rangle &= \sum_{k=-\infty}^{\infty} v(k)' \int_{-\infty}^{\infty} g(kh-\tau)u(\tau)\, d\tau \\ &= \int_{-\infty}^{\infty} \left[\sum_{k=-\infty}^{\infty} v(k)' g(kh-\tau)\right] u(\tau)\, d\tau\end{aligned}$$

so

$$y(\tau) = \sum_{k=-\infty}^{\infty} g(kh-\tau)' v(k)$$

This equation defines $(SG)^*$, the map $v \mapsto y$.

Next, we obtain the matrix representation Q of $(SG)(SG)^* : \mathbf{l}_2 \longrightarrow \mathbf{l}_2$ with respect to the standard basis. Let $\psi = SGy, y = (SG)^*v$. Then

$$\begin{aligned}\psi(k) &= \int_{-\infty}^{\infty} g(kh-\tau) y(\tau)\, d\tau \\ &= \int_{-\infty}^{\infty} g(kh-\tau) \sum_{l=-\infty}^{\infty} g(lh-\tau)' v(l)\, d\tau \\ &= \sum_{l=-\infty}^{\infty} \left(\int_{-\infty}^{\infty} g(kh-\tau) g(lh-\tau)'\, d\tau\right) v(l)\end{aligned}$$

It follows that the kl^{th}-block of Q is

$$Q_{kl} = \int_{-\infty}^{\infty} g(kh-\tau) g(lh-\tau)'\, d\tau$$

Notice that Q is (block) Toeplitz, so we can write Q_{k-l} instead of Q_{kl}.

Now we have to bring in the realization of $\hat{g}$ and determine the matrix Q explicitly. Since

$$g(t) = Ce^{tA}B1(t)$$

($1(t)$ is the unit step) we have

$$Q_k = C\int_{-\infty}^{\infty} e^{(kh-\tau)A} BB' e^{-\tau A'} 1(kh-\tau)1(-\tau)\, d\tau C'$$

Thus

$$\begin{aligned}Q_0 &= C\int_{-\infty}^{0} e^{-\tau A} BB' e^{-\tau A'}\, d\tau C' \\ &= CLC'\end{aligned}$$

Also, for $k > 0$

$$\begin{aligned} Q_k &= C\int_{-\infty}^{0} e^{(kh-\tau)A}BB'e^{-\tau A'}\,d\tau C' \\ &= Ce^{khA}LC' \end{aligned}$$

and similarly for $k < 0$

$$Q_k = CLe^{-khA'}C'$$

It follows that the transfer function of $(SG)(SG)^*$ equals

$$\cdots + \frac{1}{\lambda}Q_{-1} + Q_0 + \lambda Q_1 + \cdots$$

Finally, it is routine to check that this equals $\hat{q}$. ■

Corollary 1 *The norm of* $SG : \mathcal{L}_2 \longrightarrow \mathrm{l}_2$ *equals* $\|\hat{q}\|_\infty^{1/2}$.

Another way to express Theorem 1 is as follows. Let $\hat{w}$ be a co-spectral factor of $\hat{q}$, that is, $\hat{w}$ is stable and outer and satisfies

$$\hat{q} = \hat{w}\hat{w}^\sim$$

and let W denote the corresponding linear system. Then

$$(SG)(SG)^* = WW^*$$

so the sampled-data system SG, mapping $\mathcal{L}_2$ to l_2, and the discrete-time system W, mapping l_2 to l_2, have equal norm—the norm of W equaling $\|\hat{w}\|_\infty = \|\hat{q}\|_\infty^{1/2}$. The operator $T := W^{-1}(SG)$ is a co-isometry (its adjoint is an isometry) mapping $\mathcal{L}_2$ to l_2. It too intertwines U and V.

Corollary 2 *The norm of* $HSG : \mathcal{L}_2 \longrightarrow \mathcal{L}_2$ *equals* $\sqrt{h}\|\hat{q}\|_\infty^{1/2}$.

Proof Follows directly from the theorem and the fact that, for any ψ in l_2, the $\mathcal{L}_2$-norm of $H\psi$ equals $\sqrt{h}$ times the l_2-norm of ψ. ■

Example Consider the operator HSG with

$$\hat{g}(s) = \frac{1}{s+1} = \left[\begin{array}{c|c} -1 & 1 \\ \hline 1 & 0 \end{array}\right]$$

Then $L = 1/2$ and

$$\begin{aligned} \hat{r}(\lambda) &= \frac{e^{-h}}{2}\frac{\lambda}{1-\lambda e^{-h}} \\ \hat{q}(\lambda) &= \frac{1-e^{-2h}}{2}\frac{\lambda}{(1-\lambda e^{-h})(\lambda - e^{-h})} \\ \hat{w}(\lambda) &= \left[\frac{1-e^{-2h}}{2}\right]^{1/2}\frac{1}{1-\lambda e^{-h}} \end{aligned}$$

so

$$\begin{aligned} \|HSG\|^2 &= h\frac{1-e^{-2h}}{2}\frac{1}{(1-e^{-h})^2} \\ &= \frac{h}{2}\frac{1+e^{-h}}{1-e^{-h}} \end{aligned}$$

Thus

$$\|HSG\| \approx \begin{cases} 1 = \|G\|, & h \to 0 \\ \sqrt{h/2}, & h \to \infty \end{cases}$$

2 Norm of GH

Let's begin by observing that the adjoint of H^* is given as follows:

$$\eta = H^* y \iff \eta(k) = \int_{kh}^{(k+1)h} y(t)dt$$

Thus $H^*H = hI$; in particular, H^*H is an lti, discrete-time system.

More generally, suppose G is a bounded operator on $\mathcal{L}_2$ commuting with V. Then GH intertwines U and V:

$$V(GH) = (GH)U$$

Hence $(GH)^*(GH)$ is lti, discrete-time.

The formula for the norm of GH is much more complicated than the one for SG in the previous section. It's not necessary for G to be strictly proper, so start with

$$\hat{g}(s) = \left[\begin{array}{c|c} A & B \\ \hline C & D \end{array}\right]$$

with A stable. Define the following matrices:

$$\begin{aligned}
D_1 &= D - CA^{-1}B \\
E &= A^{-1}\left(e^{hA} - I\right) \\
C_1 &= CA^{-1} \\
C_2 &= C_1\left(I - e^{-hA}\right) \\
L_1 &= \int_0^h e^{tA'} C_1' C_1 e^{tA}\, dt \\
L_2 &= \int_0^h e^{tA'} C_1' C_2 e^{tA}\, dt \\
M &= \int_h^\infty e^{tA'} C_2' C_2 e^{tA}\, dt \\
R_0 &= hD_1'D_1 + (D_1'C_1EB) + (D_1'C_1EB)' + B'(L_1 + M)B \\
N &= D_1'C_2E + B'(L_2 + M)
\end{aligned}$$

Finally, define the discrete transfer matrices

$$\begin{aligned}
\hat{r}(\lambda) &= \left[\begin{array}{c|c} e^{hA} & B \\ \hline Ne^{hA} & 0 \end{array}\right] \\
\hat{q} &= R_0 + \hat{r} + \hat{r}^{\sim}
\end{aligned}$$

Theorem 2 *The transfer function of $(GH)^*(GH)$ is $\hat{q}$.*

Proof Letting $g(t)$ denote the impulse response function for G, we have $y = GHv$ iff

$$\begin{aligned}
y(t) &= \int_{-\infty}^{\infty} g(t-\tau)(Hv)(\tau)\, d\tau \\
&= \sum_{k=-\infty}^{\infty} \int_{kh}^{(k+1)h} g(t-\tau)\, d\tau\; v(k) \\
&= \sum_{k=-\infty}^{\infty} \int_{t-(k+1)h}^{t-kh} g(\tau)\, d\tau\; v(k)
\end{aligned}$$

Define

$$p(t) = \int_{t-h}^{t} g(\tau)\, d\tau$$

Since G is a bounded operator on $\mathcal{L}_2$, it follows that each column of the matrix $p(t)$ belongs to $\mathcal{L}_2$. We have $y = GHv$ iff

$$y = \sum_{k=-\infty}^{\infty} (V^k p)v(k)$$

It follows that the adjoint operator $(GH)^* : \mathcal{L}_2 \longrightarrow \mathrm{l}_2$ is given by

$$\psi = (GH)^* y \iff \psi(k) = \int_{-\infty}^{\infty} (V^k p)(t)' y(t)\, dt$$

Then the matrix representation Q of $(GH)^*(GH) : \mathrm{l}_2 \longrightarrow \mathrm{l}_2$ with respect to the standard basis is given by

$$Q_{kl} = \int_{-\infty}^{\infty} (V^k p)(t)'(V^l p)(t)\, dt$$

Again, Q is Toeplitz, so we can write Q_{k-l} instead of Q_{kl}.

Bringing in state-space representations, we have

$$g(t) = D\delta(t) + Ce^{tA}B1(t)$$

and so

$$\begin{aligned} p(t) &= \begin{cases} 0, & t \leq 0 \\ D + CA^{-1}\left(e^{tA} - I\right)B, & 0 < t < h \\ CA^{-1}\left(I - e^{-hA}\right)e^{tA}B, & t \geq h \end{cases} \\ &= \begin{cases} 0, & t \leq 0 \\ D_1 + C_1 e^{tA}B, & 0 < t < h \\ C_2 e^{tA}B, & t \geq h \end{cases} \end{aligned}$$

Thus

$$\begin{aligned} Q_0 &= \int_{-\infty}^{\infty} p(t)'p(t)\, dt \\ &= \int_0^h \left(D_1 + C_1 e^{tA}B\right)'\left(D_1 + C_1 e^{tA}B\right)\, dt + \int_h^{\infty} \left(C_2 e^{tA}B\right)'\left(C_2 e^{tA}B\right)\, dt \\ &= R_0 \end{aligned}$$

for $k > 0$

$$\begin{aligned} Q_k &= \int_{-\infty}^{\infty} (V^k p)(t)'p(t)\, dt \\ &= \int_{kh}^{(k+1)h} \left[D_1 + C_1 e^{(t-kh)A}B\right]' C_2 e^{tA}B\, dt + \int_{(k+1)h}^{\infty} \left[C_2 e^{(t-kh)A}B\right]' C_2 e^{tA}B\, dt \\ &= Ne^{khA}B \end{aligned}$$

and similarly for $k < 0$

$$Q_k = \left(Ne^{-khA}B\right)'$$

Finally, the transfer function of $(GH)^*(GH)$ equals

$$\cdots + \frac{1}{\lambda}Q_{-1} + Q_0 + \lambda Q_1 + \cdots$$

which in turn equals $\hat{q}$. ∎

Corollary 3 *The norm of $GH : l_2 \longrightarrow \mathcal{L}_2$ equals $\|\hat{q}\|_\infty^{1/2}$.*

As with Theorem 1, Theorem 2 has a factorization form. Let $\hat{w}$ be a spectral factor of $\hat{q}$, that is, $\hat{w}$ is stable and outer and satisfies

$$\hat{q} = \hat{w}^\sim \hat{w}$$

and let W denote the corresponding linear system. Then

$$(GH)^*(GH) = W^*W$$

so the sampled-data system GH, mapping l_2 to $\mathcal{L}_2$, and the discrete-time system W, mapping l_2 to l_2, have equal norm. Moreover, $T := (GH)W^{-1}$ is an isometry.

3 Norm of $G_1HK_dSG_2$

In this section we bring together the results in the preceding two sections to show that the system in Figure 4 has the same norm as an associated discrete-time, time-invariant system. Suppose G_1 and G_2 are continuous-time linear systems which are fdlti and stable, with G_1 causal and G_2 strictly causal. Suppose also that K_d is a discrete-time linear system, bounded from l_2 to l_2.

As before, bring in stable, discrete-time, linear time-invariant systems W_1 and W_2 satisfying the equations

$$\begin{aligned}(G_1H)^*(G_1H) &= W_1^*W_1\\ (SG_2)(SG_2)^* &= W_2W_2^*\end{aligned}$$

Then

$$\|G_1HK_dSG_2\| = \|W_1K_dW_2\|$$

Proof

$$\begin{aligned}\|G_1HK_dSG_2\|^2 &= \|(SG_2)^*K_d^*(G_1H)^*(G_1H)K_d(SG_2)\|\\ &= \|(SG_2)^*K_d^*W_1^*W_1K_d(SG_2)\|\\ &= \|W_1K_d(SG_2)\|^2\\ &= \|W_1K_d(SG_2)(SG_2)^*K_d^*W_1^*\|\\ &= \|W_1K_dW_2W_2^*K_d^*W_1^*\|\\ &= \|W_1K_dW_2\|^2\end{aligned}$$

■

The linear system $W_1K_dW_2$ is time-invariant, so its norm is readily computed.

4 Optimal sampled-data control

A direct application of the above formulas is to optimal control of sampled-data systems. Here our aim is to design a controller to provide closed-loop stability and minimize the $\mathcal{L}_2$ to l_2 (or l_2 to $\mathcal{L}_2$) induced norm of a closed-loop operator.

With reference to the sampled-data control system in Figure 1, assume G is finite-dimensional, linear time-invariant, and causal with a compatible operator matrix

$$G = \begin{bmatrix} G_{11} & G_{12} \\ G_{21} & G_{22} \end{bmatrix}$$

To bring in the notion of hybrid stability for this system, move the operators H and S in Figure 1 into the generalized plant and introduce two other discrete-time exogenous signals v_1 and v_2 to get

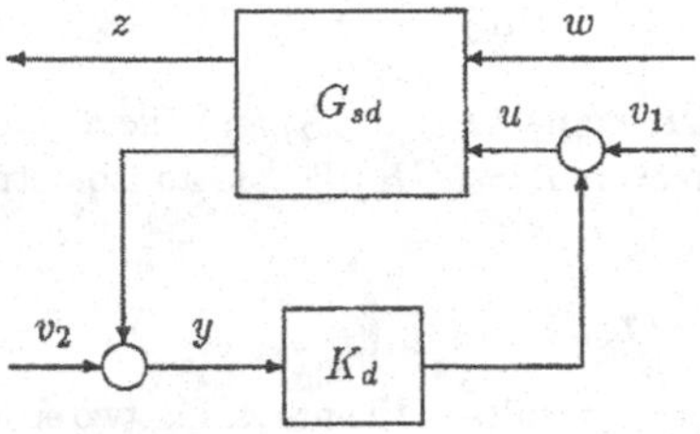

Figure 5

where u is the control sequence, y the sampled measurement, and

$$G_{sd} := \begin{bmatrix} G_{11} & G_{12}H \\ SG_{21} & SG_{22}H \end{bmatrix}$$

We use the two Hilbert spaces $\mathcal{L}_2$ and l_2 for the continuous-time and discrete-time signal spaces respectively. Thus the exogenous input w lives in $\mathcal{L}_2$ and v_1, v_2 in l_2.

The sampled-data feedback system is *hybrid stable* if the nine operators defined from w, v_1, v_2 to z, u, y in Figure 5 are all bounded on their appropriate spaces. Under this definition, a continuous-time system is *hybrid stabilizable* if there exists a digital controller to achieve hybrid stability.

For the system G to be hybrid stabilizable, it's necessary that G_{21} be strictly causal. Assume G is stabilizable (defined in Lecture 3). The discretized version of G is

$$\tilde{G} := SGH = \begin{bmatrix} SG_{11}H & SG_{12}H \\ SG_{21}H & SG_{22}H \end{bmatrix} =: \begin{bmatrix} \tilde{G}_{11} & \tilde{G}_{12} \\ \tilde{G}_{21} & \tilde{G}_{22} \end{bmatrix}$$

Assume also that $\tilde{G}$ is stabilizable. (This is true for non-pathological sampling.) It can be shown that a digital controller K_d achieves hybrid stability for G iff K_d stabilizes the discrete-time system $\tilde{G}_{22}$.

Now we can state precisely our $\mathcal{L}_2/\mathrm{l}_2$ optimal control problem as follows. Given a continuous-time plant G, design a digital controller to achieve hybrid stability and minimize the $\mathcal{L}_2$ to l_2 norm of the operator ST_{zw}. Compared with the conventional approach of dealing with the system only at its sampling instants, our treatment here is a step forward since we deal with the continuous-time input directly. We'll see that this problem can be converted into an equivalent discrete-time $\mathcal{H}_\infty$ problem.

To obviate the need for coprime factorization, we further assume that G is stable. Then the set of fdlti causal digital controllers which provide hybrid stability for G is parametrized as

$$K_d = -Q(I - \tilde{G}_{22}Q)^{-1}, \qquad \hat{q} \in \mathcal{RH}_\infty$$

where $\hat{q}$ is the discrete-time transfer matrix for Q and $\mathcal{RH}_\infty$ is with respect to the unit disk.

By this controller parametrization, it's easy to derive that

$$T_{zw} = G_{11} - G_{12} H Q S G_{21}$$

Thus our $\mathcal{L}_2/\mathrm{l}_2$ control problem is to minimize

$$\|ST_{zw}\| = \|SG_{11} - \tilde{G}_{12} Q S G_{21}\|$$

over all Q whose transfer matrix $\hat{q}$ lies in $\mathcal{RH}_\infty$. For this specification to be well-defined, we assume G_{11} too is strictly causal.

Write

$$ST_{zw} = \begin{bmatrix} I & -\tilde{G}_{12} Q \end{bmatrix} S \begin{bmatrix} G_{11} \\ G_{21} \end{bmatrix}$$

Bring in an lti, stable, discrete-time system W satisfying

$$\left(S \begin{bmatrix} G_{11} \\ G_{21} \end{bmatrix} \right) \left(S \begin{bmatrix} G_{11} \\ G_{21} \end{bmatrix} \right)^* = WW^*$$

Partition W accordingly to get

$$W = \begin{bmatrix} W_1 \\ W_2 \end{bmatrix}$$

We have the following result:

Theorem 3 *Under the foregoing assumptions, the $\mathcal{L}_2/\mathrm{l}_2$ optimal sampled-data control problem is equivalent to the following discrete-time $\mathcal{H}_\infty$ problem*

$$\inf_{\hat{q} \in \mathcal{RH}_\infty} \|W_1 - \tilde{G}_{12} Q W_2\|$$

the norm taken is the l_2-induced one.

Proof

$$\begin{aligned}
\|ST_{zw}\|^2 &= \left\| \begin{bmatrix} I & -\tilde{G}_{12} Q \end{bmatrix} S \begin{bmatrix} G_{11} \\ G_{21} \end{bmatrix} \right\|^2 \\
&= \left\| \begin{bmatrix} I & -\tilde{G}_{12} Q \end{bmatrix} S \begin{bmatrix} G_{11} \\ G_{21} \end{bmatrix} \left(S \begin{bmatrix} G_{11} \\ G_{21} \end{bmatrix} \right)^* \begin{bmatrix} I & -\tilde{G}_{12} Q \end{bmatrix}^* \right\| \\
&= \left\| \begin{bmatrix} I & -\tilde{G}_{12} \end{bmatrix} WW^* \begin{bmatrix} I & -\tilde{G}_{12} Q \end{bmatrix}^* \right\| \\
&= \left\| \begin{bmatrix} I & -\tilde{G}_{12} Q \end{bmatrix} \begin{bmatrix} W_1 \\ W_2 \end{bmatrix} \right\|^2 \\
&= \|W_1 - \tilde{G}_{12} Q W_2\|^2
\end{aligned}$$

■

Since $W_1, W_2, \tilde{G}_{12}$ all have transfer matrices in $\mathcal{RH}_\infty$, the latter optimization problem can be readily solved in the frequency domain via

$$\inf_{\hat{q} \in \mathcal{RH}_\infty} \|\hat{w}_1 - \hat{\tilde{p}}_{12} \hat{q} \hat{w}_2\|_\infty$$

using standard $\mathcal{H}_\infty$ techniques.

Example Consider a siso, single-loop, continuous-time system G with transfer matrix

$$\hat{g}(s) = \begin{bmatrix} \frac{1}{s+1} & \frac{1}{s+2} \\ \frac{1}{s+1} & \frac{1}{s+2} \end{bmatrix}$$

Then $\tilde{G}_{12} = \tilde{G}_{22}$ with transfer function easily computed as

$$\frac{\lambda(1-e^{-2h})}{2(1-\lambda e^{-2h})}$$

Since the transfer function for G_{11} is $\frac{1}{s+1}$, it follows from that

$$(SG_{11})(SG_{11})^* = W_1 W_1^*$$

where the discrete-time system W_1 has a transfer function

$$\hat{w}_1 = \left(\frac{1-e^{-2h}}{2}\right)^{1/2} \frac{\lambda}{1-\lambda e^{-h}}$$

Since $G_{11} = G_{21}$, it follows that

$$\left(S\begin{bmatrix} G_{11} \\ G_{21} \end{bmatrix}\right)\left(S\begin{bmatrix} G_{11} \\ G_{21} \end{bmatrix}\right)^* = \begin{bmatrix} W_1 \\ W_1 \end{bmatrix}\begin{bmatrix} W_1 \\ W_1 \end{bmatrix}^*$$

Thus by the preceding theorem, optimizing the $\mathcal{L}_2$ to l_2 norm of ST_{zw} leads to finding $\hat{q}$ in $\mathcal{RH}_\infty$ to minimize

$$\left(\frac{1-e^{-2h}}{2}\right)^{1/2} \left\| \frac{\lambda}{1-\lambda e^{-h}} \left(1 - \frac{\lambda(1-e^{-2h})}{2(1-\lambda e^{-2h})}\hat{q}\right)\right\|_\infty$$

This is a standard $\mathcal{H}_\infty$ problem and hence can be solved for any given sampling period h.

There's a similar dual case, the $l_2/\mathcal{L}_2$ optimal control problem.

5 Notes

This lecture is based on Chen and Francis 1990b. The first work along these lines is by Thompson, Stein, and Athans (1983). They derived an upper bound on $\|HK_dSG\|$, with G a low-pass (anti-aliasing) filter; their bound is exact in the siso case. Recently, Kabamba and Hara (1990) gave a state-space iterative procedure for computing the $\mathcal{L}_2$ induced norm in Figure 1.

Lecture 5

$\mathcal{H}_2$ Optimal Sampled-Data Control

The setup in this lecture is

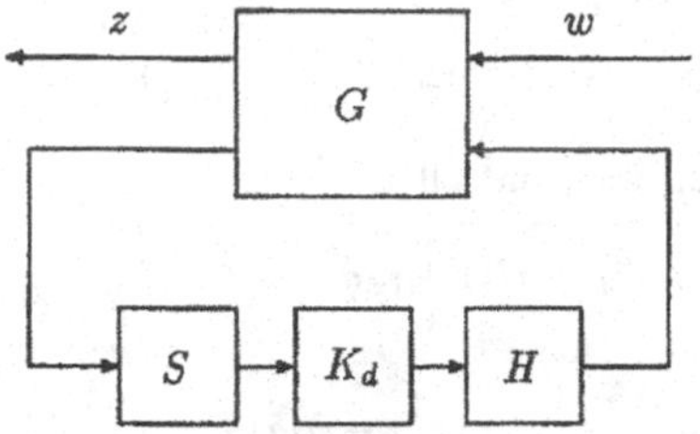

Figure 1: Sampled-data setup

The $\mathcal{H}_2$ control problem is, roughly speaking, to minimize the $\mathcal{L}_2$-norm of z for a fixed, known input w.

We begin with two examples.

1 Example 1

The simplest possible sampled-data setup is

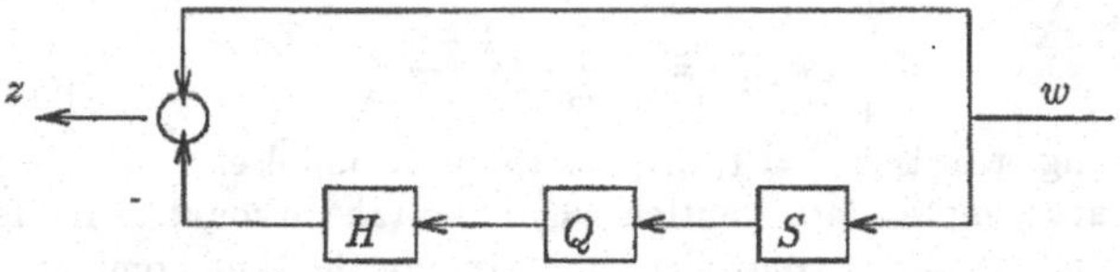

Fix a continuous w in $\mathcal{L}_2$ such that $Sw \in l_2$ and consider choosing a stable, causal Q to minimize $\|z\|_2$.

Let's temporarily make the problem even simpler by optimizing just at the sampling instants, i.e., minimizing $\|Sz\|_2$. Since

$$Sz = S(I - HQS)w = (I - Q)Sw$$

The optimal Q is obviously $Q = I$. And then

$$\begin{aligned}\|z\|_2^2 &= \|w - HSw\|_2^2 \\ &= \cdots + \int_0^h [w - w(0)]^2 + \int_h^{2h} [w - w(h)]^2 + \cdots\end{aligned}$$

The right-hand side is the minimum norm of the continuous-time z when the optimization is done only at the sampling instants.

Take r to be the unit step and $\hat{p}(s) = 1/(s+1)$. The goal is to design a stabilizing K_d to minimize $\|e\|_2$.

Again, for comparison let's first solve the simpler problem of optimizing only at the sampling instants. Define

$$P_d = SPH, \quad \rho = Sr, \quad \epsilon = Se$$

Then

$$\hat{p}_d(\lambda) = (1 - e^{-h})\frac{\lambda}{1 - e^{-h}\lambda}, \quad \hat{\rho}(\lambda) = \frac{1}{1-\lambda}$$

Parametrize all internally stabilizing controllers:

$$\hat{k}_d = \hat{q}/(1 - \hat{p}_d\hat{q}), \quad \hat{q} \in \mathcal{RH}_\infty$$

Then

$$\hat{\epsilon} = (1 - \hat{p}_d\hat{q})\hat{\rho}$$

To have ϵ in l_2 the transfer function $1 - \hat{p}_d\hat{q}$ must have a zero at $\lambda = 1$. This means that $\hat{q}$ must have the form

$$\hat{q}(\lambda) = 1 + (1-\lambda)\hat{q}_1(\lambda), \quad \hat{q}_1 \in \mathcal{RH}_\infty$$

Defining

$$\hat{\psi}(\lambda) = \frac{1}{1 - e^{-h}\lambda}$$

we get

$$\hat{\epsilon} = \hat{\psi} - \hat{p}_d\hat{q}_1$$

The unique $\hat{q}_1$ minimizing $\|\hat{\epsilon}\|_2$ is

$$\hat{q}_1(\lambda) = \frac{e^{-h}}{1 - e^{-h}}$$

and then

$$\hat{k}_d(\lambda) = \frac{1}{1 - e^{-h}}\frac{1 - e^{-h}\lambda}{1 - \lambda}$$

The resulting tracking error is $\hat{\epsilon}(\lambda) = 1$, i.e., ϵ is the unit impulse.

Now for the exact sampled-data solution where we take account of intersample behaviour. Again, parametrize all stabilizing controllers, this time in the time domain:

$$K_d = Q(I - P_dQ)^{-1}$$

Then

$$e = (I - PHQS)r$$

For e in $\mathcal{L}_2$ we need $Se \in l_2$, so Q must again have the form

$$Q = I + Q_1N, \quad \hat{n}(\lambda) = 1 - \lambda$$

Defining $v := (I - P)r$ and noting that $NSr = \delta$, the discrete-time unit pulse, we get

$$e = v - PHQ_1\delta$$

It turns out that $\hat{v}(s) = 1/(s+1)$. In this way the problem reduces to

$$\min_{Q_1} \|v - PHQ_1\delta\|_2$$

To solve this, bring in the factorization

$$(PH)^*(PH) = W^*W$$

with W, W^{-1} stable. Define $T := PHW^{-1}$ and $Q_2 = WQ_1$. The problem becomes the minimization of

$$\|v - TQ_2\delta\|_2$$

But

$$\begin{aligned} \|v - TQ_2\delta\|_2^2 &= \left\| \begin{bmatrix} T^* \\ I - TT^* \end{bmatrix} (v - TQ_2\delta) \right\|_2^2 \\ &= \|T^*v - Q_2\delta\|_2^2 + \|(I - TT^*)v\|_2^2 \end{aligned}$$

Therefore the optimal Q_2 is

$$\hat{q}_2 = \text{projection onto } \mathcal{H}_2 \text{ of } \widehat{T^*v}$$

In summary, the procedure to compute the optimal controller is as follows:

1. By spectral factorization get $\hat{w}$ such that $(PH)^*(PH) = W^*W$.
2. Compute $\hat{q}_2$ as the projection onto $\mathcal{H}_2$ of the λ-transform of $W^{-1*}(PH)^*v$.
3. Set $\hat{q}_1 = \hat{q}_2/\hat{w}$.
4. Set $\hat{q} = 1 + \hat{q}_1\hat{n}$.
5. Set $\hat{k}_d = \hat{q}/(1 - \hat{p}_d\hat{q})$.

Step 2 involves computing the λ-transform of $(PH)^*v$. There's a state-space procedure to do this, but there isn't time for its derivation.

The computations were done for $h = 0.5$ and the results are

$$\hat{w}(\lambda) = \frac{0.0581\lambda + 0.2201}{-0.6065\lambda + 1}$$

$$\hat{q}_2(\lambda) = \frac{0.8030}{-\lambda + 1.6487}$$

$$\hat{k}_d(\lambda) = \frac{-0.2601\lambda^2 + 0.8578\lambda - 0.7071}{-0.1335\lambda^2 + 0.3536\lambda - 0.2201}$$

Figure 2 shows the step response tracking error for the two cases, the discretized approximation (solid) and the exact solution (dash). Figure 3 is a plot of $\int_0^t e^2$ in the same two cases. The exact solution is about 15% better for the $\mathcal{L}_2$ criterion.

3 Problem statement

It's now time to state the $\mathcal{H}_2$ sampled-data control problem precisely. Consider the setup in Figure 1. Let T_{zw} denote the operator from w to z. Suppose T_{zw} were lti with transfer matrix $\hat{t}_{zw}(s)$. The $\mathcal{H}_2$ norm is

$$\|\hat{t}_{zw}\|_2 = \left\{ \frac{1}{2\pi} \int_{-\infty}^{\infty} \text{trace}\, [\hat{t}_{zw}(j\omega)^* \hat{t}_{zw}(j\omega)] d\omega \right\}^{1/2}$$

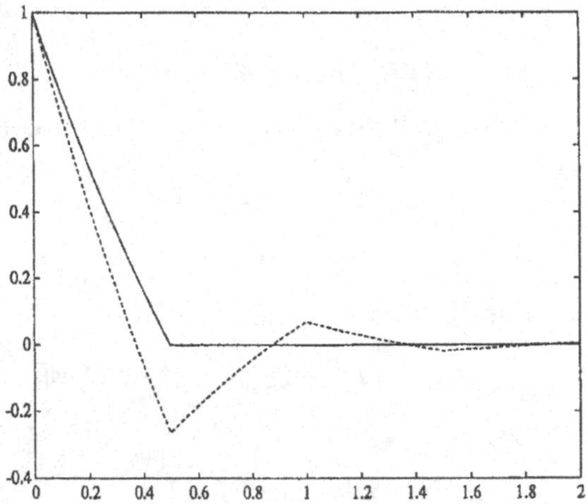

Figure 2: Step response of e: discrete approximation (solid), exact (dash)

But there's another way to represent this norm, a way which doesn't depend on time-invariance. Let $\{e_i\}$ denote the standard basis in the space of dimension dim w. Then

$$\|\hat{t}_{zw}\|_2 = \left(\sum_i \|T_{zw}\delta e_i\|_2^2\right)^{1/2}$$

The right-hand side says: apply a unit impulse at the ith input channel, find the $\mathcal{L}_2$ norm squared of the output, sum over i. Let's agree that the impulse occurs at time $t = 0-$, just prior to when the sample at $t = 0$ is taken. The derivation of this equation is routine and is therefore omitted.

This suggests the sampled-data problem, design a digital controller K_d to achieve hybrid stability and minimize

$$\left(\sum_i \|T_{zw}\delta e_i\|_2^2\right)^{1/2}$$

This problem will be dealt with in the next section by a state-space method and in the following by an operator-theoretic method.

4 State-space approach

In this section we present an approach to the $\mathcal{H}_2$ sampled-data control problem when the plant in Figure 1 is finite-dimensional. Suppose G has a minimal state-space model

$$\dot{x} = Ax + B_1 w + B_2 u \quad (1)$$

$$z = C_1 x + D_{12} u \quad (2)$$

$$y = C_2 x \quad (3)$$

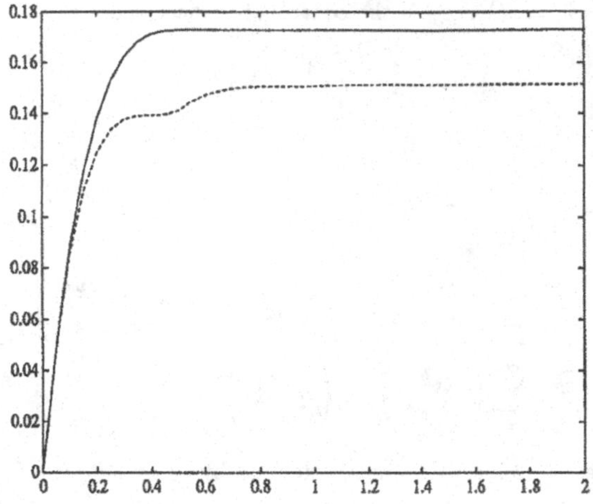

Figure 3: Step response of e: discrete approximation (solid), exact (dash)

The assumption that $D_{11} = 0$ is required for well-posedness of our $\mathcal{H}_2$ sampled-data problem. Assume the pair (A, B_2) is stabilizable and (C_2, A) is detectable. Thus G is stabilizable in the lti sense. For G to be hybrid stabilizable, we'll assume that the sampling is non-pathological. It follows that a digital controller K_d achieves hybrid stability for G iff it stabilizes G_{sd22}, the discretized G_{22}.

We start by analyzing the performance of our sampled-data system. Suppose we implement some controller K_d to provide hybrid stability. Apply input $w(t) = \delta(t)e_i$. Let z_i be the resulting output, namely, $T_{zw}\delta e_i$. Then the $\mathcal{H}_2$ performance for our sampled-data system is

$$J(K_d) := \left(\sum_i \|z_i\|_2^2\right)^{1/2} \tag{4}$$

Since the system is causal, $z_i(t) = 0$ for $t < 0$. For $t \geq 0$, we have from equations (1)-(3)

$$\begin{aligned} \dot{x} &= Ax + B_2 u, \qquad x(0) = B_1 e_i \\ z_i &= C_1 x + D_{12} u \\ y &= C_2 x \end{aligned}$$

On the interval $(kh, (k+1)h)$, since $u = H\psi$ is constant, we have

$$\begin{pmatrix} \dot{x} \\ \dot{u} \end{pmatrix} = \begin{bmatrix} A & B_2 \\ 0 & 0 \end{bmatrix} \begin{pmatrix} x \\ u \end{pmatrix}$$

$$z_i = \begin{bmatrix} C_1 & D_{12} \end{bmatrix} \begin{pmatrix} x \\ u \end{pmatrix}$$

It follows that

$$\begin{pmatrix} x(t) \\ u(t) \end{pmatrix} = e^{(t-kh)\bar{A}} \begin{pmatrix} \xi(k) \\ \psi(k) \end{pmatrix}$$

where ξ is defined by $\xi(k) = x(kh)$, ψ is the control sequence, and

$$\bar{A} := \begin{bmatrix} A & B_2 \\ 0 & 0 \end{bmatrix}$$

Thus

$$\begin{aligned}
\|z_i\|_2^2 &= \sum_{k=0}^{\infty} \int_{kh}^{(k+1)h} \|z_i(t)\|^2 \, dt \\
&= \sum_{k=0}^{\infty} \int_{kh}^{(k+1)h} \| \begin{bmatrix} C_1 & D_{12} \end{bmatrix} e^{\bar{A}(t-kh)} \begin{pmatrix} \xi(k) \\ \psi(k) \end{pmatrix} \|^2 \, dt \\
&= \sum_{k=0}^{\infty} \begin{pmatrix} \xi(k) \\ \psi(k) \end{pmatrix}' M \begin{pmatrix} \xi(k) \\ \psi(k) \end{pmatrix}
\end{aligned}$$

where M is the symmetric nonnegative matrix defined by

$$M := \int_0^h e^{\bar{A}'t} \begin{bmatrix} C_1 & D_{12} \end{bmatrix}' \begin{bmatrix} C_1 & D_{12} \end{bmatrix} e^{\bar{A}t} \, dt$$

Factor the matrix into the form

$$M =: \begin{bmatrix} C_{1d} & D_{12d} \end{bmatrix}' \begin{bmatrix} C_{1d} & D_{12d} \end{bmatrix}$$

where the number of columns in C_{1d} equals the dimension of x and similarly for D_{12d} and u. Define

$$\zeta_i(k) := C_{1d}\xi(k) + D_{12d}\psi(k)$$

Note that ζ_i is in l_2 because K_d achieves hybrid stability. Therefore the $\mathcal{L}_2$ norm of z_i equals the l_2 norm of ζ_i. Here ζ_i can be interpreted as the output of the following discrete-time system:

$$\begin{aligned}
\xi(k+1) &= A_d\xi(k) + B_{2d}\psi(k), \qquad \xi(0) = B_1 e_i && (5) \\
\zeta(k) &= C_{1d}\xi(k) + D_{12d}\psi(k) && (6) \\
\eta(k) &= C_2\xi(k) && (7) \\
\psi &= K_d\eta && (8)
\end{aligned}$$

with η being the sampled y and

$$A_d = e^{Ah} \qquad B_{2d} = \int_0^h e^{At} \, dt \, B_2$$

Let's recap: the $\mathcal{L}_2$ norm of z in Figure 1 when $w = \delta e_i$ equals the l_2 norm of ζ in (5)-(8).

Now the output of the following system is just one time unit delayed from the output in (5)-(8), and therefore the two outputs have equal l_2 norm:

$$\begin{aligned}
\xi(k+1) &= A_d\xi(k) + \delta(k)B_1 e_i + B_{2d}\psi(k), \qquad \xi(0) = 0 && (9) \\
\zeta(k) &= C_{1d}\xi(k) + D_{12d}\psi(k) && (10) \\
\eta(k) &= C_2\xi(k) && (11) \\
\psi &= K_d\eta && (12)
\end{aligned}$$

Here δ is the unit pulse. These latter equations correspond to the setup

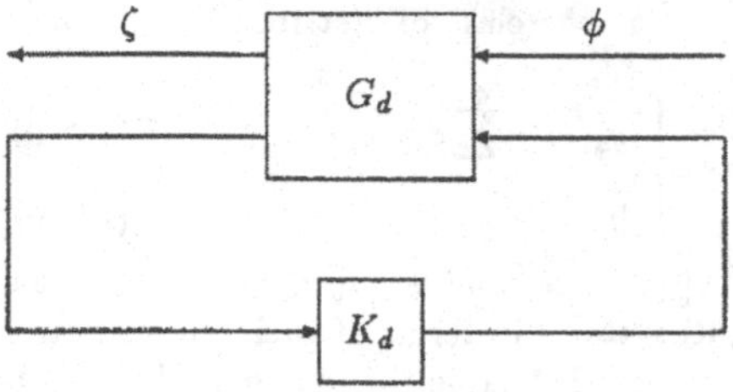

Here $\phi = \delta e_i$ and G_d has the equations

$$\begin{aligned} \xi(k+1) &= A_d\xi(k) + B_1\phi(k) + B_{2d}\psi(k) && (13)\\ \zeta(k) &= C_{1d}\xi(k) + D_{12d}\psi(k) && (14)\\ \eta(k) &= C_2\xi(k) && (15) \end{aligned}$$

Thus ζ_i is precisely the response of this system under the control $\psi = K_d\eta$ when a discrete-time impulse input $\phi = \delta e_i$ is applied. So by the discrete version of (4)

$$J(K_d) = \left(\sum_i \|\zeta_i\|_2^2\right)^{1/2} = \|\hat{t}_{\zeta\phi}\|_2$$

The right-hand side is the $\mathcal{H}_2$ norm of the discrete-time closed-loop transfer matrix from ϕ to ζ.

From our discussion before, the discrete-time system in (13)-(15) is stabilizable in the lti sense; and moreover K_d internally stabilizes G_d iff it provides hybrid stability for G.

We conclude this section by summarizing the result that has been derived.

Theorem 1 *For G given in (1)-(3), assume (A, B_2) is stabilizable, (C_2, A) is detectable, the sampling is non-pathological. Define G_d as in (13)-(15). Then the $\mathcal{H}_2$ sampled-data problem for the system (G, HK_dS) is equivalent to the $\mathcal{H}_2$ problem for the discrete-time system (G_d, K_d) in the sense that*

1. *K_d provides hybrid stability for G iff it internally stabilizes G_d; and*
2. *the $\mathcal{H}_2$ performance for the sampled-data system (G, HK_dS) equals that for the discrete-time system (G_d, K_d).*

The latter discrete-time $\mathcal{H}_2$ problem is lti, and its solution is known.

5 Operator theoretic approach

The goal in this section is to solve the $\mathcal{H}_2$ sampled-data problem using input-output, rather than state-space, methods. As a bonus, we'll be able to handle some infinite-dimensional systems, such as time-delay systems. Operator-theoretic methods were used for Examples 1 and 2.

Consider an lti plant G in the sampled-data control setup in Figure 1. To obviate the need for coprime factorization, we assume throughout this section that G is stable, i.e., $\hat{g}$ lies in $\mathcal{H}_\infty$. Since G may be infinite-dimensional, it's reasonable to allow our controllers K_d initially to be infinite-dimensional as well. We'll start with a parametrization of all infinite-dimensional, causal controllers that provide hybrid stability. Because of time constraints, the details of the derivation of this parametrization are omitted.

Let $\mathcal{A}(\mathbf{C}^{n\times m})$, or simply $\mathcal{A}$, be the class of distributions g from $(-\infty,\infty)$ to $\mathbf{C}^{n\times m}$ of the form

$$g(t) = \begin{cases} g_a(t) + \sum_{j=0}^{\infty} g_j \delta(t-t_j) & (t \geq 0) \\ 0 & (t < 0) \end{cases} \tag{16}$$

where $g_a(t) \in \mathcal{L}_1(0,\infty)$, $\sum_{j=0}^{\infty} \|g_j\| < \infty$, and $0 \leq t_0 < t_1 < \cdots$. The set $\mathcal{A}$ represents a class of infinite-dimensional, causal, lti systems in terms of their impulse responses which are bounded from $\mathcal{L}_p$ to $\mathcal{L}_p$ for every $1 \leq p \leq \infty$. Let $\hat{\mathcal{A}}$ be the image of $\mathcal{A}$ under Laplace transformation. It follows that $\hat{\mathcal{A}}$ is a proper subset of $\mathcal{H}_\infty$.

Theorem 2 *Assume $\hat{g} \in \mathcal{H}_\infty$, $\hat{g}_{22} \in \hat{\mathcal{A}}$, and the operator SG_{21} is bounded from $\mathcal{L}_2$ to $\mathbf{l}_2$. Then the set of all (lti, causal) controllers K_d which provide hybrid stability for G is parametrized as*

$$K_d = -Q(I-PQ)^{-1} = -(I-QP)^{-1}Q \qquad \hat{q} \in \mathcal{H}_\infty \tag{17}$$

where $P := SG_{22}H$.

An example when SG_{21} is bounded is where G_{21} is of the form $\hat{g}_{21}(s) = \mathrm{e}^{-\tau s}\hat{r}(s)$ with $\tau \geq 0$ and $\hat{r} \in \mathcal{RH}_\infty$ and strictly proper.

Using this controller parametrization, we can express the closed-loop operator T_{zw} in terms of the parameter Q:

$$T_{zw} = G_{11} - G_{12}HQSG_{21} \tag{18}$$

Thus our $\mathcal{H}_2$ sampled-data problem is equivalent to the following optimization problem

$$\inf_{\hat{q}\in\mathcal{H}_\infty} \sum_i \|G_{11}\delta e_i - G_{12}HQSG_{21}\delta e_i\|_2^2 \tag{19}$$

Our objective is to reduce this problem to a discrete-time, time-invariant one. To this end, we begin by studying each term in (19).

Define

$$d_i := G_{11}\delta e_i \tag{20}$$

$$\xi_i := SG_{21}\delta e_i \tag{21}$$

For the well-posedness of our $\mathcal{H}_2$ sampled-data problem, we assume from now on that $\hat{g}_{11}$ belongs to $\mathcal{H}_2$. This assumption, similar to the one made before that $D_{11} = 0$, guarantees that every d_i just defined is an $\mathcal{L}_2$ function. We'll also assume that Sg_{21}, defined in the obvious way, is an $\mathbf{l}_2$ matrix, meaning that each column of it belongs to $\mathbf{l}_2$. This assumption is similar to our requirement that $D_{21} = 0$ before. Under this assumption, each ξ_i is an $\mathbf{l}_2$ sequence.

Note that if $\hat{g}_{21}(s)$ is of the form $\mathrm{e}^{-\tau s}\hat{r}(s)$ with $\tau \geq 0$ and $\hat{r} \in \mathcal{RH}_\infty$ and strictly proper, then both of our assumptions made about G_{21} are satisfied; i.e., SG_{21} is bounded from $\mathcal{L}_2$ to $\mathbf{l}_2$ and Sg_{21} is an $\mathbf{l}_2$ matrix.

Define

$$T := G_{12}H$$

It follows that T is bounded from $\mathbf{l}_2$ to $\mathcal{L}_2$ since G_{12} is stable. The previous definitions yield

$$\|T_{zw}\delta e_i\|_2 = \|d_i - TQ\xi_i\|_2 \tag{22}$$

with $d_i \in \mathcal{L}_2$ and $\xi_i \in \mathbf{l}_2$. For any Q under consideration it's true that $TQ\xi_i$ is in the range of T, $\mathrm{Im}(T)$. To fulfill our goal, the first step is to decompose d_i into the sum of two components, one in $\mathrm{Im}(T)$, the other in $[\mathrm{Im}(T)]^\perp$. To justify this orthogonal decomposition, we require that the range of T be closed.

The following is a collection from operator theory of equivalent statements on the range of an operator being closed.

Theorem 3 *Let $\mathcal{V}_1$, $\mathcal{V}_2$ be two Hilbert spaces. Let T be a bounded linear operator from $\mathcal{V}_1$ to $\mathcal{V}_2$. Assume T is injective, i.e.,* $\mathrm{Ker}(T) = \{0\}$. *Then the following are equivalent:*

1. $\mathrm{Im}(T)$ *is closed;*
2. *there exists a bounded operator X from $\mathcal{V}_2$ to $\mathcal{V}_1$ such that $XT = I$;*
3. *T is bounded below, i.e., $\|Tx\| \geq c\|x\|$ for some $c > 0$;*
4. *T^*T has a bounded inverse on $\mathcal{V}_1$.*

The following fact, based on the above theorem, is useful in checking whether an operator has closed range.

Let $\mathcal{V}_i$, $i = 1,2,3$, be Hilbert spaces. Let $T_1 : \mathcal{V}_1 \longrightarrow \mathcal{V}_2$ and $T_2 : \mathcal{V}_2 \longrightarrow \mathcal{V}_3$ both be bounded and injective. If $\mathrm{Im}(T_1)$ *and* $\mathrm{Im}(T_2)$ *are both closed, then* $\mathrm{Im}(T_2T_1)$ *is closed.*

Return to our problem with $T = G_{12}H$. Do an inner-outer factorization $\hat{g}_{12} = \hat{g}_i\hat{g}_o$ with $\hat{g}_i$ inner and $\hat{g}_o$ outer. It's claimed that a sufficient condition for T to have closed range is that the outer factor $\hat{g}_o$ have a left inverse over $\mathcal{H}_\infty$. To see this, recall that being an inner factor, G_i is an isometry on $\mathcal{L}_2$, hence G_i has closed range. If $\hat{g}_o$ is left-invertible over $\mathcal{H}_\infty$, then the operator G_o has a left inverse on $\mathcal{L}_2$. By Theorem 3, $\mathrm{Im}(G_o)$ is closed. Finally, the fact that $\|Hx\|_2 = \sqrt{h}\|x\|_2$ holds for any x in l_2 implies that H, regarded as an operator from l_2 to $\mathcal{L}_2$, also has closed range. Invoke the just-stated fact to conclude that $\mathrm{Im}(T)$ is closed.

Hereafter, the outer factor of $\hat{g}_{12}$ will be assumed left-invertible over $\mathcal{H}_\infty$. Under this assumption, apply Theorem 3 to get that the operator T^*T from l_2 to itself has a bounded inverse.

Now consider the orthogonal projection $\Pi_{\mathrm{Im}(T)} : \mathcal{L}_2 \longrightarrow \mathrm{Im}(T)$. For any d in $\mathcal{L}_2$, $\Pi_{\mathrm{Im}(T)}d = T\eta$ where

$$\eta := (T^*T)^{-1}T^*d \tag{23}$$

To evaluate the projection onto $\mathrm{Im}(T)$, equation (23) shows that it's necessary to invert the operator T^*T. We have already encountered this operator in Lecture 4 ($(GH)^*(GH)$).

For any causal, stable, lti, and continuous-time system G, GH is a bounded operator from l_2 to $\mathcal{L}_2$. The *sampled-data gramian operator with symbol G*, denoted Ψ_G, is defined as the operator $(GH)^*(GH)$ mapping l_2 to l_2.

Lemma 1 *Let $\hat{g}$ be in $\mathcal{H}_\infty$ and let $\hat{g} = \hat{g}_i\hat{g}_o$ be an inner-outer factorization. Then*

1. *the sampled-data gramian is independent of the inner factor of its symbol, i.e., $\Psi_G = \Psi_{G_o}$; and*
2. *the sampled-data gramian Ψ_G is Toeplitz.*

Proof The first property follows from the fact that $G_i^*G_i = I$. The second property we saw in Lecure 4. ■

Since the sampled-data gramian with symbol G_{12} is Toeplitz and invertible, we can consider a spectral factorization of it by the following result.

Theorem 4 *Let Φ be a bounded self-adjoint operator from l_2 to l_2. If it is Toeplitz and has bounded inverse, then there exists some causal, Toeplitz, bounded operator Θ mapping l_2 to l_2 such that $\Phi = \Theta^*\Theta$.*

The operator Θ is a spectral factor for Φ.

By this theorem, our gramian operator has a spectral factorization

$$\Psi_{G_{12}} = \Theta^*\Theta$$

where Θ is causal, Toeplitz and bounded on l_2.

Now we are set up to derive the main result. With reference to equation (22), define

$$\Pi_{\mathrm{Im}(T)} d_i =: T\eta_i$$

to get from (23) that

$$\eta_i = \Psi_{G_{12}}^{-1} T^* d_i \tag{24}$$

Thus by (22)

$$\|T_{zw}\delta e_i\|_2^2 = \|d_i - T\eta_i\|_2^2 + \|T(\eta_i - Q\xi_i)\|_2^2$$

The first term on the right-hand side is out of control. The second term is

$$\begin{aligned} \|T(\eta_i - Q\xi_i)\|_2^2 &= \langle(\eta_i - Q\xi_i),\, T^*T(\eta_i - Q\xi_i)\rangle \\ &= \langle(\eta_i - Q\xi_i),\, \Theta^*\Theta(\eta_i - Q\xi_i)\rangle \\ &= \|\Theta(\eta_i - Q\xi_i)\|_2^2 \end{aligned}$$

We emphasize here that Θ is a discrete-time system. It is causal, lti, and stable. Note that η_i and ξ_i in the last equation are discrete-time signals. So we have arrived at an equivalent discrete-time problem.

To recap, we list the assumptions that have been made about the plant G in this section thus far:

1. $\hat{g}$ is in $\mathcal{H}_\infty$
2. $\hat{g}_{11}$ is in $\mathcal{H}_2$
3. the outer factor of $\hat{g}_{12}$ is left-invertible over $\mathcal{H}_\infty$
4. the operator SG_{21} is bounded from $\mathcal{L}_2$ to l_2 and Sg_{21} is an l_2 matrix
5. the impulse response of G_{22} lies in $\mathcal{A}$

Note from the definition of d_i in (20) that $T^*d_i = T^*g_{11}e_i$. Each column of the matrix T^*g_{11} is in l_2.

Theorem 5 *Under the five assumptions and foregoing definitions, the sampled-data $\mathcal{H}_2$ problem*

$$J^2 := \inf_{\hat{q}\in\mathcal{H}_\infty} \sum_i \|T_{zw}\delta e_i\|_2^2$$

is equivalent to the following discrete-time lti optimization problem

$$J^2 = \sum_i \|d_i - T\eta_i\|_2^2 + \inf_{\hat{q}\in\mathcal{H}_\infty} \|\Theta(\Psi_{G_{12}}^{-1}\Gamma - Q\Xi)\|_2^2$$

6 Notes

This lecture is based on Chen and Francis 1990a. The details are worked out for time-delay systems in that reference.

The first work along these lines is that of Levis, Schlueter, and Athans (1971). They looked at the linear-quadratic regulator problem with the control required to be the output of a zero-order hold; effectively, they looked at the state-feedback sampled-data $\mathcal{H}_2$ problem. They showed that there's an equivalent discrete-time problem. This observation was used also by Dorato and Levis (1971).

Theorem 2 is a modification of Zames' parametrization (1981). Theorem 3 is standard in operator theory. Theorem 4 can be found, for example, in Rosenblum and Rovnyak 1985.

Lecture 6

$\mathcal{H}_\infty$ Optimal Sampled-Data Control

The setup in this lecture is again

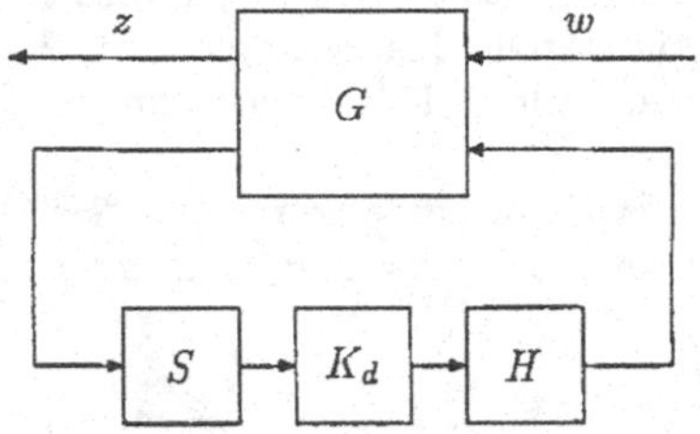

Figure 1: Sampled-data setup

The $\mathcal{H}_\infty$ control problem is to minimize the $\mathcal{L}_2$ induced norm from w to z.

A complete operator-theoretic solution to this problem is not yet available. This lecture presents the results obtained to date.

1 Lti discrete-time example

It's instructive to recall the approach in the linear time-invariant discrete-time case. So let's take the simplest example:

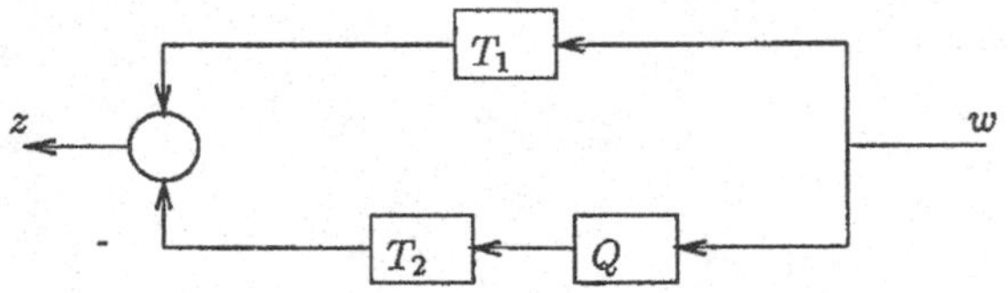

Suppose T_1, Q are fdlti and stable and $T_2 = U$, the bilateral shift on $\mathbf{l}_2$. We know that

$$\|T_1 - T_2 Q\| = \|\hat{t}_1 - \hat{t}_2 \hat{q}\|_\infty$$

and the minimum value of this norm, over all Q, equals $|\hat{t}_1(0)|$.

Let's derive this result using operator-theoretic methods. Note that T_2 is an isometry on $\mathbf{l}_2$. Bring in the subspace $\mathbf{h}_2$ of $\mathbf{l}_2$ of sequences zero for negative time. Thus $\mathbf{l}_2 = \mathbf{h}_2^\perp \oplus \mathbf{h}_2$. Also, let Π denote the orthogonal projection $\mathbf{l}_2 \longrightarrow \mathbf{h}_2^\perp$.

Now fix Q causal, i.e., $Q\mathbf{h}_2 \subset \mathbf{h}_2$. Then

$$\begin{aligned}\|T_1 - T_2 Q\| &= \|U^*(T_1 - UQ)\| \\ &= \|U^* T_1 - Q\| \\ &\geq \|\Pi(U^* T_1 - Q)|\mathbf{h}_2\| \\ &= \|\Pi U^* T_1|\mathbf{h}_2\|\end{aligned}$$

Thus

$$\min_Q \|T_1 - T_2Q\| \geq \|\Pi U^* T_1|\mathbf{h}_2\|$$

The right-hand operator,

$$\Gamma := \Pi U^* T_1|\mathbf{h}_2$$

is a Hankel operator. In fact, equality holds. This can be proved using the commutant lifting theorem of Nagy and Foias, as follows.

Let Ξ denote the shift U compressed to $\mathbf{h}_2^\perp$, i.e.,

$$\Xi := \Pi U|\mathbf{h}_2^\perp$$

Thus the following diagram commutes:

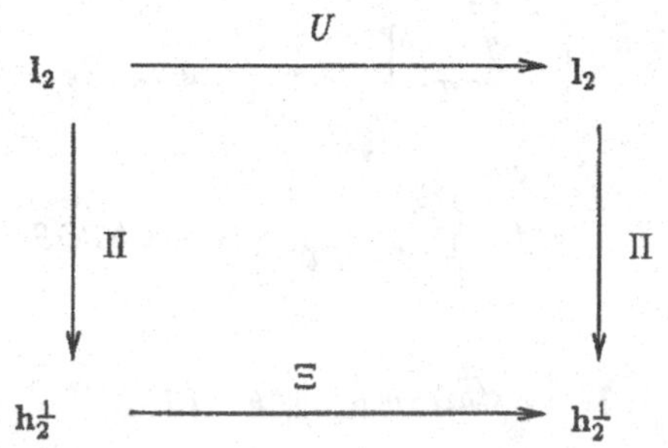

Also, the following diagram commutes:

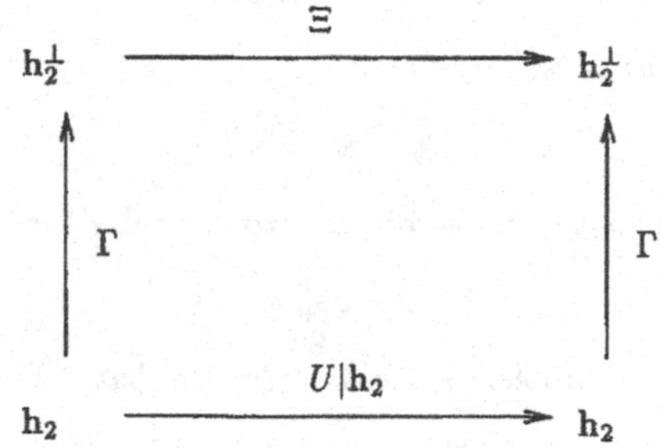

To see this, first note (using the fact that T_1 is time-invariant, and hence commutes with U) that

$$\begin{aligned} \Gamma U|\mathbf{h}_2 &= \Pi U^* T_1 U|\mathbf{h}_2 \\ &= \Pi U^* U T_1|\mathbf{h}_2 \\ &= \Pi T_1|\mathbf{h}_2 \\ &= 0 \end{aligned}$$

Secondly, to see that $\Xi\Gamma = 0$, start with the fact that $U^*\mathbf{h}_2$ is Π-invariant:

$$\begin{aligned} \Pi U^*\mathbf{h}_2 \subset U^*\mathbf{h}_2 &\implies U\Pi U^*\mathbf{h}_2 \subset \mathbf{h}_2 \\ &\implies U\Pi U^* T_1\mathbf{h}_2 \subset \mathbf{h}_2 \\ &\implies \Pi U \Pi U^* T_1|\mathbf{h}_2 = 0 \\ &\implies \Xi\Gamma = 0 \end{aligned}$$

Returning to the original problem, we note that the best constant approximation to w over the interval $[0,h]$ (in the $\mathcal{L}_2$ norm) is the mean value, $(1/h)\int_0^h w$. So we anticipate that we should be able to make the minimum value of $\|z\|_2^2$ equal to

$$\cdots + \int_0^h \left(w - (1/h)\int_0^h w\right)^2 + \int_h^{2h} \left(w - (1/h)\int_h^{2h} w\right)^2 + \cdots$$

This may not actually be achievable by a signal of the form $HQSw$. In any event, to obtain the optimal Q, first define $T := (1/\sqrt{h})H$ so that T is an isometry $\mathrm{l}_2 \longrightarrow \mathcal{L}_2$. Fix Q, stable and causal. Then

$$\|w - HQSw\|_2 = \|w - \sqrt{h}TQSw\|_2$$

Check that the following operator is an isometry:

$$\begin{bmatrix} T^* \\ I - TT^* \end{bmatrix} : \mathcal{L}_2 \longrightarrow \mathrm{l}_2 \oplus \mathcal{L}_2$$

Thus

$$\|w - HQSw\|_2^2 = \left\| \begin{bmatrix} T^* \\ I - TT^* \end{bmatrix} (w - \sqrt{h}TQSw) \right\|_2^2$$

The right-hand side equals

$$\|T^*w - \sqrt{h}QSw\|_2^2 + \|(I - TT^*)w\|_2^2$$

So the optimal Q should minimize

$$\|T^*w - \sqrt{h}QSw\|_2$$

This is a discrete-time problem in l_2. Defining

$$\psi := T^*w, \quad \phi := \sqrt{h}Sw$$

we see that the problem reduces to finding $\hat{q}$ in $\mathcal{H}_\infty$ to minimize the $\mathcal{H}_2$ norm

$$\|\hat{\psi} - \hat{q}\hat{\phi}\|_2$$

Being time-invariant, this latter problem is routine. Notice that $(1/\sqrt{h})\psi$ is the sequence

$$\left\{(1/h)\int_0^h w, (1/h)\int_h^{2h} w, \cdots\right\}$$

as anticipated above.

2 Example 2

Buoyed up by our success on Example 1, let's try a feedback problem:

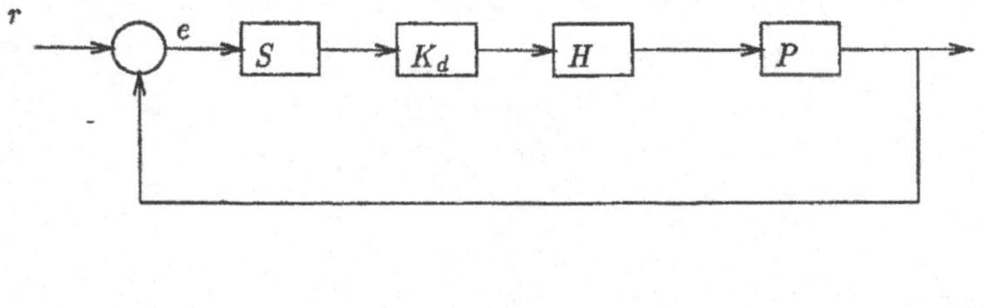

The commutant lifting theorem now guarantees the existence of an operator Γ_l having equal norm with Γ and such that the following diagram commutes:

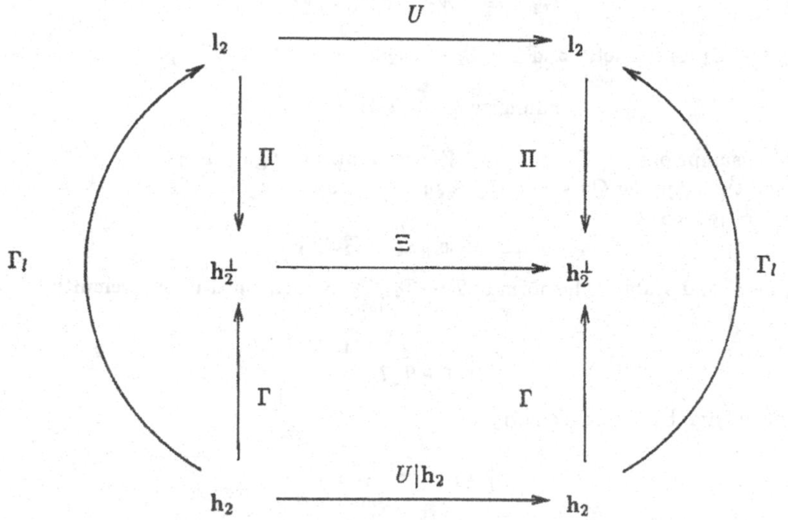

Commutativity of the outer loop says that $U\Gamma_l = \Gamma_l U|\mathbf{h}_2$. Now extend the domain of Γ_l to all of $\mathbf{l}_2$ so that $U\Gamma_l = \Gamma_l U$. This implies that Γ_l is time-invariant. Define

$$Q := U^* T_1 - \Gamma_l \tag{1}$$

Then commutativity of the left-hand semicircle, i.e., $\Pi\Gamma_l|\mathbf{h}_2 = \Gamma$, implies that $Q\mathbf{h}_2 \subset \mathbf{h}_2$, so Q is causal and stable. Finally, multiplication of (1) by U yields

$$\|T_1 - UQ\| = \|\Gamma_l\| = \|\Gamma\|$$

We have thus proved that

$$\min_Q \|T_1 - T_2 Q\| = \|\Gamma\|$$

It remains to show that

$$\|\Gamma\| = |\hat{t}_1(0)|$$

This is routine to do by expressing Γ in the frequency domain.

2 Sampled-data model-matching

Now we turn to the sampled-data model-matching $\mathcal{H}_\infty$ problem. The setup is

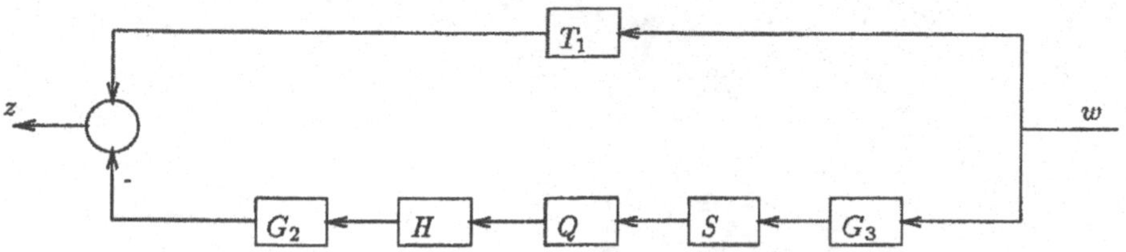

The three components T_1, G_2, G_3 are assumed fdlti, stable, and causal, with G_3 strictly causal.

As in Lecture 4 bring in the factorizations

$$G_2 H = T_2 W_2, \quad S G_3 = W_3 T_3$$

with $T_2 : \mathrm{l}_2 \longrightarrow \mathcal{L}_2$ an isometry and $T_3 : \mathcal{L}_2 \longrightarrow \mathrm{l}_2$ a co-isometry. The problem becomes

$$\text{minimize } \|T_1 - T_2 W_2 Q W_3 T_3\|$$

Under mild assumptions on G_2 and G_3, W_2 has a stable right-inverse and W_3 a stable left-inverse. Then the mapping $Q \mapsto W_2 Q W_3$ is surjective on the space of stable causal lti systems. So the problem reduces to

$$\text{minimize } \|T_1 - T_2 Q T_3\|$$

Fix Q causal and stable. The norm of $T_1 - T_2 Q T_3$ is unchanged upon premultiplying by the isometry

$$\begin{bmatrix} T_2^* \\ I - T_2 T_2^* \end{bmatrix}$$

and postmultiplying by the co-isometry

$$\begin{bmatrix} T_3^* & I - T_3^* T_3 \end{bmatrix}$$

But

$$\begin{bmatrix} T_2^* \\ I - T_2 T_2^* \end{bmatrix} T_2 Q T_3 \begin{bmatrix} T_3^* & I - T_3^* T_3 \end{bmatrix} = \begin{bmatrix} Q & 0 \\ 0 & 0 \end{bmatrix}$$

Define

$$R := \begin{bmatrix} T_2^* \\ I - T_2 T_2^* \end{bmatrix} T_1 \begin{bmatrix} T_3^* & I - T_3^* T_3 \end{bmatrix}$$

Then we have

$$\|T_1 - T_2 Q T_3\| = \left\| R - \begin{bmatrix} Q & 0 \\ 0 & 0 \end{bmatrix} \right\|$$

It follows that

$$\|T_1 - T_2 Q T_3\| \geq \|\Gamma\| \tag{2}$$

where $\Gamma : \mathbf{h}_2 \oplus \mathcal{L}_2 \longrightarrow \mathbf{h}_2^{\perp} \oplus \mathcal{L}_2$ is given by

$$\Gamma := \begin{bmatrix} \Pi & 0 \\ 0 & I \end{bmatrix} R | \mathbf{h}_2 \oplus \mathcal{L}_2$$

The commutant lifting theorem can now be used to prove equality in (2). The diagram is

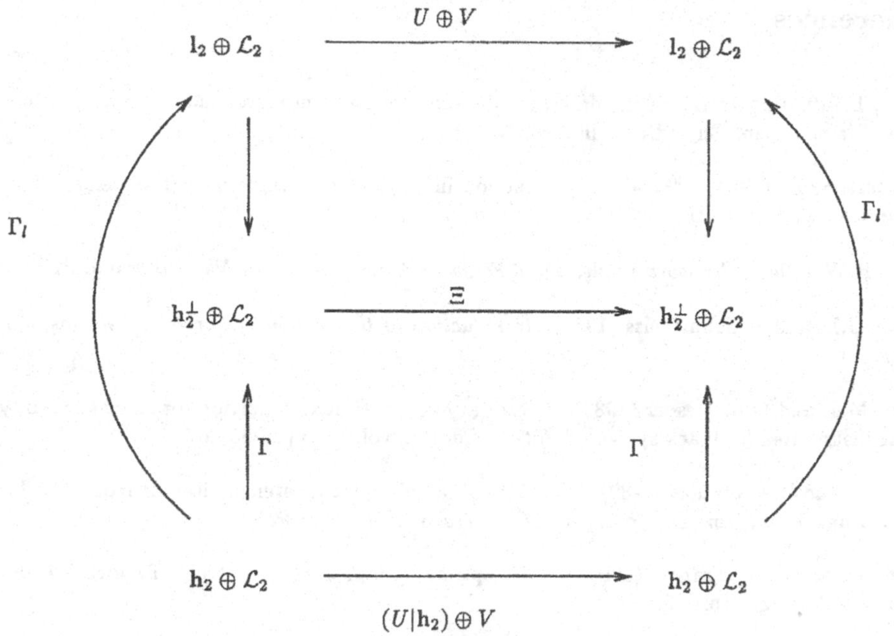

Note that this involves a mixture of discrete and continuous time. The uppermost operator, written $U \oplus V$ in the figure, is

$$\begin{bmatrix} U & 0 \\ 0 & V \end{bmatrix} \tag{3}$$

—recall (Lecture 4) that V is the operator of delay by time h. In (3) is a shift operator (of infinite multiplicity). The operator Ξ is its compression.

A procedure to compute $\|\Gamma\|$ is not known as of the time of writing.

3 Conclusion

These last four lectures have outlined an operator-theoretic approach to sampled-data control systems. There are many interesting open problems, the previous section describing just one.

4 Notes

The technique used here is based on Feintuch and Francis 1988. The commutant lifting theorem is due to Nagy and Foias (1970).

There has been recent state-space work on the $\mathcal{H}_\infty$ sampled-data problem: Hara and Kabamba (1990) show how to reduce the problem to a sequence of discrete-time lti $\mathcal{H}_\infty$ problems; Basar (1990) looks at the state-feedback problem using game theory; and Keller and Anderson (1990) present a technique to discretize an analog design which involves sampled-data $\mathcal{H}_\infty$ optimization.

References

Basar, T. (1990). "Optimum $\mathcal{H}_\infty$ designs under sampled state measurements," Technical Report, Dept. Elect. Comp. Eng., Univ. Illinois.

Bensoussan, D. (1984). "Sensitivity reduction in single-input single-output systems," *Int. J. Control*, vol. 39, pp. 321-335.

Bode, H.W. (1945). *Network Analysis and Feedback Amplifier Design*, Van Nostrand, Princeton.

Bower, J.L. and P. Schultheiss (1957). *Introduction to the Design of Servomechanisms*, Wiley, New York.

Chen, M.J. and C.A. Desoer (1982). "Necessary and sufficient condition for robust stability of linear distributed feedback systems," *Int. J. Control*, vol. 35, pp. 255-267.

Chen, T. and B.A. Francis (1989). "Stability of sampled-data systems," Report No. 8905, Dept. Elect. Eng., Univ. Toronto; to appear *IEEE Trans. Auto. Control*.

Chen, T. and B.A. Francis (1990a). "$\mathcal{H}_2$-optimal sampled-data control," Report No. 9001, Dept. Elect. Eng., Univ. Toronto.

Chen, T. and B.A. Francis (1990b). "On the $\mathcal{L}_2$-induced norm of a sampled-data system," Report No. 9002, Dept. Elect. Eng., Univ. Toronto; to appear *Systems and Control Letters*.

Desoer, C.A. and M. Vidyasagar (1975). *Feedback Systems: Input-Output Properties*, Academic Press, New York.

Dorato, P. and A.H. Levis (1971). "Optimal linear regulators: the discrete-time case," *IEEE Trans. Auto. Control*, vol. AC-16, pp. 613-620.

Doyle, J.C. and G. Stein (1981). "Multivariable feedback design: concepts for a classical modern synthesis," *IEEE Trans. Auto. Control*, vol. AC-26, pp. 4-16.

Feintuch, A. and B.A. Francis (1988). "Distance formulas for operator algebras arising in otimal control problems," in *Topics in Operator Theory*, Operator Theory: Advances and Applications, vol. 29, Birkhauser, Basel.

Francis, B.A. (1987). *A Course in $\mathcal{H}_\infty$ Control Theory*, vol. 88 in Lecture Notes in Control and Information Sciences, Springer-Verlag, New York.

Francis, B.A. and T. Georgiou (1988). "Stability theory for linear time-invariant plants with periodic digital controllers," *IEEE Trans. Auto. Control*, vol. AC-33, pp. 820-8323.

Francis, B.A. and G. Zames (1984). "On $\mathcal{H}^\infty$-optimal sensitivity theory for siso feedback systems," *IEEE Trans. Auto. Control*, vol. AC-29, pp. 9-16.

Freudenberg, J.S. and D.P. Looze (1985). "Right half-plane poles and zeros and design trade-offs in feedback systems," *IEEE Trans. Auto. Control*, vol. AC-30, pp. 555-565.

Hara, S. and P. Kabamba (1990). "On optimizing the induced norm of a sampled data system," *Proc. ACC.*

Holtzman, J.M. (1970). *Nonlinear System Theory*, Prentice-Hall, Englewood Cliffs, New Jersey.

Horowitz, I.M. (1963). *Synthesis of Feedback Systems*, Academic Press, New York.

Kabamba, P. and S. Hara (1990). "On computing the induced norm of a sampled data system," *Proc. ACC.*

Kalman, R., B.L. Ho, and K. Narendra (1963). "Controllability of linear dynamical systems," in *Contributions to Differential Equations*, vol. 1, Interscience, New York.

Keller, J.P. and B.D.O. Anderson (1990). "A new approach to controller discretization," *Proc. ACC.*

Khargonekar, P. and A. Tannenbaum (1985). "Noneuclidean metrics and the robust stabilization of systems with parameter uncertainty," *IEEE Trans. Auto. Control*, vol. AC-30, pp. 1005-1013.

Leung, G.M.H., T.P. Perry, and B.A. Francis (1989). "Performance analysis of sampled-data control systems," Report No. 8916, Dept. Elect. Eng., Univ. Toronto; to appear *Automatica.*

Levis, A.H., R.A. Schlueter, and M. Athans (1971). "On the behaviour of optimal linear sampled-data regulators," *Int. J. Control*, vol. 13, pp. 343-361.

Mees, A.I. (1981). *Dynamics of Feedback Systems*, Wiley, New York.

Sz. Nagy, B. and C. Foias (1970). *Harmonic Analysis of Operators in Hilbert Space*, North Holland.

Raggazini, J.R. and G.F. Franklin (1958). *Sampled-Data Control Systems*, McGraw-Hill, New York.

Rosenblum, M. and J. Rovnyak (1985). *Hardy Classes and Operator Theory*, Oxford University Press, New York.

Stein, E.M. and G. Weiss (1971). *Introduction to Fourier Analysis on Euclidean Spaces*, Princeton University Press, Princeton N.J.

Thompson, P.M., G. Stein, and M. Athans (1983). "Conic sectors for sampled-data feedback systems," *Systems and Control Letters*, vol. 3, pp. 77-82.

Vidyasagar, M. (1978). *Nonlinear Systems Analysis*, Prentice-Hall, New Jersey.

Willems, J.C. (1971). *The Analysis of Feedback Systems*, MIT Press, Cambridge, Mass.

Youla, D.C., H.A. Jabr, and J.J. Bongiorno, Jr., (1976). "Modern Wiener-Hopf design of optimal controllers, part II: the multivariable case," *IEEE Trans. Auto. Control*, vol. AC-21, pp. 319-338.

Zames, G. (1981). "Feedback and optimal sensitivity: model reference transformations, multiplicative seminorms, and approximate inverses," *IEEE Trans. Auto. Control*, vol. AC-26, pp. 301-320.

Zames, G. and B.A. Francis (1983). "Feedback, minimax sensitivity, and optimal robustness," *IEEE Trans. Auto. Control*, vol. AC-28, pp. 585-601.

TWO TOPICS IN SYSTEMS ENGINEERING: FREQUENCY DOMAIN DESIGN AND NONLINEAR SYSTEMS

J. William Helton*

University of California, San Diego
La Jolla, California 92093-0112

The development of a systematic theory of worst case design in the frequency domain where stability is the key consideration is developing rapidly and moving in several directions. An example of a major success in the subject is on the paradigm mixed sensitivity problem for multiport control systems. Here a beautiful and powerful theory has emerged. When one looks beyond the paradigm for new directions in the subject several areas are quickly suggested by physical considerations.

a. time varying systems,

b. non-interacting systems: sparsity patterns,

c. non-linear systems,

d. realistic performance specs.

These lectures address two of the topics. Lectures 1 and 2 concern the last topic while Lectures 3 and 4 treat nonlinear arguments. The first two lectures concern joint work with Orlando Merino while the last two are joint with Joe Ball.

For completeness we note that there are several other directions. One is infinite dimensional H^∞ control. It was not listed since philosophically it is the same as finite dimensional H^∞ control, however, mathematically it is much more difficult and so is something of a subject unto itself. Another very different direction lies in adapting H^∞ control to radically different uses, such as adaptive control.

---

* Supported in part by the Air Force Office of Scientific Research and the National Science Foundation

LECTURE 1. OPTIMAL FREQUENCY DOMAIN DESIGN

1. The fundamental H^∞ problem of control.
2. Generic motivation.
3. Explicitly solvable problems.
4. Engineering examples.

LECTURE 2. COMPUTATIONAL SOLUTIONS—

1. A computer demonstration.
2. Algorithms.
3. Diagnostics.
4. Further topics.

LECTURE 3. NONLINEAR SYSTEMS—CONTROL PROBLEM AND RESULTS

1. Introduction.
2. Preliminaries.
3. The recipe and main results.

LECTURE 4. NONLINEAR SYSTEMS—RESULTS AND METHODS

1. State feedback control.
2. Output feedback control.
3. The energy for an entire experiment.
4. The maximum principle.
5. Inner-outer factorization.

Lecture 1.

OPTIMAL FREQUENCY DOMAIN DESIGN

1. The fundamental H^∞ problem of control

A basic problem in designing stable systems is this:

At each $e^{i\theta} \in \Pi$ we are given a region $S_\theta \subset \mathbb{C}^N$ (which represents specs to be met at 'frequency' $e^{i\theta}$).

(OPT′) *Find a vector valued function f analytic in the unit disk Δ and continuous on its closure $\bar{\Delta}$, such that*

$$f(e^{i\theta}) \in S_\theta \quad \forall\, \theta.$$

Here Π stands for the unit circle $\{|z| = 1\}$ in $\mathbb{C}$ and henceforth A_N denotes the set of functions f on Π which are analytic on Δ and continuous in $\bar{\Delta}$.

In the engineering literature what we call A_N is often called H_N^∞. The distinction here is the rather technical one of continuity of functions on the boundary. Often we abbreviate A_1 by A. It is easy to put many design problems into this form, (e.g. the Horowitz templates of control) so the issue quickly becomes mathematical: computing solutions and developing a useful qualitative theory.

I always think of a picture in connection with (OPT′).

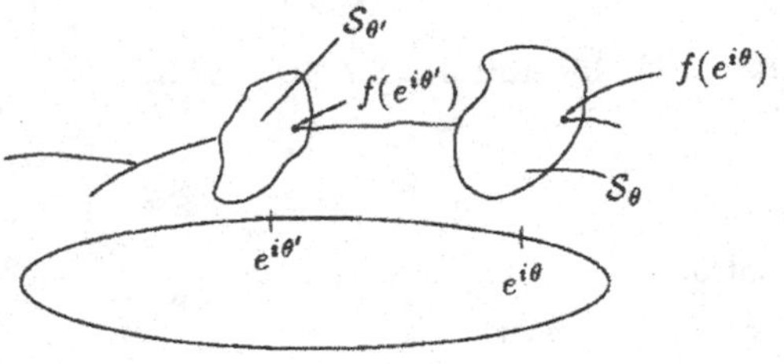

Figure L1.1.

The problem (OPT′) is very closely related to

(OPT) *Given $\Gamma \geq 0$ a map from $\Pi \times \mathbb{C}^N$ to R (which is a performance measure). Find $\gamma^* \geq 0$ and $f^* \in A_N$ which solve*

$$\gamma^* = \inf_{f \in A_N} \sup_\theta \Gamma(e^{i\theta}, f(e^{i\theta})) = \sup_\theta \Gamma(e^{i\theta}, f^*(e^{i\theta}))\,.$$

Indeed (OPT′) is a graphical version of

(OPT$_c$) *For a fixed (performance level) c find a function $f \in A_N$ with*

$$\Gamma(e^{i\theta}, f(e^{i\theta})) \leq c, \ \forall\theta.$$

(That is, f produces performance better than c). To see that (OPT′) solves (OPT$_c$) we start with Γ and denote its sublevel sets corresponding to c by

(1.1) $$\mathcal{S}_\theta(c) \triangleq \{z \in \mathbb{C}^N;\ \Gamma(e^{i\theta}, z) \leq c\}.$$

Now we take $\mathcal{S}_\theta = \mathcal{S}_\theta(c)$ in (OPT′) and see that

$$f(e^{i\theta}) \in \mathcal{S}_\theta \quad \text{if and only if} \quad \Gamma(e^{i\theta}, f(e^{i\theta})) \leq c.$$

Thus $f \in A_N$ solves (OPT′) if and only if f solves (OPT$_c$). Conversely, to go from (OPT$_c$) to (OPT′) one merely builds a *defining function* Γ for the given set $\mathcal{S}_\theta$, that is, build a Γ which satisfies

$$\begin{aligned} \Gamma(e^{i\theta},\ ) &= 1 \quad \text{on} \quad \partial\mathcal{S}_\theta \\ \Gamma(e^{i\theta},\ ) &< 1 \quad \text{inside} \quad \partial\mathcal{S}_\theta \\ \Gamma(e^{i\theta},\ ) &> 1 \quad \text{outside} \quad \partial\mathcal{S}_\theta. \end{aligned}$$

Then one solves (OPT$_1$) for that Γ.

2. Generic Motivation

The (OPT) problem is central to the design of a system where specifications are given in the frequency domain and stability is a key issue. Suppose our objective is to design a system part of which we are forced to use (in control it is called the plant) and part of which is designable

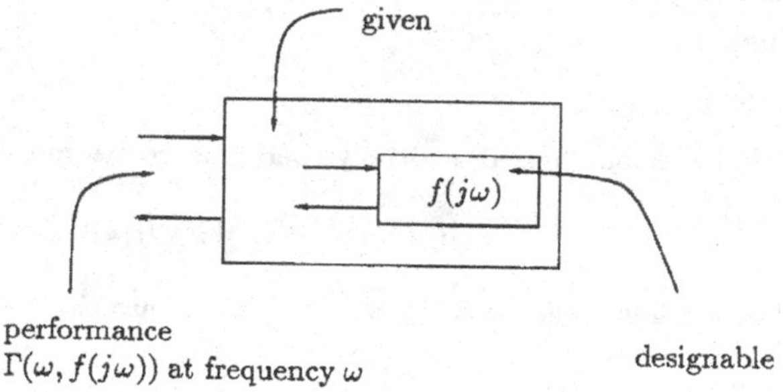

Figure L1.2.

The objective of the design is to find the admissible f which gives the best performance. The "worst case" is the frequency ω at which

$$\sup_\omega \Gamma(\omega, f(j\omega))$$

occurs. One wants to minimize this over all admissible f. The stipulation that the designable part of the circuit be stable amounts to requiring that f has no poles in the R.H.P. In other words $f \in A_N$ (R.H.P.). This is exactly the (OPT) problem for the R.H.P and, of course, conformally transforming R.H.P. to Δ transforms this problem to precisely the (OPT) problem we stated in

§1. However, in discussing applications we shall stay with the R.H.P. and $j\omega$-axis formulation of problems as is conventional in engineering. . Even when parts of the system other than the designable part are in H^∞ one can frequently reparametrize to get (OPT). Consequently (OPT) arises in a large class of problems.

Indeed the (OPT) problem is so basic that I am fond of calling it *the fundamental H^∞ problem of control.* This sits in distinction to the fundamental problem of H^∞-control. I don't know what that problem is.

3. Explicitly Solvable Problems. When each $\mathcal{S}_\theta(c)$ is a "disk" such as

$$\mathcal{S}_\theta(c) = \left\{ (z_1, z_2, \ldots, z_N) \in \mathbb{C}^N \colon \sum_{\ell=1}^{N} p_\ell(e^{i\theta}) |K_\ell(e^{i\theta}) - z|^2 \leq c \right\}$$

for some complex valued functions K_ℓ and positive valued functions p_ℓ in $\mathbb{C}^N$ the problem has an *explicit solution.* Other solvable (OPT) problems are easy to recognize since so few of them exist. They are (OPT)

(i) when each $\mathcal{S}_\theta(c)$ is a disk on $\mathbb{C}^N$

(ii) when each $\mathcal{S}_\theta(c)$ is a matrix ball

(iii) when each $\mathcal{S}_\theta(c)$ is a ball of symmetric matrices $M = M^T$

(iv) when each $\mathcal{S}_\theta(c)$ is a ball of antisymmetric matrices $M = -M^T$.

A function Γ each of whose sublevel sets $\mathcal{S}_\theta(c)$ are disks in this sense will be called *quasicircular.*

There is a huge literature on solutions to these problems. See books by Ball-Gohberg-Rodman **[BGR]**, Dym **[Dym]**, Foias-Frazho **[FF]**, Francis **[Fr]**, Helton **[H3]**, Nagy-Foias **[NF]**, Young **[Yng]**.

4. Engineering Examples

Example 1. The famous "mixed sensitivity" performance measure of control is

$$\tilde{\Gamma}(\omega, T) \triangleq W_1(j\omega)\,|T-1|^2 + W_2(j\omega)\,|T|^2 \tag{1.2}$$

where W_1 weights low frequencies and W_2 weights high frequencies. The *basic H^∞ control problem* is

> *Find a compensator which produces an internally stable system with acceptable performance (mixed sensitivity) over all frequencies.*

This converts directly to an (OPT) problem over functions T analytic in the R.H.P. (denoted $A(R.H.P.)$) which meet interpolation conditions

$$T(\xi_k) = r_k \quad k = 1, 2, \ldots, m \tag{INT}$$

imposed by the engineering system one wants to control. Then we get the mathematical statement:

Basic H^∞ Control Problem: Find such a T which gives a certain performance

$$\tilde{\Gamma}(\omega, T(j\omega)) \leq c\,.$$

This is the (OPT_c) problem except for the interpolation constraints.

The interpolation constraints are easily dealt with by a reparametrization in function space. For example, if $m = 2$, then express T as

$$T(\xi) = \frac{1}{(1+\xi)^2}\left[r_1\frac{(\xi-\xi_2)}{\xi_1-\xi_2}(1+\xi_1)^2 + r_2\frac{(\xi-\xi_1)}{\xi_2-\xi_1}(1+\xi_2)^2\right] + \frac{(\xi-\xi_1)(\xi-\xi_2)}{(1+\xi)^2}H(\xi) \tag{1.3}$$

where H is in A(R.H.P.). Clearly T sweeps through the desired class (INT) as H sweeps through A(R.H.P.). Abbreviate (1.3) to $T = a + bH$ and substitute into $\tilde{\Gamma}$ to define

$$\Gamma(\omega, H) \triangleq \tilde{\Gamma}(\omega, a + bH)\,.$$

Then (OPT_c) for Γ is equivalent to the basic control problem.

The graphical version of this problem goes as follows. First note via simple algebra that the sublevel sets $S_\omega(c)$ of $\tilde{\Gamma}$ are always disks and that they vary in a pattern like this:

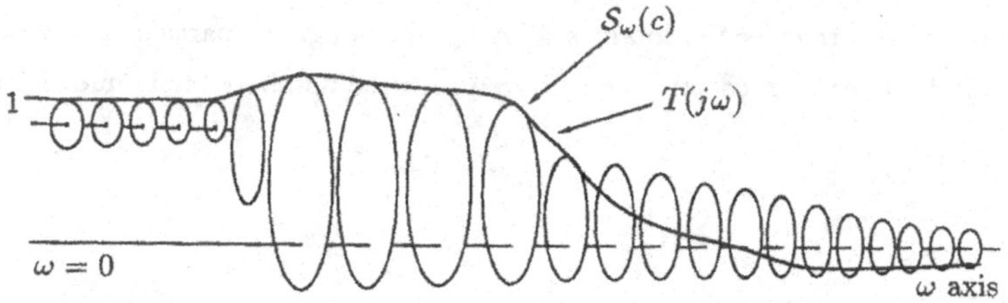

Figure L1.3.

Thus finding T in A meeting (INT) whose values on the $j\omega$ axis lie in $S_\omega(c)$ is equivalent to a basic "disk" problem of the explicitly solvable type.

Historically, the algebraic formulation (1.1) and graphical formulation Fig. L1.3 together with effective proposals for solution were done independently in 1983 by Kwaakernak [**Kw**] and Helton [**H1**], respectively. Also Doyle [**Dy1**] simultaneously gave a different approach which also was both physically correct and an effective computationally. The original paper of Zames and Francis [**ZF**] took $W_2 = 0$.

I might insert a caveat here to any practically oriented individual. Physically one is not given W_1 and W_2. The basic H^∞ control problem is a great abbreviation of the design process. In my opinion very little intelligent discussion about design can be carried out at this level of abbreviation.

Indeed a salty comment of mine along these lines pertains to a debate which persists in the H^∞ control community. The issue is whether or not a control theory will ever exist which in practice sets the weights W_1, W_2 once and for all. Opponents contend that in an industrial setting enormous tuning of the weights W_1, W_2 must occur. I don't wish to take sides here on the outcome. What impresses me about this debate is less the arguments of one side or the other, but the fact that it has not evolved significantly in the last five years. I (somewhat tongue in cheek) attribute this lack of progress to the physical imprecision of talking primiarly in terms of mixed sensitivity (IIb.1) and W_1, W_2. The W_1, W_2 are in fact derivable (with explicit formulas) from more primitive specifications such as tracking error, gain-phase margins, bandwidth constraints, etc. (c.f. **[DF]**, **[H1]**, **[BHMer]**). If the H^∞ culture commonly used this or an even more precise language possibly the debate on tuning could advance to a higher plane.

It seems to me that another practical issue might be worth emphasizing (since it is frequently treated incorrectly). It is an outgrowth of the fact that the basic control problem for many plants (ones with a pole or zero on the jw axis) is ill conditioned. This is because the basic control problem has not been correctly posed; the difficulty lies in the fact that the usual notion of internal stability:

$$(1.4) \qquad T, (I+PC)^{-1}, (I+PC)^{-1}P, C(I+PC)^{-1} \quad \text{are in} \quad H^\infty$$

while philosophically correct is too loose practically speaking. In particular a designer must have specified initially enough constraints to have produced numbers (or bounded functions)

$$M_1, \ M_2, \ M_3, \ M_4$$

so that

$$(1.5) \qquad |T| \le M_1, \ |(I+PC)^{-1}| \le M_2, \ |C(I+PC)^{-1}| \le M_3, \ |(I+PC)^{-1}P| \le M_4,$$

through the entire R.H.P. or even on a slightly larger region. Note that (1.4) is just a strengthened form of (1.3), in that (1.3) says that these functions must be bounded in the R.H.P., but does not say by how much. The point is that we must a priori say what these bounds are.

As mentioned before, forgetting constraints (1.4) becomes deadly at a jw-axis pole or zero of $\mathbb{P}$. For example, when $P(jw_0) = 0$ a naive H^∞ solution produces compensators with $|C(jw_0)|$ of arbitrarily large size. Fortunately, adding constraints (1.4) to the standard H^∞ control solution is easy to do using a function space reparameterization like (1.2). The interested reader is referred to Part I **[BHMer]** which does an example thoroughly; also **[H2]** mentions this tersely.

Exceptions to this occur in common engineering practice and they must be treated by explicitly building the exceptions into the mathematics of the control problem. For example, if the compensator is an integrator $\frac{1}{s}$ times something else $\tilde{c}(s)$, then the compensator $c(s) = \frac{\tilde{c}(s)}{s}$ certainly is not bounded at $s = 0$. In practice $c(s)$ cannot have a true pole at 0, and so there is a bound it must satisfy. However, for design purposes typically one models $c(s)$ as having a true pole at 0. With the current level of software one treats cases like this specially. One assumes that there will be a pole at 0, then does all of the optimization in terms of $\tilde{c}(s)$, a function which must be uniformly bounded at 0.

Example 2. Smoothing. For MIMO problems the mixed sensitivity function Γ is not smooth because it contains matrix norms. (However, it is quasicircular.) To approximate Γ by a smooth function one merely needs to approximate $\| \ \|_{M_{mn}}$ by a smooth approximate, such as

$$\|M\|_{M_{mn}} = [\mathrm{tr}\,(M^*M)^{2^k}]^{1/2^k} .$$

Small k should produce good approximations for most engineering purposes.

Example 3. Power mismatch (cf. [H4,5], [Y], [YS]). This area actually was the predecessor of H^∞ control. A paradigm problem is that of choosing the source impedance ρ_s (expressed here in the scattering formalism) to produce the amplifier in Figure L1.4 having maximum gain. The transistor is assumed to be unilateral and its scattering parameters S_{ij} are given, as is the load scattering parameter ρ_L.

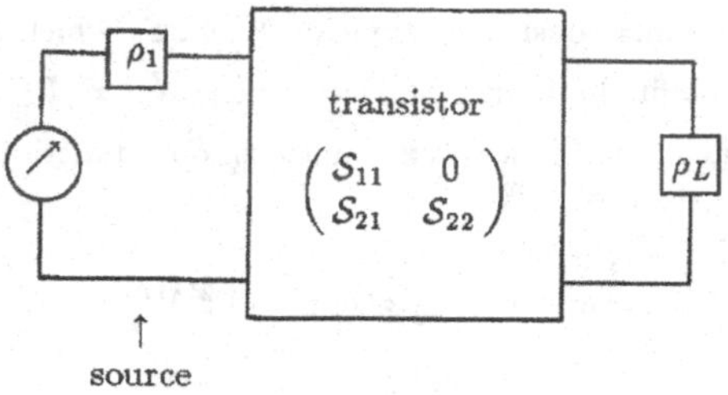

Figure L1.4.

The gain (normalized) for this amplifier can be computed to be

$$G = |S_{21}|^2 \, \frac{(1-|\rho_S|^2)}{|1-S_{11}\rho_S|^2} \, \frac{1-|\rho_L|^2}{|1-S_{22}\rho_L|^2} .$$

It is common practice in a narrowband (fixed ω) design to plot the level sets of G as a function of the unknown ρ_S. One uses this graphical aid in selecting a suitable ρ_S. To do a broadband design using H^∞ methods one can produce an (OPT) problem by choosing Γ in several ways. For example,

$$\Gamma(\omega, \rho_S) = \frac{1}{G} W_1 + W_2 |\rho_S|^2$$

where W_1, W_2 are weight functions (of frequency) which select the frequency band where one requires good performance. Here W_2 is chosen to insure $|\rho_S| \le 1$ for all frequencies. Note at frequencies where $W_1 = 1$ and $W_2 = 0$ the sublevel sets $\mathcal{S}_\omega(c)$ of Γ are just the disks of the type seen in Figure L1.5.

Smith Chart

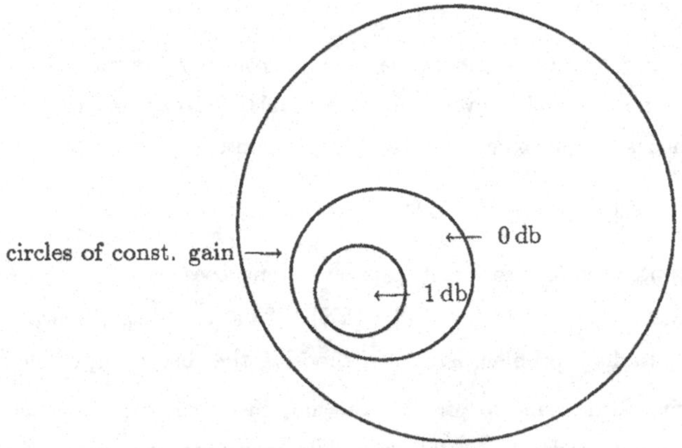

Figure L1.5

Example 4. Two competing constraints typically yield S_θ which are intersections of two disks, etc. Algebraically these fit the form $\Gamma(\omega, z) = \max\{\Gamma_1(\omega, z), \Gamma_2(\omega, z)\}$ where Γ_1 and Γ_2 are two given performance functions. Since max is not a smooth function, Γ will not be smooth. However, it is approximated by

$$\Gamma = \max\{\Gamma_1, \Gamma_2\} \approx (\Gamma_1^{2^k} + \Gamma_2^{2^k})^{1/2^k}$$

well even when k is small.

Example 5. Plant uncertainty naturally leads to (OPT) problems with very complicated Γ. Our formulation is to start with a performance measure $\tilde{\Gamma}$ which depends on what one believes the plant P to be at frequency ω and the choice T of the designable parameter at ω. The basic design optimization problem is:

(UNCOPT) $$\inf_{T \in A_N^\infty} \sup_{\omega} \sup_{p \in R_\omega} \tilde{\Gamma}(\omega, p, T(j\omega)).$$

Here R_ω denotes the range of values p at frequency ω which you believe the plant $P(j\omega)$ might actually take. For this problem "tightening the specs" amounts to calculating the "tightened" performance measure

(UNC) $$\Gamma(\omega, T) = \sup_{p \in R_\omega} \tilde{\Gamma}(\omega, p, T).$$

After this is done solving the full (UNCOPT′) problem is equivalent to (OPT).

(OPT) $$\inf_{f \in A_N} \sup_{\omega} \Gamma(\omega, f(j\omega))$$

Thus (OPT) *is the pure* H^∞ *part of* (UNCOPT) (which is another good reason for calling (OPT) the fundamental H^∞ problem of control).

Plant uncertainty when treated in this way simply amounts to a methematization of the age old engineering adage:

In the presence of uncertainty tighten the specs

The maximization in (UNC) is time consuming and is a subject unto itself (the structured singular value (s.s.v.) and environs). Consequently, one certainly expects that the most effective numerical algorithms at the $k+1^{st}$ iteration will update current guess f^k and $P^k(j\omega)$ by doing a (UNC) step increase $\tilde{\Gamma}(\omega, P, f^k(j\omega))$ for each ω, then an f step to decrease $\|\Gamma\|_\infty$, then another P step, then another f step, etc.

Doyle's μ-synthesis (after substantial reparameterization and a compromise) is such an algorithm. It is very natural since the function $\Gamma(\omega, p, z)$ of (UNCOPT) is in many control applications quasicircular for fixed p. Thus coordinate optimization produces a sequence $P^k \in L^\infty$ and $f^k \in H^\infty$ where the update to f^k comes by solving a quasicircular (OPT) problem. One could imagine an infinite variety of choices between spending a long time on each p maximization step before going to an f step or vice versa. Consequently, it is clear that to begin a systematic study of the H^∞ plant uncertainty problem (UNCOPT) one should analyze two extreme situations. The *first* is where one does the p maximization completely; this is (UNC). It is in principle studied by the s.s.v. school. The *second* is (OPT) for the "tightened" performance measure Γ. Ultimately, one expects that wisdom gained from studying these two extreme cases will combine to form a mature theory of (UNCOPT).

Lecture 2.

COMPUTER SOLUTIONS

The first lecture was designed to generate interest in the OPT problem and bring us to the question: How do we solve it? That is the topic of this lecture. A glib answer to the question is: Run a computer program.

1. A computer demonstration

Orlando Merino and I have a program for solving OPT which is easy to use and easy to modify for performing experiments. It runs out of Mathematica, a very powerful package which does both algebraic and numerical operations. Also we have a Fortran version of the program but it is harder to use. If you are interested in trying our program please write or better yet send e-mail to:

helton@osiris.ucsd.edu

Here is how one uses our program called **anopt** on simple examples: Your inputs are

Example 1. Solve the problem

$$\inf_{f \in A} \sup_{\theta} \left| 0.8 + \left(\frac{1}{e^{i\theta}} + f(e^{i\theta}) \right)^2 \right|^2$$

This is an OPT problem with Γ given by $\Gamma(e, z) = \left| 0.8 + (\frac{1}{e} + z)^2 \right|^2$ where e is an abbreviation for $e^{i\theta}$.

The main steps in solving it with our program are to enter Mathematica, call our package `Anopt.m`, to type in the formula for Γ and label it g, initial guess f_0, and finally to run the program. For our example this looks like

<<Anopt.m

$g = \text{Abs}\,[0.8 + (1/e + z[1])^2]^2$

$f0 =$ InitialGuess $[\{0\}, 128]$

Anopt$[g, f0, 12, 10 \wedge (-10)]$

Here the `Anopt` arguments are: `Anopt`[Γ, initial guess, maximum number of iterations, stopping tolerance]. Also other arguments are: `InitialGuess` [algebraic formula for the initial guess, gridsize].

The program *runs* and produces a table as you watch the screen, one line per iteration. The table lists the perfomance level, step size, how close $\Gamma(e^{i\theta}, f^k(e^{i\theta}))$ is to being frequency independent, and another diagnostic called Wno.

it #	sup $\Gamma[e, f_k(e)]$	norm $[f_k - f_{(k-1)}]$	Sup−Inf	Wno
0	0.3240000000000000E+01	********	0.32E+01	0.0E+00
1	0.1345600000000000E+01	0.40E+00	0.64E+00	0.0E+00
2	0.1025813183350302E+01	0.14E+00	0.59E−01	0.0E+00
3	0.1000349531551161E+01	0.14E−01	0.73E−03	0.0E+00
4	0.1000000079892810E+01	0.20E−03	0.16E−06	0.0E+00
5	0.1000000043437466E+01	0.61E−07	0.86E−07	0.0E+00
6	0.1000000010172161E+01	0.29E−07	0.21E−07	0.0E+00
7	0.1000000010172162E+01	0.48E−15	0.21E−07	0.0E+00

0
NORMAL EXIT
0Number of Iterations : 7
Opt. value sup{$\Gamma(f)$}: 1.00000
Diagnostic (Sup−Inf) 0.211003E−07

The $e^{i\theta}$ grid values of the analytic function that solves OPT are stored in the Mathematica variable denote **Solution**.

The full power of Mathematica applies. For example, a little function

Plot3DVector[Solution,0.3]

which we wrote gives a picture of the curve

$$\{(\theta, f^*(e^{i\theta})) \colon 0 \le \theta \le 2\pi\}$$

Figure L2.1.

Here .3 is a factor that sets scales. Also the plotting features of Mathematica can help us understand the performance function Γ itself. Here is a plot of sublevel sets:

$$\mathcal{S}_0(c) = \{z \colon \Gamma(1, z) \le c\}\ .$$

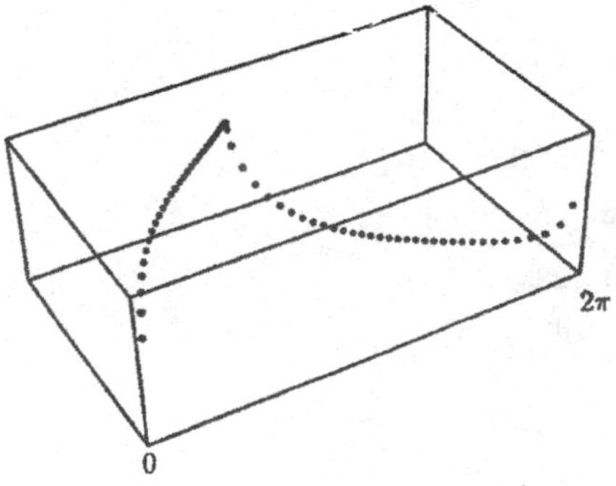

Figure L2.2.

Here .3 is a factor that sets scales. Also the plotting features of Mathematica can help us understand the performance function Γ itself. Here is a plot of sublevel sets:

$$\mathcal{S}_0(c) = \{z\colon\, \Gamma(1, z) \leq c\}\,.$$

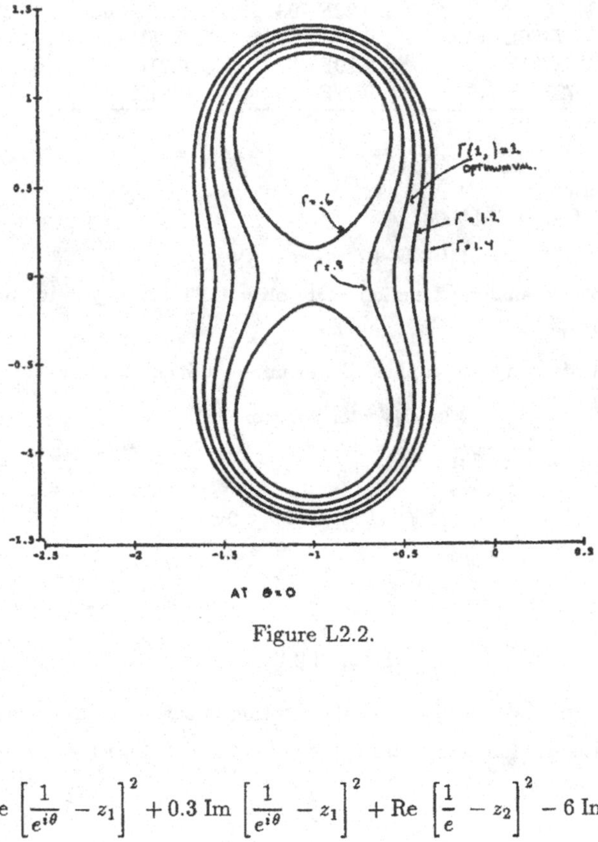

Figure L2.2.

Example 2.

Solve

$$\inf_{f \in A_2} \sup_{\theta} \left\{ \mathrm{Re}\left[\frac{1}{e^{i\theta}} - z_1\right]^2 + 0.3\ \mathrm{Im}\left[\frac{1}{e^{i\theta}} - z_1\right]^2 + \mathrm{Re}\left[\frac{1}{e} - z_2\right]^2 - 6\ \mathrm{Im}\left[\frac{1}{e^{i\theta}} - z_2\right]^2 \right\}.$$

The `Anopt` run looks like:

<<Anopt.m

$g = \mathrm{Re}\ [1e\ - z[1]]^2 + 0.3\ \mathrm{Im}\ [1/e - z[1]]^2$
$+\mathrm{Re}\ [1/e\ - z[2]]^2 + 6\ \mathrm{Im}\ [1/e\ - z[2]]^2$

$f0 = \mathrm{InitialGuess}\ [\{0, 0\}, 128]$

$\mathrm{Anopt}[g, f0, 12, 10 \wedge (-10)]$

The output is:

it #	sup $\Gamma[e, f_k(e)]$	norm $[f_k - f_{k-1})]$	Sup−Inf	Wno
0	0.6300000000000000E+01	********	0.43E+01	0.15E+02
1	0.2388810617317908E+01	0.21E+01	0.73E−04	0.27E+02
2	0.2356068189969852E+01	0.29E+00	0.19E−01	0.94E+00
3	0.2348112231884711E+01	0.85E−01	0.57E−06	0.13E+00
4	0.2348074079343865E+01	0.10E−01	0.20E−04	0.23E+01
5	0.2348065839256768E+01	0.26E−02	0.88E−10	0.44E−02
6	0.2348065797007293E+01	0.34E−03	0.22E−07	0.79E−03
7	0.2348065787866543E+01	0.87E−04	0.55E−12	0.15E−03
8	0.2348065787817682E+01	0.11E−04	0.26E−10	0.27E−04
9	0.2348065787807094E+01	0.29E−05	0.58E−14	0.50E−05
10	0.2348065787807040E+01	0.40E−06	0.36E−13	0.67E−06
11	0.2348065787807029E+01	0.55E−07	0.53E−14	0.70E−07
12	0.2348065787807029E+01	0.63E−08	0.53E−14	0.25E−08

0

0MAXIMUM NUMBER OF ITERATIONS EXCEEDED

0Number of iterations:	12
0pt. value sup $\{\Gamma(f)\}$:	2.34807
Diagnostic (sup − inf)	0.532907E−14
Diagnostic (Wno V)	0.247667E−08

The grid values of f^* are stored in **Solution**.

2. Various computational efforts

This section is hardly complete and makes only a few remarks. The subject naturally divides into numerics for

(a) Γ which are quasi-circular (sublevel sets are disks).

(b) General Γ.

The first subject contains the numerics of H^∞ control and so is a huge field within the engineering (not the numerical analysis) community. I shall not discuss it.

Numerical efforts on OPT for general Γ and related problems are carried out by various groups using very different methods.

(1) Peak point methods—Mayne-Polak-Salucidean, Fan-Tits

(2) Linear programming—Streit, Boyd, Daleh, Pearson.

(3) Convex programming—Boyd.

(4) μ-synthesis—Doyle, the Honeywell group, Chu, Lenz, etc. (solves UNCOPT)

(5) Quasi-circular gradient Newton—Merino-Helton.

(6) Frequency dependent, conformal mapping ($N = 1$)—Sideris.

Codes are available from several of these groups, including Fan-Tits, Streit, Boyd, Helton-Merino. Efforts (1), (2) and (3) are carried out independently of numerical efforts in classical H^∞ control while (4), (5), and (6) iterate classical H^∞ control solutions.

We shall not describe all of these numerical approaches here but we mention one general class of algorithms here called quasicircular iteration methods. The methods 4, 5, and 6 on our list are of this type. Recall from §Ia that OPT can be solved "exactly" when the sublevel sets $S_\theta(c) = \{z\colon \Gamma(e^{i\theta}, z) \le c\}$ are disks in $\mathbb{C}^N$ (quasicircular Γ). When the sublevel sets are not disks, no "exact" method is known. Since the problem is highly nonlinear, there is probably no other way to solve it than to do it approximately. Here iterative methods are an important choice. The general class of algorithms for solving OPT iteratively works as follows.

Quasicircular Iteration

Given a function Γ and f^0 in A_N, to produce an update $f^1 \in A_N$,

(1) Obtain a quasicircular approximation $\tilde{\Gamma}(e^{i\theta}, z)$, to $\Gamma(e^{i\theta}, f^0(e^{i\theta}) + z)$.

(2) Solve OPT for $\tilde{\Gamma}$ and denote its solution by h.

(3) Find t that minimizes $\sup_\theta \Gamma(e^{i\theta}, f^0(e^{i\theta}) + th(e^{i\theta}))$.

(4) Set $f^1 = f^0 + th$.

One characteristic of quasicircular Iteration Algorithms and other similar methods is that (computationally efficient) techniques known for solving the quasicircular case are taken advantage of.

An analysis of quasicircular algorithms appears in **[HMer1]** as well as a list of detailed algorithms.

3. Diagnostics

An extremely valuable tool in computation especially with iterative methods is a collection of *diagnostics* which tell if a function f^* is close to a solution or not close to a solution. The diagnostics we know for OPT are based directly on pure mathematics theorems which we state below. We emphasize that these diagnostics apply to OPT itself and so can be used with any computer program (provided Γ is smooth). The first diagnostic is

(DIAG I)
$$\sup_\theta \Gamma(e^{i\theta}, f^k(e^{i\theta})) - \inf_\theta \Gamma(e^{i\theta}, f^k(e^{i\theta}))$$

The second diagnostic is based on the functions

$$a_j(e^{i\theta}) \triangleq \frac{\partial \Gamma}{\partial z_j}(e^{i\theta}, f^k(e^{i\theta})) \,.$$

The diagnostic is

(DIAG II)
$$\sum_{j=1}^{N} \|P_{H^{2\perp}}(a_j/a_{j_0})\|_{L^2} \,.$$

which is easy to compute. This diagnostic becomes valid when wno $a_{j_0} = 1$, something which is easy to check on the computer. At optimum both (DIAG I) and (DIAG II) should equal zero to nearly machine precision.

These diagnostics have already been illustrated in the tables which our computer examples printed out. In both examples the column labeled $\sup_\theta - \inf_\theta$ equals DIAG I. In Example 2 the column labeled wno equals DIAG II. In Example 1, the column labeled wno is something simpler than DIAG II but is consistent with it. This is because the $N = 1$ case is simpler than the $N > 1$ case.

The diagnostics are based directly on a mathematics theorem which characterizes local solutions to OPT. We now state this theorem and from it the reader can see how (DIAG I) and (DIAG II) arise. This gives a nice example of how theory of OPT can have direct practical consequences.

A *local solution* $\in A_N$ is a solution to OPT in some neighborhood $V \subset A_N$ of f. It is a *strict local solution* if it is unique in V (for some V). The function f is a (*strict*) *directional solution* if for each $h \in A_N$ we have that f is a (strict) local solution to OPT when the set of admissible functions is restricted to $\{f + th,\ t \in \mathbb{R}\}$.

Of course, every (strict) local solution is a (strict) directional solution. The converse is true in finite dimensional spaces, but not in general infinite dimensional spaces. The next theorem gives a complete characterization of directional solutions, and therefore, necessary conditions for local solutions.

THEOREM L2.1. [**HMer2**] *Let Γ be of class C^3 and f^* in A_N be such that the $\mathbb{C}^N$ valued function $a(e^{i\theta}) \triangleq \frac{\partial \Gamma}{\partial z}(e^{i\theta}, f^*(e^{i\theta}))$ is never 0 on $\mathbb{T}$. If f^* is a directional solution to* OPT, *then,*

(I) *$\Gamma(e^{i\theta}, f^*(e^{i\theta}))$ is constant in $e^{i\theta}$.*

(II) *There exist $F \in H^1_N$ and λ: $\mathbb{T} \to \mathbb{R}^+$ measurable such that $a(e^{i\theta}) = e^{i\theta}\lambda(e^{i\theta})F(e^{i\theta})$, for almost all $e^{i\theta} \in \mathbb{T}$.*

(III) *For each $h \in N(a) = \{h\colon a(e^{i\theta})^t h(e^{i\theta}) = 0 \text{ for all } e^{i\theta}\}$,*

$$\sup_\theta \{\overline{h(e^{i\theta})^t} A(e^{i\theta}) h(e^{i\theta}) + \operatorname{re} h(e^{i\theta})^t B(e^{i\theta}) h(e^{i\theta})\} \geq 0\,.$$

Here $A(e^{i\theta})$, $B(e^{i\theta})$ are the $N \times N$ matrix valued functions with (ℓ, j) entries given by $\frac{\partial^2 \Gamma}{\partial \bar{z}_\ell \partial z_j}(e^{i\theta}, f^*(e^{i\theta}))$ and $\frac{\partial^2 \Gamma}{\partial z_\ell \partial z_j}(e^{i\theta}, f^*(e^{i\theta}))$ respectively.

Conversely, conditions (I), (II), *and* (III) *with strict inequality are sufficient for f^* to be a strict directional solution.*

A condition that together with (I), (II) becomes sufficient (but not necessary) for strict local solutions has been treated in [**HMer2**]. This condition is automatically satisfied if $\Gamma(e^{i\theta}, z)$ is strictly convex in z for each $e^{i\theta} \in \mathbb{T}$.

The case $N = 1$ is special and deserves some comments. Observe that in this case the set $N(a)$ in (III) is trivial. Therefore (III) is automatically satisfied, and (I), (II) become necessary and sufficient for directional optimality, and for local optimality as well. Another observation is that when $N = 1$, the condition (II) has been classically known as the requirement that the winding number of the function a about zero is positive.

In the general case ($N \geq 1$), Conditions (II) and (III) are not completely satisfactory in that they involve unknown functions λ and F (in (II)) and the space $N(a)$ (in (III)). This is particularly important for use of the conditions in computer programs. A practical version of condition III appears in **[HMer2]**. A practical alternative versions of (II) which requires only knowledge of the function $a(\cdot)$ is

THEOREM L2.2. [**HMer1**] *If j_0 in $\{1, N\}$ is such that a_{j_0} is never 0 and has* wno $a_{j_0} = 1$, *then* II *of Theorem* L2.1 *is equivalent to*

$$\textit{For each } j = 1, \dots, N, \quad \textit{the function } \frac{a_j}{a_{j_0}} \textit{ is analytic .}$$

The hypothesis on the winding number in Theorem L2.2 above seems to be restrictive, but this is deceiving because of

THEOREM L2.3. [**Mer**] *If $N = 1$, for generic Γ,* wno $a = 1$.

Theorem L2.3 above applies also when $N > 1$ to each a_j. To see this all one has to do is to fix all complex variables except for z_{j_0}, to obtain a (scalar) problem OPT for which $a = a_j$. While this does not prove the $N > 1$ version of Theorem L2.3 it suggests that it is true. Computer experiments strongly bear this out.

Now it is easy to understand DIAG I and DIAG II. They check to see if conditions I and II of Theorem L2.1 are close to holding DIAG I checks (I) of the theorem, namely, how flat performance is, while DIAG II checks the Theorem L2.2 version of (II). Automatically we check wno a_j before invoking DIAG II. Thus if it appears on the screen it is valid.

Intuitively informative is the case when $a(\cdot)$ has entries that extend meromorphically to a neighborhood of the unit disk in $\mathbb{C}$, then condition (II) of Theorem L2.1 becomes

THEOREM L2.4. [**H6**] *Assume the hypotheses to Theorem* L2.1. *If for each $j = 1, \dots, N$, the function a_j extends meromorphically to a neighborhood of the disk, then* (II) *in Theorem* L2.1 *is equivalent to:*

> *The number of common zeros of the a_j's inside the disk minus*
>
> *the total number of poles of the a_j's inside the disk is positive.*

Here multiplicity must be counted.

Thus Theorems L2.2 and L2.4 above give formulations of (II) which are basically self-contained. An equation like II appears in the literature for the first time in Lempert's (c.f. [**L**]) characterization of extremals for the Kobayashi distance, a special case of Theorem L2.1.

We conclude by directing the reader to another application of Theorem L2.1. In [**HMer2**] and [**HMer3**] we use it to show that H^∞ coordinate descent algorithms almost *never* reach the true optimum.

4. Further topics

For iterative algorithms the *order of convergence* is a critical indicator of performance. This is a highly mathematical subject and is analyzed in **[HMer1]**. Our SISO algorithm is second order convergent, our MIMO algorithm is first order convergent, and we are at the moment experimenting with ways to increase its order of convergence.

Connections with other branches of mathematics are described in **[H7]**. In particular there are strong connections with the field of several complex variables, and issues involving optimal analytic disks **[W]**, **[K]**.

References for Lectures 1 and 2

[BGR] J. Ball, I. Gohberg and L. Rodman. Book, to appear in Operator Theory: Advances and Applications series, Birkhäuser, 1990.

[BHM] J. Bence, J. W. Helton and D. Marshall, H^∞ optimization, Proc., Conference on Decision and Control, Athens, 1986.

[BHMer] J. Bence, J. W. Helton and O. Merino, A primer in H^∞ control, documentation for software: UCSD, 1985.

[DFT] J. C. Doyle, B. A. Francis and A. Tannenbaum. Book, to appear.

[Dym] J. C. Doyle, Synthesis of robust controllers, Proc., IEEE Conference on Decision and Control, December, 1983, pp. 109–114.

[FF] C. Foias and A. Frazho, "The Commutant Lifting Approach to Interpolation Problems," in *Operator Theory: Advances and Applications*, vol. OT44, 1990.

[Fr] B. A. Francis, *A course in H^∞ control theory*, Lecture Notes in Control and Information Sci., vol. 88, Springer-Verlag, 1986.

[H1] J. W. Helton,An H^∞ approach to control, IEEE Conference on Decision and Control, San Antonio, December 1983

[H2] J. W. Helton, Worst case analysis in the frequency domain: an H^∞ approach to control, IEEE Trans. Auto. Control, AC-30 (1985), 1154–1170.

[H3] J. W. Helton, et.al, *Operator Theory, Analytic Functions, Matrices, and Electrical Engineering*, in Regional Conference Series in Mathematics, No. 68, Amer. Math. Soc., 1987.

[H4] J. W. Helton, Non-Euclidean functional analysis and electronics, Bull. Amer. Math. Soc., **7** (1982), 1–64.

[H5] J. W. Helton, Broadband gain equalization directly from data, IEEE Trans. Circuits and Systems, CAS-28 (1981), 1125–1137.

[H6] J. W. Helton, Optimization over spaces of analytic functions and the Corona problem, **13** (1986), 359–375.

[H7] J. W. Helton, Optimal frequency domain design vs. an area of several complex variables, pleniary address. MTNS, 1989.

[HMer1] J. W. Helton and O. Merino, A novel approach to accelerating Newton's method for sup-norm optimization arising in H^∞-control, preprint.

[HMer2] J. W. Helton and O. Merino, Conditions for Optimality over H^∞, preprint, pp.1–65.

[HMer3] J. W. Helton and O. Merino, Conditions for optimality over H^∞, American Conference on Control, 1990.

[K] S. Kranz, *Function Theory of Several Complex Variables*, Wiley, New York, 1982.

[Her] R. Herman, book, to appear.

[Kw] Kwaakernak, Robustness optimization of linear feedback systems, Proc., Conference on Decision and Control, December 1983.

[L] L. Lempert, Complex geometry in convex domains, Proceedings Internation Congress of Mathematics, Berkeley 1986.

[MH] D. Marshall and J. W. Helton, Frequency domain design and analytic selections, Indiana Journal of Math, 1990.

[Mer] O. Merino, Optimization over spaces of analytic functions, UCSD Thesis, 1988.

[NF] B. Sz.-Nagy and C. Foias, *Harmonic Analysis of Operators on Hilbert Space*, American Elsevier, New York, 1970.

[Y] D. C. Youla, A new theory of broadband impedance matching, IEEE CAS, 1964.

[YS] D. C. Youla and M. Saito, Interpolation with positive real functions, J. Franklin Inst., **284** (1967), 77–108.

[ZF] G. Zames and B. A. Francis, Feedback and minimax sensitivity, Advanced Group for Aerospace Research and Development NATO Lectures Notes No. 117, Multivariable Analysis and Design Techniques.

[Yng] N. Young, *An Introduction to Hilbert Space*, Cambridge University Press, 1988.

Lecture 3

NONLINEAR SYSTEMS: CONTROL PROBLEM AND RESULTS

The next two lectures address aspects of the nonlinear time–invariant H^∞ control problem in the discrete time setting. A recipe is presented which is shown to generate a solution of the H^∞ "model matching" problem for stable plants and which we conjecture generates a genuine solution in more general circumstances. This appears to be the canonical generalization to the nonlinear case of the Doyle-Glover-type equations (discrete time) for the linear model matching problem. The work is joint with Joe Ball and it appears in **[BH2]**, **[BH4]**, **[BH3]**, **[BH5]**; also see **[BH1]**. Ciprian Foias also has a lecture on this topic in this volume and he was very helpful to us in the early stages of this work. Another promising approach is to use power series expansions (see **[F]**).

1. INTRODUCTION

For mathematical convenience all problems are presented in discrete time. Therefore, for $\mathbb{C}$ any real Hilbert space (usually finite dimensional), we denote by $\ell^{2+}_{\mathbb{C}}$ the space of $\mathbb{C}$–valued sequences $\vec{u} = \{\vec{u}(n)\}_{n\geq 0}$ which are norm square–summable. The following basic problem of H^∞-control theory is described by Figure L3.1 (see [Fr]).

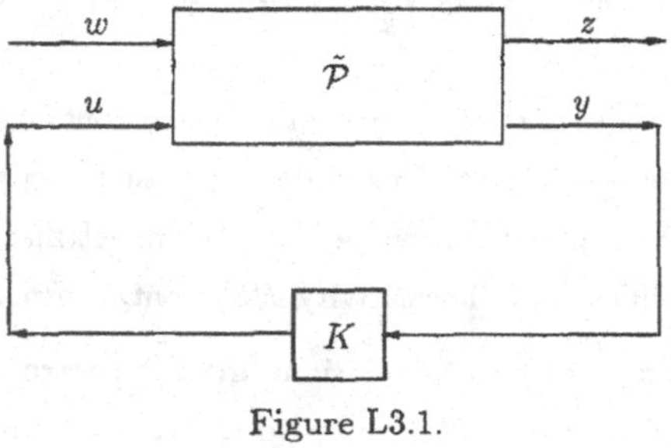

Figure L3.1.

(CTR). *Given a system $\mathcal{P}$ find a feedback K which produces an internally stable system so that the input–output (IO) map $T_{zw} : \vec{w} \to \vec{z}$ is contractive, that is it satisfies*

$$\|T_{zw}(\vec{w})\|_{\ell^{2+}_{Z}} \leq \|\vec{w}\|_{\ell^{2+}_{W}}.$$

Here $\mathcal{P}$ and K are assumed to be causal time–invariant IO maps. The input signals $\vec{w}$ and $\vec{u}$ for $\mathcal{P}$ are assumed to have values in finite dimensional input spaces W and U while the output signals $\vec{z}$ and $\vec{y}$ are assumed to have values in output spaces Z and Y. We demand that the system be well–posed so that the closed loop IO map T_{zw} is well–defined, causal, time–invariant

IO map. We do not insist that $\mathcal{P}$ or K be linear but do require that they take 0 to 0. In the linear case *internal stability* of the closed loop system can be defined in a couple of ways (see [Fr]). In terms of IO maps, it means that the output signal $\vec{z}$ as well as the internal signals $\vec{y}$ and $\vec{u}$ are in ℓ_Z^{2+}, ℓ_Y^{2+} and ℓ_U^{2+} respectively for any choice of ℓ_W^{2+}–input $\vec{w}$, even in the presence of ℓ^{2+}–perturbations of the internal signals. A state space formulation of internal stability is the state for the composite closed loop system in Figure L3.1 evolves stably from any initial state whenever any ℓ_W^{2+} signal is put into the system. These two definitions are equivalent in the linear case if we assume that we are working with minimal state space realizations for $\mathcal{P}$ and for K. For the nonlinear case we shall be working with a state space representation for $\mathcal{P}$

(1.1)
$$\mathcal{P}: \quad \begin{aligned} \dot{x} &= F(x, w, u) \\ z &= G_1(x, w, u) \\ y &= G_2(x, w, u) \end{aligned}$$

and shall seek a state space representation for K

(1.2)
$$K: \quad \begin{aligned} \dot{\xi} &= f(\xi, y) \\ u &= g(\xi, y) \end{aligned}$$

so that the resulting closed loop state space system is stable in one of the senses described in section 2.

An ambitious goal is to give a recipe which produces a controller which solves (CTR) in general and which is computationally implementable at least for systems having only "mild" nonlinearities. We succeed in giving the recipe for the model matching problem for stable plants. For a large class of linear mixed sensitivity H^∞–control problems the solution reduces to the usual maximum entropy or central solutions in H^∞–control. We shall give theorems which describe the range of validity of the recipe to a reasonable extent. Elsewhere **[BH3]**, **[BH5]** we work out in detail the example of a plant consisting of a linear system followed by a memoryless function h. We find that at least one instance where the recipe indeed leads to a solution is the case where $\|h(x)\|^2$ is strictly convex. On the other extreme if h is a saturation nonlinearity then serious difficulties arise.

Recently a number of papers (see **[B]**, **[LAKG]**, **[PS]**, **[T]**, **[YS]**) have explored the connections of H^∞–control with differential game theory (see **[BO]**). Various formulas here can be interpreted that way. Indeed the equation for the unknown energy function e which we present is just a Hamilton-Jacobi-Bellman-Isaacs equation for a particular differential game. The work in (**[BH1]**, **[BH2]**, **[BH3]**) originally motivated by a search for a nonlinear Beurling–Lax theory with applications in fact led to a rediscovery of some of these ideas in differential game theory. Our contribution here is two–fold. First, we remark that one can associate to a system of differential equations many differential games depending on information structure and order of play in the game. Selecting which games are useful for solving the H^∞-control problem is a tricky business. We believe that we have done it here correctly for stable plants. Secondly, we adapt the machinery from differential games to the setting of the infinite horizon nonlinear H^∞-control problem where stability is a crucial consideration. Indeed it is not clear if differential games helps much in understanding the basic recipe of this paper. Our constructions are very clear from passive circuit considerations and exactly why this particular game (the FICTR in Lecture 4) corresponds to (CTR) is still not philosophically clear to us. It just works.

2. PRELIMINARIES

Now we give some definitions. Let us suppose that we are given an IO map $\mathcal{P} : \vec{w} \to \vec{y}$ which is modeled by state space equations

$$\tag{2.1} \begin{aligned} \bar{x} &= F(x, w) \\ y &= G(x, w) \end{aligned}$$

Thus if the initial state is x, the output sequence $\vec{y}$ is generated by the input sequence $\vec{w}$ according to the recursion

$$\tag{2.2} \begin{aligned} \vec{x}(n+1) &= F(\vec{x}(n), \vec{w}(n)), \qquad \vec{x}(0) = x \\ \vec{y}(n) &= G(\vec{x}(n), \vec{w}(n)). \end{aligned}$$

We assume that the element 0 of the state space X is an equilibrium point; then $0 = F(0,0)$, $0 = G(0,0)$. Call the system *stable* (resp. *bounded input stable* BIstable) provided that for any input sequence $\vec{w}$ in ℓ^{2+}_W (resp. in $\ell^{\infty +}_W$) and any initial state the resulting sequence $\vec{x}$ of states is uniformly bounded.

Now suppose that we have specified an energy function $e : X \to \mathbb{R}^+$ for which $e(x) = 0 \iff x = 0$. We say that the system (2.1) is *weakly asymptotically e-stable* if for each initial state x there is some choice of input sequence $\vec{w}_x \in \ell^{2+}_W$ for which the resulting sequence of states $\vec{x}$ generated by (2.2) has the property that

$$\tag{2.3} \lim_{x \to \infty} e(\vec{x}(n)) = 0.$$

Similarly we say that (2.1) is *asymptotically e-stable* if (2.3) holds for *any* choice of input sequence $\vec{w} \in \ell^{2+}_W$. Next we say that (2.1) is simply *e-stable* if at least

$$\sup_n e(\vec{x}(n)) < \infty$$

for any choice of input sequence $\vec{w} \in \ell^{2+}_W$. We call e a Lyapunov function if it is positive except at $x = 0$, and proper that is $e(x(k)) \to \infty \Rightarrow \|x(k)\| \to \infty$. Thus for a Lyapunov function e-stability (resp. e-asymptotic stability) force true stability (resp. asymptotic stability).

Internal stability of the multicomponent system also comes in several versions corresponding to the types of stability just described. The basic requirement is that the system in Figure L3.1 be stable in whatever sense required *regardless of initial condition* on the state. This is the sense in which these lectures use the term internal stability.

We say that the system (2.1) is *input-output* (IO) *passive* if for any $\vec{w} \in \ell^{2+}_W$ and for $m = 0, 1, \ldots$ we have

$$\sum_{k=0}^{m} \|\vec{y}(k)\|^2 \leq \sum_{k=0}^{m} \|\vec{w}(k)\|^2$$

where the output sequence $\vec{y}$ is generated from (2.2) with $x = 0$. Alternatively we may express this by

$$\|P_m \vec{y}\|^2_{\ell^{2+}_Y} \leq \|P_m \vec{w}\|^2_{\ell^{2+}_W}$$

where for any sequence $\vec{x}$ in ℓ^{2+} we define

$$[P_m \vec{x}](k) = \begin{cases} \vec{x}(k), & 0 \leq k \leq m \\ 0, & k > m \end{cases}$$

Ultimately we would like a solution of (CTR) for which T_{zw} is IO passive. An excellent reference on various notions of passivity is the article **[CW]** by Chua and Wyatt.

3. THE RECIPE AND MAIN RESULTS

Now we turn to the recipe for solving (CTR). We assume that we are given a state space representation as in (1.1) for the plant $\mathcal{P}$. In addition we shall assume that G_2 is independent of u:

(3.1) $$y = G_2(x, w)$$

and, for each fixed x that G_2 is a diffeomorphism as a function of w. Thus there is a smooth function $G_2^I(x, \cdot)$ such that

(3.2) $$\begin{aligned} y &= G_2(x, G_2^I(x, y)) \\ w &= G_2^I(x, G_2(x, w)). \end{aligned}$$

The main unknown is a function $e : X \to \mathbb{R}^+$ on the state space; intuitively we think of $e(x)$ as representing how much potential energy is in the state x and the main equation is an energy balance inequality.

Recipe for (CTR)

1. Find a function $e : X \to \mathbb{R}^+$ satisfying

(ENGY) $$e(x) \geq \max_w \min_u \{e(F(x, w, u)) + \|G_1(x, w, u)\|^2 - \|w\|^2\}$$

as well as

$$e(x) = 0 \quad \text{if and only if} \quad x = 0.$$

With this choice of the function e define

$$Q(x,w,u) \triangleq e(F(x,w,u)) + \|G_1(x,w,u)\|^2 - \|w\|^2$$

2. Compute

$$u^*(x,w) \triangleq \arg\min_u Q(x,w,u)$$

3. Use as feedback law the dynamic compensator $\vec{u} = K(\vec{y})$ given by state space equations

$$\begin{aligned} \tilde{\xi} &= F(\xi, G_2^I(\xi,y), u^*(\xi, G_2^I(\xi,y))) \\ u &= u^*(\xi, G_2^I(\xi,y)) \end{aligned} \qquad \blacksquare$$

It is also possible to give continuous time analogs of these recipes. In the continuous time setting, the state space representation for the plant has the form

$$\mathcal{P}: \quad \begin{aligned} \dot{x} &= F(x,w,u) \\ z &= G_1(x,w,u) \\ y &= G_2(x,w) \end{aligned}$$

and we seek a state space representation for the compensator K of the form

$$K: \quad \begin{aligned} \dot{\xi} &= f(\xi,y) \\ u &= g(\xi,y) \end{aligned}$$

If in the recipes (ENGY) is replaced by

$$\text{(ENGY} - \text{CT)} \qquad 0 \geq \max_w \min_u \{\nabla e(x) \cdot F(x,w,u) + \|G_1(x,w,u)\|^2 - \|w\|^2\}$$

we obtain continuous time analogs of all the results stated here explicitly only for discrete time.

We now state our results concerning the validity of the recipes. We expect more definitive results will follow in time.

An assumption we need at the moment is

$$\text{(MM)} \qquad G_2(x,w,u) = G_2(w).$$

and as before G_2 is an invertible function. It is satisfied for the model matching problem. We shall call e a *Lyapunov function* provided that

(1) $e(x) \geq 0$ for all x,

(2) $e(x) = 0$ if and only if $x = 0$, and

(3) if $x(n) \to \infty$, then $e(x(n)) \to \infty$.

We have the following result:

Theorem 3.1. *Let $\mathcal{P}$ be a BIStable plant with state space equations of the form*

$$\begin{aligned}\dot{x} =& F(x,w,u)\\ z =& G_1(x,w,u)\\ y =& G_2(x,w),\end{aligned}$$

Assume $(x,w) \to u^*(x,w)$ is continuous and the compensator K is constructed as in the Recipe CTR *where we have found an energy function e which is a Lyapunov function. Suppose in addition $\mathcal{P}$ is in the special form* (MM)*. Then the closed loop output–feedback configuration with compensator K as in Recipe* CTR *is BIstable, K is BIstable, and so the system is bounded input internally stable.*

It is possible that internal stability result in Theorem L3.1 in fact holds under the weaker conditions that the system with input variable $\begin{bmatrix} u \\ y \end{bmatrix}$ and output variable $\begin{bmatrix} z \\ w \end{bmatrix}$ given by state space equations be stable. This is discussed further in Lecture 4 during the proof of Theorem L3.1.

Proofs of the results mentioned in this section will be discussed in the next section. We also mention that if assumption (3.1) fails, we replace assumption (3.2) with the assumption that the function $\tilde{G}_2(x,w) \triangleq G_2(x,w,u^*(x,w))$ be a diffeomorphism as a function of w for each fixed x, i.e. that there is a function $\tilde{G}_2^I(x,y)$ for which

$$(3.2') \qquad \begin{aligned} y =& \tilde{G}_2(x, \tilde{G}_2^I(x,y))\\ w =& \tilde{G}_2^I(x, \tilde{G}_2(x,w)).\end{aligned}$$

In Lecture 4 we give a formal recipe extending CTR under this assumption. Also we analyze it to some extent.

Lecture 4.

NONLINEAR SYSTEMS: METHODS AND RESULTS

1. STATE-FEEDBACK CONTROL

We first consider a special case of the general output feedback problem, a ***state space*** version of the problem where here it is assumed that the sensor whose output is fed into the compensator $\check{K}$ can measure the present value of the state x and the input w; this amounts to the special case of the general problem where the compensator K is required to be memoryless and the second output y of $\mathcal{P}$ is assumed to be $\begin{bmatrix} x \\ w \end{bmatrix}$. This leads to the *full information control problem* (see Figure L4.1):

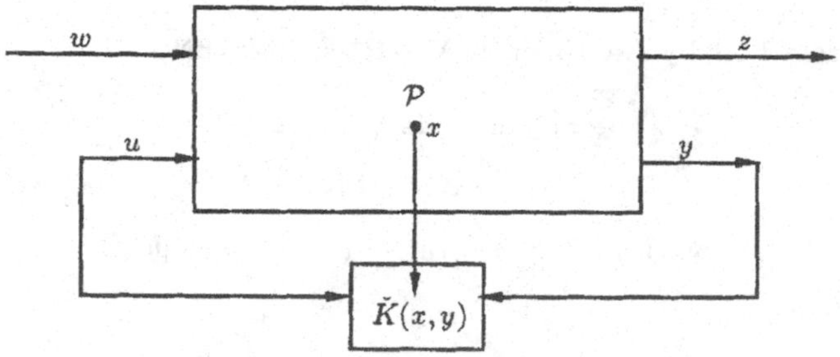

Figure L4.1.

(FICTR). *Solve (CTR) for the special case where $\check{K}$ is required to be memoryless and y is assumed to be given by* $y = \begin{bmatrix} x \\ w \end{bmatrix}$.

For this problem we have the formal recipe, which we state without assumptions on G_2.

Recipe for (FICTR)

(1) Find $e\colon X \to R^+$ which solves $e(x) \geq \max_w \min_u Q_e(x,w,u)$ where $Q_e = e(F(x,w,u)) + \|z\|^2 - \|w\|^2$.

(2) Now that e is known find $u^*(x,w) = \text{minimizer}_u\, Q_e(x,w,u))$.

(3) Suppose that

$$y = G_2(x, w, u^*(x,w)) \tag{1.1}$$

is for each x invertible as a map of $w \to y$ and define $\check{K}$ by the implicit formula

$$\check{K}y = u^*(x,w)$$

where y and w satisfy (1.1). For example, if G_2 does not depend on u, then

$$\check{K}y = u^*(x, G_2^I(x, y))$$

Theorem L4.1. *Suppose a function $e : X \to \mathbb{R}^+$ exists and $u^*(x,w)$ is defined as in the Recipe (FICTR). Then the closed loop transfer function T_{zw} as in Figure L4.1 with $u = u^*(x,w)$ is input–output passive, i.e.*

$$\|P_M T_{zw}(\vec{w})\|^2_{\ell_Z^{2+}} \le \|P_M \vec{w}\|^2_{\ell_W^{2+}}$$

for $M = 0, 1, 2, \ldots$. Consequently

$$\|T_{zw}(\vec{w})\|^2_{\ell_Z^{2+}} \le \|\vec{w}\|^2_{\ell_W^{2+}}$$

for all $\vec{w} \in \ell_W^{2+}$. Moreover, the state space system in Figure L4.1 constructed by Recipe FICTR is e–stable.

Proof of Theorem L4.1 By assumption $e : X \to \mathbb{R}^+$ satisfies (ENGY), so

$$e(x) \ge \max_w \min_u Q(x, w, u)$$

where

$$Q(x,w,u) = e(F(x,w,u)) + \|G_1(x,w,u)\|_Z^2 - \|w\|_W^2.$$

Hence for any fixed x and w,

$$e(x) \ge \min_u Q(x,w,u) = Q(x,w,u^*(x,w)).$$

Plugging in the definition of Q gives

$$e(x) \ge e(F(x,w,u^*)) + \|G_1(x,w,u^*)\|^2 - \|w\|^2 \tag{1.2}$$

for all $x \in X$ and $w \in W$ where $u^* = u^*(x,w)$ is given as in the Recipe CTR or FICTR.

Now we analyze the key energy balance equation (1). We first show that the Figure L4.2 system is IO passive. Fix $\vec{w}$ in ℓ^{2+} an input string in Figure L4.2. The first step is to prove

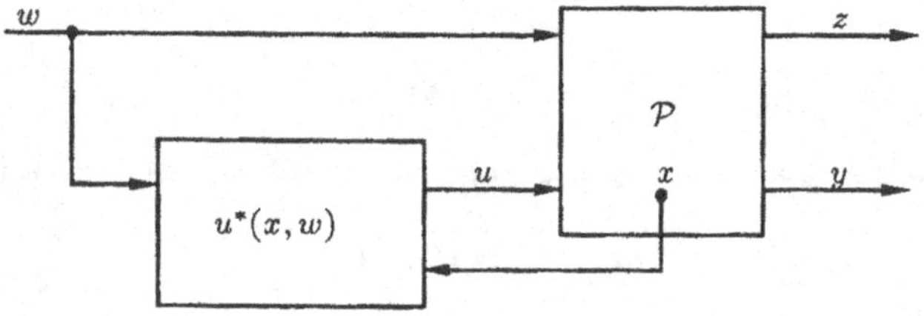

Figure L4.2.

Lemma L4.2. *The inequality (1.2) plus positivity of e implies that the system is IO passive and e-stable.*

Proof. The state equations for Figure L4.2 are:

$$\vec{x}(n+1) = F(\vec{x}(n)), \vec{w}(n), u^*(\vec{x}(n), \vec{w}(n))), \qquad \vec{x}(0) = 0$$
$$\vec{z}(n) = G_1(\vec{x}(n), \vec{w}(n), u^*(\vec{x}(n), \vec{w}(n))).$$

Suppose that $\vec{w}$ in ℓ^{2+} is an input string to Figure L4.2, then (1) together with (2) simply says that

$$e(\vec{x}(k)) - e(\vec{x}(k+1)) \geq \|\vec{z}(k)\|^2 - \|\vec{w}(k)\|^2$$

for all k. Summing from $k = 0$ to $k = M$ gives

$$e(\vec{x}(0)) - e(\vec{x}(M+1)) \geq \sum_{k=1}^{M} \|\vec{z}(k)\|^2 - \sum_{k=1}^{M} \|\vec{w}(k)\|^2. \tag{1.3}$$

In words, this says

state energy at time 0— state energy at time $M + 1 \geq$ energy through z— energy in through w.

Since $x(0) = 0$ we have $e(\vec{x}(0)) = 0$ and by assumption $e(x) \geq 0$ for all x. We conclude that

$$\sum_{k=0}^{M} \|\vec{z}(k)\|^2 \leq \sum_{k=0}^{M} \|\vec{w}(k)\|^2$$

whenever $\vec{z} = T_{zw}(\vec{w})$. This shows that the IO operator $\tilde{T}_{zw}$ for Figure L4.2 is input–output passive as asserted, since

$$\|P_M T_{zw}(\vec{w})\|^2_{\ell^{2+}} \leq \|P_M \vec{w}\|^2_{\ell^{2+}} .$$

To prove e–stability of the system in Figure L4.2, observe that (1.3) implies that

$$e(\vec{x}(0)) + \sum_{k=0}^{M} \|\vec{w}(k)\|^2 \geq \sum_{k=0}^{M} \|\vec{z}(k)\|^2 + e(\vec{x}(n+1))$$

so

$$e(\vec{x}(0)) + \|\vec{w}\|^2_{\ell^{2+}} \geq e(\vec{x}(n+1)) \quad \forall n .$$ ■

To complete the proof of Theorem L4.1 requires observing that the state equations

$$\vec{x}(n+1) = F(\vec{x}(n), \vec{w}(n), \check{K}\vec{y}(n))$$
$$\vec{z}(n) = G_1(\vec{x}(n), \vec{w}(n), \check{K}\vec{y}(n))$$

for Figure L4.1 actually equal the state equations for Figure L4.2. Thus inputting $\vec{w}$ to Figure L4.1 produces the same sequence of states as inputting $\vec{w}$ to Figure L4.2. The lemma implies that Figure 1 with $\check{K}$ given by FICTR is passive and stable. ■

2. Output feedback control

First we state the natural extension of Recipe CTR from Lecture 3 to G_2 which depend on u.

Recipe CTR

1. Follow Recipe CTR of Lecture 3 for two steps, thus producing $u^*(x,w)$.
2. Define the compensator K as follows: Assume $w \to G_2(\xi, w, u^*(\xi,w)) \triangleq Y$ is invertible for each ξ and let $w(\xi, Y))$ denote the inverse map. The state equations for K are

$$\tilde{\xi} = F(\xi, w(\xi,y), u^*(\xi, w(\xi,y)))$$
$$u = u^*(\xi, w(\xi,y)) .$$

Proof of Theorem L3.1. A key observation here is that Figure L3.1 with K given by Recipe CTR and Figure L4.1 with $\check{K}$ given by Recipe FICTR have the same IO operator T_{zw}. When this is proved we have that the Recipe FICTR control law produces a passive, e-stable system. It still remains to check internal stability.

The first step is to determine T_{zw} for the closed loop system of Figure L3.1.
Its state equations are:

For each fixed x, ξ let $y(x, \xi, w)$ denote the solution to

$$y = G_2(x, w, u^*(\xi, w(\xi, y))) \tag{2.1}$$

then

$$\begin{aligned} \tilde{x} &= F(x, w, u^*(\xi, w(\xi,y))) \\ \tilde{\xi} &= F(\xi, w(\xi,y), u^*(\xi, w(\xi,y))) \\ z &= G_1(x, w, u^*(\xi, w(\xi,y))) \end{aligned} \tag{2.2}$$

where $y = y(x, \xi, w)$. Suppose that at time k we have equality of the states $x = \xi$. Then (2.2) tells us that at time $k+1$, the states $\tilde{x}$ and $\tilde{\xi}$ are

$$\tilde{x} = F(x, w, u^*(x, w(x,y)))$$
$$\tilde{\xi} = F(x, w(x,y), u^*(x, w(x,y)))$$

where $y = y(x, x, w)$ satisfies

$$y = G_2(x, w, u^*(x, w(x,y))) \tag{2.3}$$

However, $w(x,y)$ satisfies

$$y = G_2(x, w(x,y), u^*(x, w(x,y))) .$$

By the $w \to y$ invertibility assumption on G_2 we have that $w = w(x,y)$. Consequently (2.2) becomes two copies of

$$\tilde{x} = F(x, w, u^*(x, w))$$

and

$$z = G_1(x, w, u^*(x, w)) .$$

These are the state equations for Figure L4.2 which we have already seen are equivalent to those for Figure L4.1. This proves passivity of the closed loop control system.

For stability of this system it is essential to assume that $\mathcal{P}$ is stable. Also we put heavy restrictions on $G_2(x, w, u)$, namely, that it does not depend on x; hopefully this is not essential and can be removed.

First recall that Theorem L4.1 showed that the system in Figure L4.2 is e-stable where (by hypothesis) e is a Lyapunov function. This implies that for any input $\vec{w}$ in ℓ_W^{2+} its states $\vec{\xi}(k)$ are uniformly bounded in k, since if $\vec{\xi}(k) \to \infty$, then $e(\vec{\xi}(k)) \to \infty$ by the Lyapunov property and contradicts e-stability. That is

$$\tilde{\xi} = F(\xi, w, u^*(\xi, w)) \tag{2.4}$$

produces uniformly bounded $\vec{\xi}(k)$ for any $\vec{w}$ in ℓ_W^{2+} (for any initial $\vec{\xi}(0)$).

Next we turn to stability of the output feedback system of Figure L3.1. Its state equations are (2.2). If we input the sequence $\vec{w}$ in ℓ_W^{2+}, then the resulting sequence $\vec{y}$ is

$$\vec{y}(k) = G_2(\vec{w}(k), u^*(\vec{\xi}(k), \vec{w}(k))) \tag{2.5}$$

and so it is not influenced by the state $\vec{x}(k)$ of $\mathcal{P}$. By the definition of $w(\xi, y)$ we have

$$w(\xi, y) = w(\xi, G_2(w, u^*(\xi, w))) = w$$

which applies at time k to give $w(\vec{\xi}(k), \vec{y}(k)) = \vec{w}(k)$. Now the string $\vec{y}$ feeds into K and gives

$$\begin{aligned} \vec{\xi}(k+1) &= F(\vec{\xi}(k), \vec{w}(\vec{\xi}(k), \vec{y}(k)), u^*(\vec{\xi}(k), w(\vec{\xi}(k), \vec{y}(k))) \\ &= F(\vec{\xi}(k), \vec{w}(k), u^*(\vec{\xi}(k), \vec{w}(k))) . \end{aligned} \tag{2.6}$$

This formula for the states $\vec{\xi}(k)$ depends on the initial value $\vec{\xi}(0) = \xi_0$, but *not* on the initial value $x(0) = x_0$. Since the ξ equation of (2.2) is effectively (2.6), we see that the states $\vec{\xi}(k)$ it generates are uniformly bounded (for any initial x_0, ξ_0).

Finally we analyze $\vec{x}(k)$. It is determined by

$$\vec{x}(k+1) = F(\vec{x}(k), \vec{w}(k), u^*(\vec{\xi}(k), \vec{w}(k))) .$$

The continuity of $u^*(\ ,\ )$ implies that $u^*(\vec{\xi}(k), \vec{w}(k))$ is uniformly bounded, consequently the BIStability of F implies that $\vec{x}(k)$ is uniformly bounded (independent of the initial conditions $\vec{x}(0)$ and $\vec{\xi}(0)$). Theorem L3.1 is proved. ∎

3. The energy for an entire experiment

There is a completely different layer of structure parallel to the one we have described so far in Lectures 3 and 4. It adds more precision to results, but may be even harder for computation. Instead of working with the function e on states, we could work with a function $\mathcal{E}$ on $X \times \ell_W^{2+} \times \ell_U^{2+}$ which equals the energy used by the system starting in state x, when strings $\vec{w}$ and $\vec{u}$ are fed into the system. First we give definitions and then we show how $\mathcal{E}$ and e are related.

Denote by $\mathcal{F}_x^{\mathcal{P}}$ the input–output map associated with the system $\mathcal{P}$, i.e. if $\begin{bmatrix} \vec{w} \\ \vec{u} \end{bmatrix}$ is a sequence of inputs in $\ell_W^+ \oplus \ell_U^+$, then $\begin{bmatrix} \vec{z} \\ \vec{y} \end{bmatrix} = \begin{bmatrix} \mathcal{F}_{x1}^{\mathcal{P}}(\vec{w}, \vec{u}) \\ \mathcal{F}_{x2}^{\mathcal{P}}(\vec{w}, \vec{u}) \end{bmatrix} = \mathcal{F}_x^{\mathcal{P}}(\vec{w}, \vec{u})$ is the associated sequence of outputs generated with initial state equal to x defined recursively by

$$\begin{aligned} \vec{x}(n+1) &= F(\vec{x}(n), \vec{w}(n), \vec{u}(n)), \qquad \vec{x}(0) = x \\ \vec{z}(n) &= G_1(\vec{x}(n), \vec{w}(n), \vec{u}(n)) \\ \vec{y}(n) &= G_2(\vec{x}(n), \vec{w}(n), \vec{u}(n)) \end{aligned}$$

Let $\mathcal{D}_x \subset \ell_W^{2+} \oplus \ell_V^{2+}$ be the set of ℓ^{2+}–sequences $\begin{bmatrix} \vec{w} \\ \vec{u} \end{bmatrix}$ for which the associated output sequence $\vec{z}$ is also norm–square–summable:

$$\mathcal{D}_x = \left\{ \begin{bmatrix} \vec{w} \\ \vec{u} \end{bmatrix} \in \ell_W^2 \oplus \ell_V^2 : \mathcal{F}_{x1}^{\mathcal{P}}(\vec{w}, \vec{u}) \in \ell_Z^{2+} \right\}.$$

We now provide a link between critical points for two different types of energy functions. In general, if $x \in X$ and $\vec{x} \in \ell_X^2$, by $(x, \vec{x})$ we denote the sequence

$$(x, \vec{x})(k) = \begin{cases} x, & k = 0 \\ \vec{x}(k-1), & k \geq 1. \end{cases}$$

Proposition L4.2. *Define a function*

$$\mathcal{E} : \left\{ (x, \vec{w}, \vec{u}) : x \in X, \begin{bmatrix} \vec{w} \\ \vec{u} \end{bmatrix} \in \mathcal{D}_x \right\} \to \mathbb{R}$$

by

$$\mathcal{E}(x, \vec{w}, \vec{u}) = \|\mathcal{F}_{x1}^{\mathcal{P}}(\vec{w}, \vec{u})\|_{\ell_Z^{2+}}^2 - \|\vec{w}\|_{\ell_W^2}^2$$

and suppose that for each $x \in X$ there is a choice of isolated critical pint $(\vec{w}_x^*, \vec{u}_x^*)$ for $\mathcal{E}(x, \cdot, \cdot)$, depending smoothly on x, in the interior of $\mathcal{D}_x$ which is a local max-min point for $\mathcal{E}(x, \cdot, \cdot)$:

$$\mathcal{E}(x, \vec{w}_x^*, \vec{u}_x^*) = \max_{\vec{w} \in W_x} \min_{u \in U_x} \mathcal{E}(x, \vec{w}, \vec{u})$$

where W_x and U_x are neighborhoods of $\vec{w}_x^*$ and $\vec{u}_x^*$ in ℓ_W^{2+} and ℓ_U^{2+} respectively[1]. *Define a function $e : X \to \mathbb{R}$ by*

$$e(x) = \mathcal{E}(x, \vec{w}_x^*, \vec{u}_x^*).$$

Then e satisfies (ENGY) with equality locally, i.e., for each $x \in X$ the function

$$Q(x,w,u) = e(F(x,w,u)) + \|G_1(x,w,u)\|^2 - \|w\|^2$$

has a critical point (w_x^*, u_x^*) which is a local max.–min. for $Q(x,\cdot,\cdot)$

$$Q(x, w_x^*, u_x^*) = \max_{w \in Q_w} \min_{u \in Q_u} Q(x,w,u)$$

(Q_w and Q_u are open neighborhoods of w_x^* and u_x^* respectively) such that

$$e(x) = Q(x, w_x^*, u_x^*).$$

Moreover, the critical point $(\vec{w}_x^*, \vec{u}_x^*)$ for $\mathcal{E}(x,\cdot,\cdot)$ and the critical point (w_x^*, u_u^*) for $Q(x,\cdot,\cdot)$ are connected in the following way:

(3.1) $$\vec{w}_x^* = (w_x^*, \vec{w}^*_{F(x,w_x^*,u_x^*)})$$

(3.2) $$\vec{u}_x^* = (u_x^*, \vec{u}^*_{F(x,w_x^*,u_x^*)})$$

Remark

Equations (3.1) and (3.2) tell us how to get $\begin{bmatrix} w_x^* \\ u_x^* \end{bmatrix}$ from $\begin{bmatrix} \vec{w}_x^* \\ \vec{u}_x^* \end{bmatrix}$. Conversely, under the assumption of Proposition L4.3 we can recover $\begin{bmatrix} \vec{w}_x^* \\ \vec{u}_x^* \end{bmatrix}$ from the function $x \to \begin{bmatrix} w_x^* \\ u_x^* \end{bmatrix} \in W \times U$ as the solution of the recurrence equations

$$\vec{x}^*(k+1) = F(\vec{x}^*(k), \vec{w}_x^*(k), \vec{u}_x^*(k)), \qquad \vec{x}^*(0) = x$$

$$\begin{bmatrix} \vec{w}_x^*(k) \\ \vec{u}_x^*(k) \end{bmatrix} = \begin{bmatrix} w_{\vec{x}^*(k)} \\ u^*_{\vec{x}^*(k)} \end{bmatrix}$$

In addition, if we only assume that e is known and that the function $Q(x,w,u) = e(F(x,w,u)) + \|G_1(x,w,u)\|^2 - \|w\|^2$ has a max.–min. point (w_x^*, u_x^*) for each x, the above recursion generates a candidate for a max.–min. point for $\mathcal{E}(x, \vec{w}, \vec{u})$. A stability constraint natural for the problems here is that the resulting sequence $\begin{bmatrix} \vec{w}_x^* \\ \vec{u}_x^* \end{bmatrix}$ be in $\ell_W^2 \oplus \ell_U^2$.

---

[1] A more intuitive assumption is:
For each $x, \vec{w}$ there is an open set $V_{x,\vec{w}} \subset \ell_U^{2+}$ so that $\mathcal{F}_{x_1}(\vec{w}, \vec{u}) \in \ell^{2+}$. In other words, for each $x, \vec{w}$, there is a stabilizing open set of input strings $\vec{u}$ in ℓ_U^{2+} which "passively stabilize" the system. It is primarily this assumption plus some added "frills" about the existence of a $\max_{\vec{w}} \min_{\vec{u}}$ which the proposition tells us give state feedback making the system in Figures L3.1 and L3.2 passive.

Proposition L4.3 is very much of a differential games nature (see [**BO**]). The proof of Proposition L4.3 is probably not in the game theory literature, since the hard part of the proof involves going from the infinite dimensional problem on ℓ^{2+} to the finite dimensional ENGY equations of the recipe. Typical game theory proofs apply only to finite duration games (where critical points are usually saddlepoints) so the structure is a bit different.

We are now in a position to state the string–max–min assumption needed for our stability result.

(STRMXMN). *The solution* $e : X \to \mathbb{R}^+$ *of the (ENGY) equation is of the form*

$$e(x) = \mathcal{E}(x, \vec{w}_x^*, \vec{u}_x^*)$$

where $\mathcal{E}, \vec{w}_x^*, \vec{u}_x^*$ *are as in Proposition L4.3.*

Under this assumption we obtain rigorously the following stability result for the controller constructed in the Recipe FICTR and Recipe CTR.

Theorem L4.5. *Let* $\mathcal{P}$ *be as in (FICTR), assume that (STRMXMN) holds and that the FI feedback* $u = u^*(x, w)$ *is constructed as in the Recipe FICTR (where equality holds in (ENGY)). Then the closed loop–system in Figure L4.1 is IO passive and is weakly asymptotically e–stable. Under the* (MM) *assumption if e is a Lyapunov function, then the output feedback system in Figure L3.1 built by Recipe CTR is internally stable in a weak asymptotic sense.*

4. The maximum principle

There is an equivalent form of the condition (ENGY) on e when we insert an equality. It is the Hamiltonian form of the equations often called the maximum principle. One has not only state vectors $x(k)$ evolving in time but co-state vector $p(k)$ evolving in time. For statements of this formulation of (ENGY) see [**BO**, Th. 1, Ch. 6]. Sometimes those are easier to use in solving examples explicitly for an H^∞ control example see [**BH5**].

5. Inner-outer factorization

A closely related subject is that of factoring systems. In linear H^∞ control factoring a transfer function (easily obtained from $\mathcal{P}$) as a product of a J-inner and a J-outer function gives a linear fractional parametrization of all compensators K solving CTR. For model matching Joe Ball and I have a reasonably successful extension of this to nonlinear $\mathcal{P}$. The reader is referred to [**BH1**], [**BH2**], [**BH3**]. Here we merely mention that the Recipe for J-inner-outer factorization is an extension of Recipe CTR or FICTR, in that one must construct e and $u^*(x, w)$ and once they are found proceed by solving the equation:

$$\left(J\begin{pmatrix} r \\ s \end{pmatrix}, \begin{pmatrix} r \\ s \end{pmatrix}\right) + e(x) = \left(J\gamma_x\begin{pmatrix} r \\ s \end{pmatrix}, \gamma_x\begin{pmatrix} r \\ s \end{pmatrix}\right) + e(\hat{F}^\times\left(x, \gamma_x\begin{pmatrix} r \\ s \end{pmatrix}\right)) \tag{5.1}$$

for the map $\gamma_x\binom{W}{U} \longrightarrow \binom{W}{U}$ which meets the initial condition

$$\gamma_x \begin{pmatrix} 0 \\ 0 \end{pmatrix} = \begin{pmatrix} w_x \\ u_x \end{pmatrix}$$

where w_x, u_x are the solutions to (ENGY) with equality. Here $F^{\times}(x,y) = F(x, G^I(x,y))$. This is a hard step and after one has solved for γ_x the rest of the construction for inner and outer factors is easier. Solving (5.1) is a Morse Theory type of problem. This is directly added to the equations of Recipe FICTR and so is considerably harder. Articles **[BH1–3]** give a theory of these equations which is fairly satisfactory. Actually solving them in practice might be very hard.

References for Lectures 3 and 4

[B] T. Basar, A dynamic games approach to controller design: disturbance rejection in discrete time, in Proc. 28th Conf. on Dec. and Control, Tampa, 1989, pp. 407–414.

[BO] T. Basar and G. J. Olsder, Dynamic Non–Cooperative Game Theory, Math. Sci. and Eng. Series Academic Press (New York), 1977.

[BH1] J. A. Ball and J. W. Helton, Shift invariant manifolds and nonlinear analytic function theory, Integral Equations and Operator Theory II (1988), 615–725.

[BH2] J. A. Ball and J. W. Helton, Factorization of nonlinear systems; toward a theory for nonlinear H^∞ control, IEEE Conference on Decision and Control, Austin, TX (1988), pp. 2376–2381.

[BH3] J. A. Ball and J. W. Helton, Inner–outer factorization of nonlinear operators, preprint.

[BH4] J. A. Ball and J. W. Helton, H^∞ control for nonlinear plants: connections with differential games, IEEE Conference on Decision and Control, Tampa, Florida (1989), pp. 956–962.

[BH5] J. A. Ball and J. W. Helton, H^∞ control for stable nonlinear plants (preprint).

[CW] L. O. Chua and J. L. Wyatt, et.al, Energy concepts in the state-space theory of nonlinear N-ports: losslessness, IEEE Trans. Circuits and Syst., **CAS-29** (July 1982), 417–30.

L. O. Chua and J. L. Wyatt, et.al, Energy concepts in the state-space theory of nonlinear N-ports: passivity, IEEE Trans. Circuits and Syst., **CAS-28** (January 1981), 48–61.

[DGKF] J. C. Doyle, K. Glover, P. P. Khargonekar and B. A. Francis, State–space solutions to standard H_2 and H_∞ control problems, IEEE Trans. Auto. Control 34 (1989), 831–847.

[F] C. Foias, Lectures in this volume.

[Fr] B. A. Francis, A Course in H_∞ Control, Springer–Verlag, 1987.

[LAKG] D. J. N. Limebeer, B. D. O. Anderson, P. P. Khargonekar and M. Green, A game theoretic approach to control for time varying systems, preprint.

[PS] G. P. Papavassilopoulos and M. G. Safonov, Robust control via game theoretic methods, in Proc. 28th IEEE Conf. on Dec. and Control, Tampa, 1989, pp. 382–387.

[T] G. Tadmor, The standard H_∞ problem and the maximum principle: the general linear case, Univ. Texas at Dallas Technical Report 192, 1989.

[YS] I. Yaesh and V. Shaked, Game theory approach to optimal linear estimation in the minimum H^∞-norm sense, in Proc. 28th IEEE Conf. on Dec. and Control, Tampa, 1989, pp. 421–425.

Hearty thanks are in order to Neola Crimmins for producing this manuscript.

THE POLYNOMIAL APPROACH TO $\mathcal{H}_\infty$-OPTIMAL REGULATION

Huibert Kwakernaak
University of Twente
Department of Applied Mathematics
P. O. Box 217, 7500 AE Enschede, The Netherlands

Abstract

After a review of the "polynomial approach" to the solution of the single-input-single-output minimum sensitivity and mixed sensitivity problems, a new solution is presented of the standard $\mathcal{H}_\infty$-optimal regulation problem based on polynomial J-spectral factorization. Besides a parametrization of all suboptimal compensators an explicit expression for all *optimal* compensators is obtained.

CONTENTS

1 Introduction

These notes review some of the main results and recent progress in the "polynomial" approach to $\mathcal{H}_\infty$-optimal regulation of linear systems. In this approach, systems are represented by rational transfer functions, and polynomials and polynomial matrices are resorted to when it comes to computations. $\mathcal{H}_\infty$ optimality is established by the "equalizer principle" in the earlier work and J-spectral factorization in the more recent results.

Section 2 deals with the minimum sensitivity problem for single-input-single-output systems. This is the simplest $\mathcal{H}_\infty$-optimal feedback problem. It is shown how the equalizer principle simply and straightforwardly leads to a generalized eigenvalue problem. The minimum sensitivity problem was originally considered by Zames (1981) and triggered a true avalanche of work on $\mathcal{H}_\infty$-optimization in control. The present treatment of the minimum sensitivity problem is based on Kwakernaak (1985).

Section 3 first explains that the minimum sensitivity paradigm is inadequate for feedback control system design. Control system robustness and frequency response shaping are used to motivate an extension of the minimum sensitivity problem that is known as the *mixed sensitivity problem.* The terminology is from Verma and Jonckheere (1984) but the problem was first considered by Kwakernaak (1983). The solution for the SISO case is treated in some detail. It is based on the equalizer principle and follows the lines of Kwakernaak (1986) adapted to the SISO case.

An important conclusion from this work on the minimum and mixed sensitivity problems is that solutions of $\mathcal{H}_\infty$-optimization problems are characterized by a cancelation phenomenon.

In Section 4 the *standard* $\mathcal{H}_\infty$-optimal regulator problem is treated. This problem was defined by Francis and Doyle (1987) and is described by Francis (1987). It is a generalized regulation problem that encompasses many $\mathcal{H}_\infty$-optimal control problems of interest as special cases. Most of the results of this section are new, although the work was precursed by other "polynomial" attempts to solve the standard problem (Kwakernaak, 1987; Boekhoudt, 1988; Kwakernaak, 1990). Part of the results reported here originate from joint work with G. Meinsma. The present solution is based on J-spectral factorization. Obviously these results are related to other work on $\mathcal{H}_\infty$-optimization relying on J-specral factorization, in particular that of Green (1989).

The crucial result in the current approach is a rational matrix inequality that characterizes suboptimal solutions of the optimal regulation problem, that is, solutions such that the ∞-norm that is to be minimized does not exceed a given number. By using rational J-spectral factorization it is not difficult to find explicit formulas for all suboptimal compensators, and, in particular, all stabilizing suboptimal compensators. Subsequently, the rational J-spectral factorization is reduced to two polynomial J-spectral factorizations. These two polynomial J-spectral factorizations are the counterparts of the two Riccati equations that arise in recent work on the state space approach to the $\mathcal{H}_\infty$-optimal regulation problem (Doyle *et al.*, 1989; Glover and Doyle, 1989).

The numerical details of polynomial J-spectral factorization do not seem to have been investigated much. We present an adaptation of Callier's symmetric extraction method (Callier, 1985) that was developed for ordinary polynomial spectral factorization.

It turns out that the cancelation phenomenon that characterizes $\mathcal{H}_\infty$-optimal solutions (as opposed to *suboptimal* solutions) manifests itself in the form of a singularity in the J-spectral factorization. This singularity may be identified and exploited to find exact optimal solutions. A further adaptation of the symmetric extraction procedure leads to an explicit formula for all $\mathcal{H}_\infty$-optimal compensators.

To obtain $\mathcal{H}_\infty$-optimal compensators numerically, first a simple line search procedure is employed to delimit the minimal value of the ∞-norm that is to be optimized. Each step of the search involves the solution of a linear matrix polynomial equation and two polynomial J-spectral factorizations. The indicator for the line search is whether the compensator is stabilizing or not. Following the initial search the minimal value of the ∞-norm and the corresponding compensators are computed exactly by a root finding procedure that uses the singularity phenomenon as an indicator.

The tone of these notes is tutorial. Many details are missing but the exposition is reasonably self-contained. No deep mathematics are used. Familiarity with the polynomial matrix fraction description of linear multivariable systems such as found in Kailath

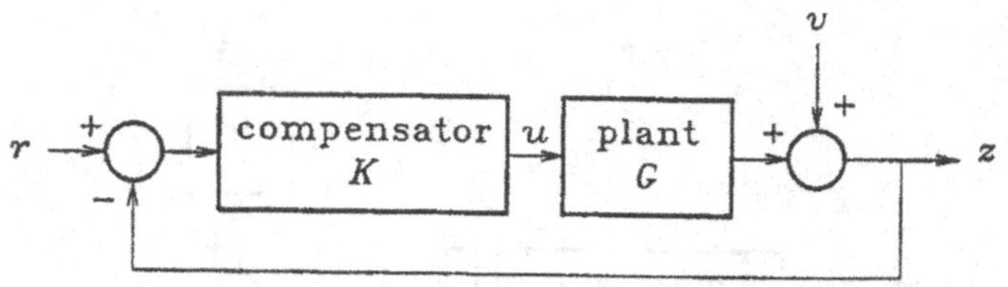

Figure 1: A basic SISO feedback configuration

(1980) is needed for Section 4. Other useful material on polynomial matrices and their application in system and control theory is available in Kučera (1979) and Callier and Desoer (1982).

2 The SISO Minimum Sensitivity Problem

We begin by considering the problem of minimizing the sensitivity function of a single-input single-output feedback system. This is the simplest $\mathcal{H}_\infty$ optimal regulation problem. Fig. 1 shows the basic configuration.

A SISO plant with given transfer function G is connected in feedback with a compensator with transfer function K. It is desired to design the feedback system such that first of all it is stable. In addition, the effect of the disturbance v on the output z of the feedback system should be as small as possible. We confine ourselves to the case where the transfer functions G and K are rational, that is, both the plant and the compensator are finite-dimensional linear time-invariant systems.

Ignoring the reference input r, we see that the closed-loop system satisfies the "signal balance equation"

$$z = v - GKz, \tag{1}$$

where z and v are the Laplace transforms of the output and disturbance, respectively. Assuming that $1 + GK$ is not identical to zero, we solve for z as

$$z = Sv, \tag{2}$$

where

$$S = \frac{1}{1+GK} \tag{3}$$

is the *sensitivity function* of the closed-loop system. The extent to which disturbances affect the output z of the closed-loop system is determined by the behavior of $S(j\omega)$, $\omega \in \mathrm{R}$, as a function of the frequency ω. The smaller $|S(j\omega)|$ is, the less disturbances at the frequency ω affect the output.

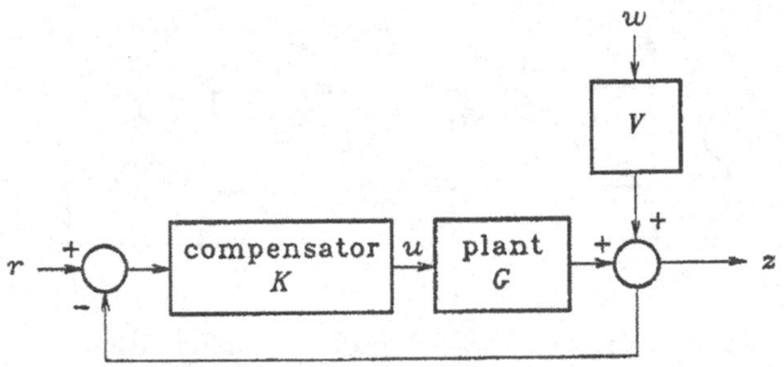

Figure 2: Basic configuration with shaping filter for the disturbance.

2.1 $\mathcal{H}_\infty$ optimization

One way of measuring how small is the sensitivity of the closed-loop system to disturbances is to consider the ∞-norm

$$\|S\|_\infty := \sup_{\omega \in \mathbb{R}} |S(j\omega)| \tag{4}$$

of the sensitivity function S.

The minimum sensitivity problem is the problem of determining the compensator K such that (1) the closed-loop system is stable, and (2) the ∞-norm $\|S\|_\infty$ of the sensitivity function is *minimal*.

This problem formulation is too naive to be practical, for the following reason. Physical plants have the property that as the frequency ω increases to infinity, the frequency response of the plant $G(j\omega)$ decreases to zero. Since the compensator is to be implemented as a physical system, it also should have a frequency response function $K(j\omega)$ that decreases to zero or at least remains bounded as ω increases. Combination of these two observations leads to the conclusion that as ω increases to ∞, the sensitivity function $S(j\omega)$ approaches the value 1. Hence, for high frequencies it is not useful to attempt to make $|S(j\omega)|$ smaller than 1. It is important to make $|S(j\omega)|$ small for those frequencies that dominate in the disturbance.

This leads to considering the minimization of the $\mathcal{H}_\infty$-norm of the *weighted* sensitivity function $V(j\omega)S(j\omega)$, $\omega \in \mathbb{R}$, where the frequency-dependent weighting function V is chosen to be large at those frequencies where the disturbance is large.

In Fig. 2 the weighting function V is represented as a *shaping filter* for the disturbance. The function SV is the transfer function from the external signal w to the control system output z.

2.1.1 Definition (The minimum sensitivity problem.) The minimum sensitivity problem is the problem to find a (rational) compensator transfer function K such that

1. the closed-loop system is stable,

2. the ∞-norm $\|SV\|_\infty$ of the weighted sensitivity function is minimal.

■

In Subsection 2.3 it will be explained what "stable" is understood to be. Compensators K such that the closed-loop system is stable are called ***stabilizing*** compensators.

2.2 The equalizer principle

Our approach to the solution of $\mathcal{H}_\infty$ optimal regulation problems is based on the *equalizer principle*. This idea has been borrowed from statistical decision theory. In statistical decision theory the equalizer principle is used to obtain minimax decision rules from Bayesian decision rules that have constant risk functions. Transposed to control theory, the equalizer principle serves to find $\mathcal{H}_\infty$ optimal compensators from compensators that minimize a quadratic criterion and have constant loss functions.

The idea of the equalizer principle may be explained quite simply. Suppose that we have found a stabilizing compensator K_o such that the square of the magnitude $|S_o(j\omega)V(j\omega)|^2$ of the resulting loss function S_oV is *constant*, say equal to λ_o^2. Such a compensator is said to be *equalizing*. If we take λ_o nonnegative, then we have $\|S_oV\|_\infty = \lambda_o$.

Suppose that *in addition* K_o minimizes a quadratic criterion of the form

$$\int_{-\infty}^{\infty} \Phi(j\omega)|S(j\omega)V(j\omega|^2\, d\omega, \tag{5}$$

where $\Phi(j\omega)$, $\omega \in \mathsf{R}$, is some nonnegative real function that is not identical to zero. Then it is easy to see that K_o minimizes $\|SV\|_\infty$ with respect to all stabilizing compensators.

Assume that it does not, that is, there is another compensator K_* such that for the resulting loss function S_*V we have $\|S_*V\|_\infty < \|S_oV\|_\infty = \lambda_o$. Then clearly

$$|S_*(j\omega)V(j\omega)| < \lambda_o \quad \text{for all} \quad \omega \in \mathsf{R}. \tag{6}$$

It follows that

$$\begin{aligned}\int_{-\infty}^{\infty} \Phi(j\omega)|S_*(j\omega)V(j\omega)|^2\, d\omega &< \int_{-\infty}^{\infty} \Phi(j\omega)\lambda_o^2\, d\omega \\ &= \int_{-\infty}^{\infty} \Phi(j\omega)|S_o(j\omega)V(j\omega)|^2\, d\omega,\end{aligned} \tag{7}$$

which contradicts the fact that K_o minimizes (5). Hence, the hypothesis that there exists a compensator that achieves a smaller value of $\|SV\|_\infty$ than K_o does is false.

2.2.1 Summary (Equalizer principle for the SISO minimum sensitivity problem.) Suppose that the stabilizing compensator K_o is equalizing, that is,

$$|S_o(j\omega)V(j\omega)|^2 = \lambda_o^2 \tag{8}$$

for some real constant λ_o. Then if there exists a nonnegative real function $\Phi(j\omega)$, $\omega \in \mathsf{R}$, that is not identical to zero such that K_o minimizes

$$\int_{-\infty}^{\infty} \Phi(j\omega)|S(j\omega)V(j\omega|^2\, d\omega \tag{9}$$

with respect to all stabilizing compensators, K_o minimizes $\|SV\|_\infty$ with respect to all stabilizing compensators. •

2.3 Equalizing compensators

We continue our discussion of the minimum sensitivity problem by seeing if we can actually find equalizing compensators. The question whether there also exists a function Φ such that (5) is minimized is ignored for the moment.

To convert the question of finding equalizing compensators to an algebraic problem we write the transfer functions G and K of the plant and compensator and the weighting function V in the explicit rational forms

$$G = \frac{N}{D}, \quad V = \frac{A}{B}, \quad K = \frac{Y}{X}. \tag{10}$$

N, D, A and B are given polynomials, and X and Y polynomials to be determined. Without loss of generality we assume that N and D are coprime, that is, have no common polynomial factors, and that the plant has no hidden unstable modes. Also A and B may be assumed to be coprime. Since only the behavior of $|V(j\omega)|$, $\omega \in \mathrm{R}$, is important, there is no loss in generality in assuming that both A and B have all their roots in the closed left-half complex plane. Finally, also the compensator is assumed to have no hidden unstable modes.

With (10) we may write

$$SV = \frac{V}{1+GK} = \frac{\frac{A}{B}}{1+\frac{NY}{DX}} = \frac{ADX}{(DX+NY)B} = \frac{ADX}{\chi B}. \tag{11}$$

The polynomial

$$\chi := DX + NY \tag{12}$$

is the *closed-loop characteristic polynomial* of the feedback system. The roots of χ are the poles of the closed-loop system, that is, the set of poles of all all transfer functions that may be associated with the closed-loop system consists of the roots of χ. If χ has all its roots in the open left-half complex plane, the feedback system is stable.

Equalizing compensators may now be found by considering the equation

$$V^{\sim}VS^{\sim}S = \lambda^2, \tag{13}$$

with λ a real number. Here if S is any rational function, $S^{\sim}$ is the *adjoint* defined by

$$S^{\sim}(s) = S(-s). \tag{14}$$

With (11) and (12) we may rewrite (13) in the form

$$A^{\sim}AD^{\sim}DX^{\sim}X = \lambda^2\chi^{\sim}\chi B^{\sim}B. \tag{15}$$

We deal with this equation by factorization. Since we wish to find stabilizing compensators, we are only interested in solutions of this equation such that the polynomial χ is *Hurwitz,* that is, has all its roots in the left-half complex plane. Actually, we want χ to be *strictly Hurwitz,* that is, with all its roots in the *open* left-half plane, because this ensures both asymptotic and BIBO stability. For the time being, though, we are content if χ is Hurwitz.

It is always possible to factor

$$D^{\sim}D = \hat{D}^{\sim}\hat{D}, \quad X^{\sim}X = \hat{X}^{\sim}\hat{X}, \tag{16}$$

where the polynomials $\hat{D}$ and $\hat{X}$ both are Hurwitz. This is called (polynomial) *spectral factorization.* Then (15) may be solved with χ Hurwitz if we let

$$\lambda\chi B = A\hat{D}\hat{X}, \tag{17}$$

because A and B are Hurwitz by assumption. Next, factor the polynomials N, D, X, and Y as

$$D = D_-D_+, \quad N = N_-N_+, \quad X = X_-X_+, \quad Y = Y_-Y_+, \tag{18}$$

where the subscript $-$ means that the polynomial factor is Hurwitz, and the subscript $+$ that the polynomial has all its roots in the *open* right half-plane (i.e., is *strictly anti-Hurwitz.*) Then we may take

$$\hat{D} = D_-D_+^{\sim}, \quad \hat{N} = N_-N_+^{\sim}. \tag{19}$$

Substituting $\chi = DX + NY$ into (17) we thus obtain

$$\lambda B(D_-D_+X_-X_+ + N_-N_+Y_-Y_+) = AD_+^{\sim}D_-X_+^{\sim}X_-. \tag{20}$$

Note that the first term on the left and the term on the right both contain a factor D_-. This implies that also the second factor on the left has this factor, which can only be Y_-, Thus, we take $Y_- = D_-$. Next, observe that the left-hand side has a factor B, which by assumption is Hurwitz. On the right-hand side this factor may only be contained in X_-. Thus, we have $X_- = BX'_-$, with X'_- to be determined. Substituting Y_- and X_- and canceling common factors we thus have

$$\lambda(D_+BX'_-X_+ + N_-N_+Y_+) = AD_+^{\sim}X_+^{\sim}X'_-. \tag{21}$$

The first term on the left and the term on the right of this equation both have a factor X'_-. Hence, the second term on the left also contains this factor, which can only be N_-. Thus, we take $X'_- = N_-$. After cancelation, what remains may be written as

$$\frac{1}{\lambda}AD_+^{\sim}X_+^{\sim} = D_+BX_+ + N_+Y_+. \tag{22}$$

If we can solve this polynomial equation for strictly anti-Hurwitz polynomials X_+ and Y_+, then the compensator

$$K = \frac{D_-Y_+}{BN_-X_+} \tag{23}$$

is both stabilizing and equalizing. The closed-loop characteristic polynomial generated by this compensator is given by

$$\chi = DX + NY = D_-N_-(D_+BX_+ + N_+Y_+) = \frac{1}{\lambda}D_-N_-AD_+^{\sim}X_+^{\sim}. \tag{24}$$

The sensitivity function corresponding to equalizing compensators that satisfy (22) is easily found to be

$$S = \lambda \frac{BD_+X_+}{AD_+^\sim X_+^\sim}. \tag{25}$$

Before studying the polynomial equation (22) in some detail, we formulate the following result:

2.3.1 Assertion (Improvability of the solution of the SISO minimum sensitivity problem.) For any given real λ the SISO minimum sensitivity problem has a solution such that

$$\|SV\|_\infty \leq |\lambda| \tag{26}$$

if and only if the polynomial equation

$$\frac{1}{\lambda} AD_+^\sim X_+^\sim = D_+BX_+ + N_+Y' \tag{27}$$

has a solution X_+, Y' such that the compensator

$$K = \frac{D_-Y'}{BN_-X_+} \tag{28}$$

stabilizes the closed-loop system. •

What we have shown so far is that if the polynomial equation (27) has a solution such that X_+ is strictly anti-Hurwitz then the corresponding compensator is stabilizing and equalizing such that $\|SV\|_\infty = |\lambda|$. This proves the sufficiency of the condition of 2.3.1. Its necessity is shown later.

2.4 Solution of the polynomial equation

We discuss the solution of the linear polynomial equation (27). To emphasize the dependence of its solution on λ, we rewrite the equation as

$$\frac{1}{\lambda} AD_+^\sim X_{+,\lambda}^\sim = D_+BX_{+,\lambda} + N_+Y'_\lambda. \tag{29}$$

For $\lambda = \infty$ the equation reduces to

$$0 = D_+BX_{+,\infty} + N_+Y'_\infty. \tag{30}$$

Inspection shows that it has the solution

$$X_{+,\infty} = N_+, \quad Y'_\infty = -D_+B. \tag{31}$$

Denote

$$\deg(D_+) = d, \quad \deg(B) = b, \quad \deg(N_+) = n, \tag{32}$$

where deg indicates the degree of a polynomial. Then,

$$\deg(X_{+,\infty}) = n, \quad \deg(Y'_\infty) = d + b. \tag{33}$$

We investigate whether we can find solutions to (29) that have these same degrees. To this end, we need introduce the additional assumption that the rational weighting function $V = A/B$ is *proper*, that is, the degree a of its numerator A is less than or equal to the degree b of the denominator B. Then the degree of the left-hand side of (29), which equals $a + d + n$, is less than or equal to the degree $b + d + n$ of the two terms on the right-hand side. Expanding the polynomial equation and term-by-term identification of the coefficients of terms with like powers in the indeterminate variable s hence results in $d+b+n$ equations in the $n+1$ unknown coefficients of $X_{+,\lambda}$ and the $d+b+1$ unknown coefficients of Y'_λ. Since the equation is homogeneous, the single degree of freedom that remains may be used, say, to normalize one of the coefficients. It is easy to see that the coefficient equations are *linear*, and, hence, may easily be solved. We consider a simple example.

2.4.1 Example (Minimum sensitivity.) Suppose that the plant transfer function G and the weighting function V are given by

$$G(s) = \frac{s-1}{(s-2)(s-3)}, \qquad V(s) = 1. \tag{34}$$

Since $D_+(s) = (s-2)(s-3)$, $N_+(s) = s - 1$, and $B = 1$, we have

$$X_{+,\infty}(s) = -1 + s, \quad Y'_\infty(s) = -6 + 5s - s^2. \tag{35}$$

This shows that the degree of $X_{+,\lambda}$ is one and that of Y'_λ two. Writing

$$X_{+,\lambda}(s) = x_0 + x_1 s, \quad Y'_\lambda(s) = y_0 + y_1 s + y_2 s^2, \tag{36}$$

and expanding the polynomial equation

$$\begin{aligned} &\frac{1}{\lambda}(6 + 5s + s^2)(x_0 - x_1 s) \\ &\quad = (6 - 5s + s^2)(x_0 + x_1 s) + (-1 + s)(y_0 + y_1 s + y_2 s^2), \end{aligned} \tag{37}$$

that follows from (29), we obtain a set of four linear equations in the five unknown coefficients x_0, x_1, y_0, y_1, and y_2. Solution results in

$$x_0 = 1 + \frac{6}{\lambda}, \quad x_1 = -(1 - \frac{6}{\lambda}), \tag{38}$$

$$y_0 = 6(1 - \frac{1}{\lambda})(1 + \frac{6}{\lambda}), \quad y_1 = -5(1 - \frac{2}{\lambda})(1 - \frac{3}{\lambda}), \quad y_2 = (1 + \frac{1}{\lambda})(1 - \frac{6}{\lambda}), \tag{39}$$

which may be multiplied by any nonzero real constant. Hence, the equalizing compensator is given by

$$K(s) = \frac{Y'_\lambda(s)}{X_{+,\lambda}(s)} = \frac{6(1 - \frac{1}{\lambda})(1 + \frac{6}{\lambda}) - 5(1 - \frac{2}{\lambda})(1 - \frac{3}{\lambda})s + (1 + \frac{1}{\lambda})(1 - \frac{6}{\lambda})s^2}{(1 + \frac{6}{\lambda}) - (1 - \frac{6}{\lambda})s}. \tag{40}$$

Note that any constant multiplied into the coefficients may be canceled.

The closed-loop characteristic polynomial that results from this compensator is according to (24)

$$\chi(s) = \frac{1}{\lambda}(s+2)(s+3)\left[(1+\frac{6}{\lambda}) + (1-\frac{6}{\lambda})s\right]. \tag{41}$$

We take a closer look at this solution. The closed-loop characteristic polynomial χ has two fixed roots -2 and -3, and a root

$$\pi_\lambda = -\frac{1+\frac{6}{\lambda}}{1-\frac{6}{\lambda}} \tag{42}$$

that depends on λ. For $|\lambda|$ sufficiently large, in particular for $|\lambda| > 6$, all roots of χ are in the open left-half complex plane, so that the closed-loop system is strictly stable. As λ decreases from $+\infty$, for $\lambda = 6$ the closed-loop pole π_λ moves to the right-half plane via infinity. As λ increases from $-\infty$ for $\lambda = -6$ the closed-loop pole moves into the right-half plane via the origin.

This shows that 6 is the smallest value of $|\lambda|$ for which a stabilizing compensator is obtained. Inspection of (40) reveals that also the compensator transfer function has the pole π_λ. For $\lambda = 6$, the compensator transfer function reduces to

$$K(s) = \frac{10 - \frac{5}{3}s + 0\cdot s^2}{2 - 0\cdot s} = \frac{5}{6}(6-s), \tag{43}$$

which shows that the pole π_λ has *canceled* in the compensator transfer function. Accordingly, the closed-loop characteristic polynomial reduces to $\frac{1}{3}(s+2)(s+3)$, which has both its roots in the open left-half plane, and hence results in a closed-loop system that is asymptotically stable. Similarly, for $\lambda = -6$ we have for the compensator transfer function

$$K(s) = \frac{0 - 10s + \frac{5}{3}s^2}{0 - 2s} = \frac{5}{6}(6-s). \tag{44}$$

Again, the pole π_λ has canceled, and the *same* compensator is obtained, with the same closed-loop poles that are strictly in the open left-half plane.

For this compensator, $\|SV\|_\infty = 6$. It is the equalizing stabilizing compensator that results in the smallest ∞-norm for SV. It is tempting to believe that it is $\mathcal{H}_\infty$ optimal. •

2.5 Reduced solutions

In the preceding subsection we concluded that if the rational function V is proper, the polynomial equation

$$\frac{1}{\lambda}D_+^\sim X_{+,\lambda}^\sim = D_+BX_{+,\lambda} + N_+Y_\lambda' \tag{45}$$

has a solution $X_{+,\lambda}$, Y_λ' of degrees n and $d+b$, respectively, which may be obtained by solving a set of linear coefficient equations. For $|\lambda| = \infty$ the solution is given by $X_{+,\infty} = N_+$, $Y_\infty' = -D_+B$. We see from (24) that those closed-loop poles that depend on λ consist of the roots of $X_{+,\lambda}^\sim$. For $|\lambda| = \infty$ these are the roots of $N_+^\sim$, which lie strictly in the open left-half plane.

As λ decreases from ∞ or increases from $-\infty$, these closed-loop poles describe a locus in the complex plane. Contemplation of (45) and a little thought reveal that as λ approaches zero, the roots of $X^{\sim}_{+,\lambda}$, and, hence, those closed-loop poles that depend on λ, approach the roots of N_+. Since these roots are located in the open *right-half* complex plane, for some value of λ between $\pm\infty$ and 0 all the roots that change with λ cross over from the left- to the right-half plane. They migrate either through infinity, or through some point on the imaginary axis.

If $X^{\sim}_{+,\lambda}$ has a root on the imaginary axis, then $X_{+,\lambda}$ has this same root (because $X_{+,\lambda}$ has real coefficients). Inspection of (45) shows that if $X^{\sim}_{+,\lambda}$ and $X_{+,\lambda}$ have a common root, then also Y'_λ has this root. If the common polynomial factor corresponding to this root is canceled from $X_{+,\lambda}$ and Y'_λ, the resulting polynomials are still a solution of (45). We call this a *reduced* solution.

Similarly, if a root of $X_{+,\lambda}$ passes through infinity to the right-half complex plane, the leading coefficient of both $X_{+,\lambda}$ and $X^{\sim}_{+,\lambda}$ vanishes. As a result, also the leading coefficient of Y'_λ vanishes, and again we obtain a reduced solution. The reduced solution this time corresponds to "cancelation" of a root at ∞.

Each time a closed-loop poles passes from the left- to the right-half plane a reduced solution is found. The values of λ for which this occurs may be found by considering the polynomial equation (45) where we now look for reduced solutions, that is, solutions with degrees

$$\deg(X_{+,\lambda}) = n-1, \quad \deg(Y'_\lambda) = d+b-1. \tag{46}$$

Expansion of the polynomial equation now results in $d+b+n$ homogeneous linear equations in the $d+b+n$ unknown coefficients of $X_{+,\lambda}$ and Y'_λ. These equations only have a solution for those values of λ for which the determinant of the coefficient matrix vanishes. The determination of those values of λ actually is a generalized eigenvalue problem. We do not explore this in detail, but illustrate what happens by an example.

2.5.1 Example (Reduced solutions.) We continue the previous example. Taking $\deg(X_{+,\lambda}) = n-1 = 0$, and $\deg(Y'_\lambda) = d+b-1 = 1$, we let

$$X_{+,\lambda}(s) = x_0, \quad Y'_\lambda(s) = y_0 + y_1 s. \tag{47}$$

Expansion of the polynomial equation

$$\frac{1}{\lambda}(6+5s+s^2)x_0 = (6-5s+s^2)x_0 + (-1+s)(y_0+y_1 s) \tag{48}$$

leads to a set of linear equations that may be arranged as

$$\begin{bmatrix} 6(1-\frac{1}{\lambda}) & -1 & 0 \\ -5(1+\frac{1}{\lambda}) & 1 & -1 \\ 1-\frac{1}{\lambda} & 0 & 1 \end{bmatrix} \begin{bmatrix} x_0 \\ y_0 \\ y_1 \end{bmatrix} = 0. \tag{49}$$

This set of equations has a nonzero solution if and only if the determinant of the coefficient matrix is zero. Since this determinant equals $2 - \frac{12}{\lambda}$, a nonzero solution only exists if $\lambda = \lambda_o = 6$. For this value of λ the set of equations has the solution $x_0 = 6$, $y_0 = 30$, $y_1 = -5$ (or any multiple). This shows that we have the reduced solution

$$X_{+,\lambda_o}(s) = 6, \quad Y'_{\lambda_o}(s) = 30 - 5s, \tag{50}$$

which agrees with what we found in Example 2.4.1.

In the minimum sensitivity problem of this example we obtain only one reduced solution. In general, there are several, namely, one corresponding to each root of $X_{+,\lambda}$ that migrates from a root of N_+ to one of $N_+^\sim$. •

2.6 Optimality of reduced solutions

Compensators that correspond to the largest value of $|\lambda|$ for which a reduced solution to the polynomial equation (45) exists are stabilizing, because after cancelation of the closed-loop pole that migrates into the right-half plane all *remaining* roots of $X_{+,\lambda}^\sim$ lie strictly in the open left-half complex plane. Among all stabilizing *equalizing* compensators these compensators minimize weighted sensitivity. It is tempting to believe that these compensators minimize weighted sensitivity among *all* stabilizing compensators.

According to the equalizer principle, equalizing compensators minimize sensitivity if they minimize the quadratic criterion

$$\int_{-\infty}^{\infty} \Phi(j\omega)|S(j\omega)V(j\omega)|^2 \, d\omega \tag{51}$$

for some nontrivial nonnegative function Φ. We construct such a function for reduced compensators by first finding sufficient and necessary conditions for the minimization of (51), and then determining Φ such that these conditions are satisfied.

2.6.1 Assertion (Sufficient and necessary conditions for the minimization of (51).) Let Φ be a nonzero rational function that is nonnegative on the imaginary axis. Then, sufficient and necessary conditions for the minimization of (51) with respect to all stabilizing compensators are that

1. $\Phi S^\sim SV^\sim V$ is strictly proper, and
2. $\Phi S^\sim DNV^\sim V$ has all its poles in the closed left-half complex plane.

•

2.6.2 (Proof.) Let U be a 2×2 unimodular polynomial matrix (i.e., a polynomial matrix whose determinant is a nonzero constant) such that

$$[D \quad N]U = [1 \quad 0]. \tag{52}$$

Such a matrix exists if D and N are coprime, that is, have no common factors. In particular, we may write

$$U = \begin{bmatrix} U_{11} & -N \\ U_{21} & D \end{bmatrix}. \tag{53}$$

The compensator $K = Y/X$ is determined by the two polynomials X and Y. Equivalently, we may characterize the compensator by the two poynomials R and Q defined by

$$\begin{bmatrix} X \\ Y \end{bmatrix} = U \begin{bmatrix} R \\ Q \end{bmatrix}. \tag{54}$$

The closed-loop characteristic polynomial may now be written as

$$\chi = DX + NY = [D \quad N] \begin{bmatrix} X \\ Y \end{bmatrix} = [D \quad N]\, U \begin{bmatrix} R \\ Q \end{bmatrix} = R. \tag{55}$$

Hence, the closed-loop system is stable if and only if the polynomial R is Hurwitz. Substituting $X = U_{11}R - NQ$ and $Y = U_{21}R + DQ$ we find

$$S = \frac{DX}{DX + NY} = DU_{11} - DN\frac{Q}{R} = DU_{11} - DNF, \tag{56}$$

where F is the rational function $F = Q/R$. Similarly, we find for the compensator

$$K = \frac{Y}{X} = \frac{U_{21}R + DQ}{U_{11}R - NQ} = \frac{U_{21} + DF}{U_{11} - NF}. \tag{57}$$

It is clear that we may obtain all stabilizing compensators by varying R and Q such that R is Hurwitz. Equivalently, we may let F vary over all rational functions that have their poles in the (closed) left-half complex plane. The representations (56) and (57) for S and K in terms of F are called the *Youla-Bongiorno-Jabr parametrizations* of S and K.

Through the representation (56) we may now consider the minimization of (51) with respect to all F that have their poles in the left-half complex plane. The sufficient and necessary conditions for the minimization of (51) follow from elementary methods in Wiener-Hopf optimization (Youla *et al.*, 1976). •

We are now ready to construct Φ. If an equalizing compensator is to satisfy the optimality condition, from (25) and 2.6.1 we need

$$\Phi \cdot \lambda \frac{B^{\sim} D_{+}^{\sim} X_{+,\lambda}^{\sim}}{A^{\sim} D_{+} X_{+,\lambda}} \cdot D_{+} D_{-} N_{+} N_{-} \frac{A^{\sim} A}{B^{\sim} B} = \lambda \Phi \frac{D_{+}^{\sim} X_{+,\lambda}^{\sim} D_{-} N_{+} N_{-} A}{X_{+,\lambda} B} \tag{58}$$

to have all its poles in the left-half complex plane. The factor $X_{+,\lambda}$ has all its roots in the open right-half complex plane, and therefore must be canceled. Hence, we take the numerator of Φ as $X_{+,\lambda}^{\sim} X_{+,\lambda}$, where the extra factor $X_{+,\lambda}$ has been included to make Φ nonnegative on the imaginary axis. To make sure that condition 1 of 2.6.1 is satisfied, we take the denominator of Φ as $N_{+}^{\sim} N_{+}$. Contemplation of (58) reveals that this choice maximizes the degree of the denominator of Φ while keeping all poles of (58) in the left-half complex plane. Thus,

$$\Phi = \frac{X_{+,\lambda}^{\sim} X_{+,\lambda}}{N_{+}^{\sim} N_{+}} \tag{59}$$

ensures that condition 2 of 2.6.1 is satisfied. Condition 1 holds if and only

$$\Phi S^{\sim} S V^{\sim} V = \lambda^{2} \frac{X_{+,\lambda}^{\sim} X_{+,\lambda}}{N_{+}^{\sim} N_{+}} \tag{60}$$

is strictly proper. Because for *non*-reduced solutions $\deg(X_{+,\lambda}) = n = \deg(N_+)$, condition 1 is not satisfied for non-reduced solutions. Since for reduced solutions the degree of $X_{+,\lambda}$ drops, only for such solutions Φ may be constructed such that the conditions of 2.6.1 are satisfied. To be sure that the reduced equalizing compensator stabilizes the closed-loop system, we need take the reduced solution corresponding to the *largest* value of $|\lambda|$ for which a reduced solution exists.

2.6.3 Summary (Solution of the minimum sensitivity problem.) The minimum sensitivity problem is solved by the compensator

$$K = \frac{D_- Y'_{\lambda_o}}{BN_- X_{+,\lambda_o}}, \tag{61}$$

where X_{+,λ_o} and Y'_{λ_o} satisfy the polynomial equation

$$\frac{1}{\lambda} A D_+^\sim X_{+,\lambda}^\sim = D_+ B X_{+,\lambda} + N_+ Y'_\lambda \tag{62}$$

for the largest value λ_o of $|\lambda|$ for which a reduced solution exists. •

It can be shown that the optimal solutions of the minimum sensitivity problem are *unique*.

We can now prove the necessity of the condition of 2.3.1. For $|\lambda| > \lambda_o$ the solution $X_{+,\lambda}$ of (62) is strictly Hurwitz (because none of its roots crosses the imaginary axis). At $|\lambda| = \lambda_o$ a root of $X_{+,\lambda}$, crosses over into the right-half plane. $X_{+,\lambda}$ can never become Hurwitz again as $|\lambda|$ decreases, because otherwise a stable closed-loop system would exist whose weighted sensitivity is less than λ_o.

2.7 Minimum complementary sensitivity and the regulability number

In conclusion of this section on the minimum sensitivity problem we consider the minimization of the *complementary sensitivity function.* In the notation of the block diagram of Fig. 1 the complementary sensitivity function is defined as

$$T = \frac{GK}{1+GK}. \tag{63}$$

It is the transfer function from the external input r to the output z, and is called the complementary sensitivity function because

$$T + S = 1. \tag{64}$$

In the next section it is seen that T plays a role in characterizing the robustness of the closed-loop system against plant perturbations. The smaller T is, the more robust the system.

The problem of minimizing $\|VT\|_\infty$, with

$$V = \frac{A}{B} \tag{65}$$

a weighting function, may be solved in a way that is completely analogous to the solution of the minimum weighted sensitivity problem. Comparison of

$$VT = \frac{VKG}{1+KG} = \frac{ANY}{(DX+NY)B} \tag{66}$$

with the corresponding expression (11) for the minimum sensitivity problem shows that the roles of D and N on the one hand and X and Y on the other are reversed. The solution of the minimum weighted complementary sensitivity problem may be phrased as follows.

2.7.1 Summary (Solution of the minimum complementary sensitivity function.) The minimum weighted complementary sensitivity problem is solved by the compensator

$$K = \frac{BD_-Y'_{\lambda_o}}{N_-X_{+,\,\lambda_o}}, \tag{67}$$

where Y'_{λ_o} and $X_{+,\,\lambda_o}$ satisfy the polynomial equation

$$\frac{1}{\lambda}AN_+^\sim(Y'_\lambda)^\sim = D_+X_{+,\,\lambda} + N_+BY'_\lambda \tag{68}$$

for the largest value λ_o of $|\lambda|$ for which a reduced solution exists. •

The equations that solve the minimum sensitivity and the minimum complementary sensitivity function are symmetric. It turns out that there is an interesting relation between the solutions of the *unweighted* versions of the two problems.

2.7.2 Summary (Relation between the solutions of the unweighted minimum sensitivity and complementary sensitivity problems.) Suppose that the plant G has both zeros and poles in the open right-half complex plane. Then

$$\min_{K \text{ stabilizing}} \|S\|_\infty = \min_{K \text{ stabilizing}} \|T\|_\infty. \tag{69}$$

•

2.7.3 Proof (Relation between the solutions of the unweighted minimum sensitivity and complementary sensitivity problems.) To solve the minimum unweighted sensitivity problem we need consider the equation

$$\frac{1}{\lambda}D_+^\sim X_+^\sim = D_+X_+ + N_+Y'. \tag{70}$$

Substituting $D_+X_+ = \frac{1}{\lambda}D_+^\sim X_+^\sim - N_+Y'$ into the left-hand side we obtain after rearrangement

$$-\frac{1}{\lambda}N_+^\sim(Y')^\sim = (1-\frac{1}{\lambda^2})D_+X_+ + N_+Y'. \tag{71}$$

This equation is very similar to the one we need solve for the solution of the minimum unweighted complementary sensitivity function. Contemplation reveals that a reduced solution of (70), which also satisfies (71), generates a reduced solution of the equation for the mimimum complementary sensitivity problem. Hence, the reduced solution of

(70) for the largest possible value of $|\lambda|$ generates a reduced solution for the minimum complementary sensitivity problem for the largest possible value of $|\lambda|$. This proves that the minimum ∞-norms are equal. •

The result 2.7.2 means that for a plant that has both right-hand plane poles and zeros the minimum ∞-norm of the unweighted sensitivity function and that of the unweighted complementary sensitivity function are *equal.* They are not achieved by the same compensator, though. For such plants, the minimum norm of the two sensitivity functions may be considered as a measure of the extent to which the system may be controlled by feedback.

We therefore define this norm as the *regulability number* of the plant, denoted as κ. Its meaning is as follows. Whatever stabilizing compensator is used to regulate the plant, the magnitude of either sensitivity function is never uniformly less than κ. In fact, typically either magnitude *exceeds* κ over some frequency region. Hence, κ is a measure for the inherent limitations in regulating the plant. The larger κ, the more difficult it is to regulate the plant. It may be proved that $\kappa > 1$.

For plants that have either all their poles or all their zeros in the left-half complex plane, or both, 2.7.2 does not hold, and the definition of the regulability number does not make sense. The situation is as follows.

1. *Right-half plane poles but no right-half plane zeros.* $\|S\|_\infty$ may be made 0 by infinite gain-feedback. This same feedback achieves $\|T\|_\infty = 1$, which is the best that can be accomplished for this type of system.

2. *Right-half plane zeros but no right-half plane poles.* The best control is no control: $K = 0$ achieves $\|S\|_\infty = 1$ and $\|T\|_\infty = 0$.

3. *No right-half plane poles and zeros.* $\|S\|_\infty$ can be made 0 by infinite-gain feedback and $\|T\|_\infty$ is made 0 by using no control at all.

We illustrate the significance of the regulability number by an example. Another interesting example involving a multiple inverted pendulum system may be found in Kwakernaak and Westdijk (1985).

2.7.4 Example (Regulability number.) Consider the plant with transfer function

$$G(s) = \frac{s-\zeta}{s-\pi}, \tag{72}$$

with ζ and π positive real numbers. The result that we obtain holds for *any* plant with a single right-half plane pole and a single right-half plane zero, because left-half plane poles and zeros do not affect the regulability number. To determine minimum unweighted sensitivity we need solve

$$\frac{1}{\lambda}(-s-\pi)X_+^\sim = (s-\pi)X_+ + (s-\zeta)Y'. \tag{73}$$

For $\lambda = \infty$, both X_+ and Y' have degree 1, so that for the reduced solution (obtained for $\lambda = \lambda_o$) both polynomials have degree 0, that is, are constants. Substitution of $s = \zeta$ yields

$$\frac{1}{\lambda_o}(-\zeta - \pi)X^{\sim}_{+,\lambda_o} = (\zeta - \pi)X_{+,\lambda_o}. \tag{74}$$

Because X_{+,λ_o} is a constant it may be canceled from the equation and we find

$$\lambda_o = \frac{\pi + \zeta}{\pi - \zeta}, \qquad \kappa = \left|\frac{\pi + \zeta}{\pi - \zeta}\right|. \tag{75}$$

The regulability number κ is completely determined by the right-half plane pole and zero of the plant. If the pole and zero are close, κ is large. This exposes very clearly the well-known difficulty in regulating plants with near-canceling pole-zero pairs in the right-half plane. •

3 The SISO Mixed Sensitivity Problem

As we have seen, the solution of the minimum sensitivity problem has the property that

$$|S(j\omega)V(j\omega)| = \lambda, \tag{76}$$

with λ a nonnegative constant. It follows that

$$|S(j\omega)| = \frac{\lambda}{|V(j\omega)|}. \tag{77}$$

This shows that the magnitude of the sensitivity function is inversely proportional to the magnitude of the weighting function V. Hence, the *shape* of $|S|$ directly follows from the shape of V, which is a design parameter, and can be chosen at will. The *level* at which $|S|$ takes it values is determined by λ, which follows from the solution of the optimization problem.

In spite of this attractive property, the minimum sensitivity problem is not useful for practical control system design. The reasons may be itemized as follows.

- The optimal compensator has the same pole excess as the *inverse* of the plant transfer function. This follows by inspection of the compensator transfer function (23) with the degrees of Y_+ and X_+ given by (46). Since physical plants always are proper and usually strictly proper, the optimal compensator transfer function is at best proper, and typically not proper.

- If the plant has zeros or poles on the imaginary axis, the optimal closed-loop system has these as closed-loop poles. This follows by inspection of the optimal closed-loop characteristic polynomial as given by (24).

Compensators that have improper transfer functions make the plant saturate. Closed-loop systems with poles on the imaginary axis are unstable from a practical point of view.

In this section we study a modification of the minimum-sensitivity problem that is known as the *mixed sensitivity problem.* This problem formualtion allows consideration of

1. performance optimization, and

2. robustness optimization.

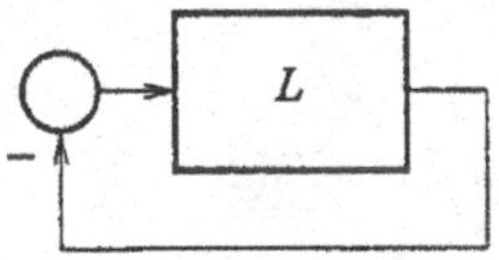

Figure 3: A SISO feedback system

3.1 Robustness of SISO closed-loop systems

A feedback system is *robust* if it does not lose its useful properties when the plant or compensator transfer functions change. The principal property the closed-loop system should preserve is of course *stability*. Hence, we first discuss *stability robustness*.

Consider the feedback configuration of Fig. 3. L is the transfer function of the series connection of the plant and the compensator. We investigate the stability of the feedback system when the loop gain is perturbed from its nominal value L_o to the actual value L. Writing

$$L = \frac{N}{D}, \quad L_o = \frac{N_o}{D_o}, \tag{78}$$

with D, N, D_o, and N_o polynomials, we introduce the following assumptions:

3.1.1 Assumptions (Robustness analysis.)

1. The nominal closed-loop system is stable, that is, all its closed-loop poles are stricly in the left-half complex plane.

2. The nominal and perturbed loop gains L_o and L are proper.

3. The nominal denominator polynomial D_o and the perturbed denominator polynomial D have the same numbers of roots in the open right-half plane, and no roots on the imaginary axis.

•

The perturbed and nominal closed-loop characteristic polynomials are given by

$$\chi = D + N, \quad \chi_o = D_o + N_o, \tag{79}$$

respectively. Consider the rational function

$$\sigma = \frac{\chi D_o}{\chi_o D} = \frac{(D+N)D_o}{(D_o+N_o)D}, \tag{80}$$

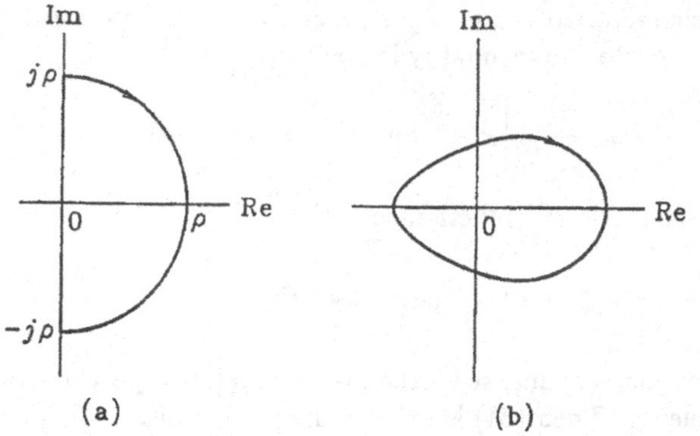

Figure 4: Left: a D-contour
Right: the image of the contour under a biproper rational map.

which we introduce to study the relation between the nominal and perturbed closed-loop characteristic polynomials. First we note that by the assumption that both L and L_o are proper σ is biproper, that is, both σ and its inverse are proper.

Suppose now that the argument s of σ traverses the D-contour with radius ρ of Fig. 4(a), and let us look at the corresponding contour that the *image* of σ traverses in the complex plane, as illustrated in Fig. 4(b). According to the *principle of the argument* of complex function theory, the number of times the image encircles the origin equals the number of zeros of σ inside the D-contour minus the number of its poles inside the D-contour. It follows that if the image of σ does *not* encircle the origin, $(D+N)D_o$ and $(D_o+N_o)D$ have the same numbers of roots inside the D-contour.

Now assume that ρ is so large that the D-contour encloses *all* right-half plane roots of $D+N$, D_o, D_o+N_o, and D. Then because by assumption D and D_o have the same numbers of roots inside the contour, it follows that if the image of σ does not encircle the origin, the nominal closed-loop characteristic polynomial D_o+N_o and the perturbed closed-loop polynomial $D+N$ have the same numbers of right-half plane roots. Since by the assumption that the nominal closed-loop system is stable the polynomial D_o+N_o has *no* right-half plane roots, neither does $D+N$ if and only if the image of σ does not encircle the origin.

To establish a sufficient condition that guarantees that σ does not encircle the origin we rewrite it in the form

$$\sigma = \frac{(D+N)D_o}{(D_o+N_o)D} = \frac{1+\frac{N}{D}}{1+\frac{N_o}{D_o}} = \frac{1+L}{1+L_o} = 1 + \frac{L-L_o}{1+L_o}. \tag{81}$$

Obviously, if

$$\left|\frac{L-L_o}{1+L_o}\right| < 1 \tag{82}$$

on the D-contour, the image of σ cannot encircle the origin. Because σ is biproper, the condition that (82) holds on the D-contour is easily seen to be equivalent to the condition that it holds on the imaginary axis, that is

$$\left|\frac{L(j\omega)-L_o(j\omega)}{1+L_o(j\omega)}\right| < 1 \quad \text{for all } \omega \in \mathbb{R}. \tag{83}$$

We use (83) to characterize a class of perturbations that do not affect closed-loop stability. To this end, we rewrite the inequality in the form

$$\left|\frac{(L-L_o)/L_o}{w} \cdot w\frac{L_o}{1+L_o}\right| < 1 \quad \text{on the imaginary axis,} \tag{84}$$

where w is a given (rational) function such that

$$\left|\frac{(L-L_o)/L_o}{w}\right| < 1 \quad \text{on the imaginary axis.} \tag{85}$$

Thus, w is a function that measures the maximal relative perturbation $(L-L_o)/L_o$ as a function of frequency. Then (84) is satisfied if

$$\left|w\frac{L_o}{1+L_o}\right| \leq 1 \quad \text{on the imaginary axis.} \tag{86}$$

Rewriting (85) and (86) in terms of ∞-norms we summarize as follows.

3.1.2 Summary (Doyle's condition for robust stability.) (Doyle, 1979.) Under the assumptions 3.1.1 the closed-loop system is stable for all perturbations such that

$$\left\|\frac{(L-L_o)/L_o}{w}\right\|_\infty < 1. \tag{87}$$

if

$$\|wT_o\|_\infty \leq 1. \tag{88}$$

•

Here

$$T = \frac{L}{1+L} \tag{89}$$

is the *complementary sensitivity function* introduced earlier, and T_o is the *nominal* complementary sensitivity function. The condition (88) may be proved to be not only sufficient but also necessary.

The result 3.1.2 implies that the smaller the ∞-norm $\|T_o w\|_\infty$ is, the larger the perturbations are that may be admitted without making the feedback system unstable. This may be recognized as follows. Suppose that for a given function w we have

$$\|T_o w\|_\infty = \delta, \tag{90}$$

with δ some positive number. Then the closed-loop system remains stable for all perturbations satisfying

$$\left\|\frac{(L-L_o)/L_o}{w}\right\|_\infty < \frac{1}{\delta}. \tag{91}$$

Thus, the smaller δ is, the larger the perturbations that are allowed.

This makes it tempting to define *robustness optimization* as the problem of minimizing the weighted complementary sensitivity $\|T_o w\|_\infty$. The following example illustrates, however, that 3.1.2 may not give a very realistic impression of the allowable perturbations.

3.1.3 Example (Application of Doyle's criterion.) Consider a feedback system with loop gain

$$L(s) = \frac{1}{s\theta + \varepsilon}, \tag{92}$$

with θ a known positive parameter and ε an uncertain parameter. In compliance with condition 3 of 3.1.1 the nominal value ε_o of ε may be either positive or negative but not zero.

It is easy to see that the closed-loop characteric polynomial of the perturbed system is

$$s\theta + (1 + \varepsilon). \tag{93}$$

Hence, the perturbed closed-loop system is stable if and only if

$$\varepsilon > -1. \tag{94}$$

Let us now see what the robustness criterion 3.1.2 says. The nominal complementary sensitivity function is

$$T_o(s) = \frac{1}{\theta s + 1 + \varepsilon_o}. \tag{95}$$

We have $\|T_o w\|_\infty \leq 1$ if we take

$$w(s) = \frac{1}{T_o(s)} = \theta s + 1 + \varepsilon_o. \tag{96}$$

The relative perturbation is given by

$$\frac{L(s) - L_o(s)}{L_o(s)} = \frac{\varepsilon_o - \varepsilon}{\theta s + \varepsilon}. \tag{97}$$

It follows that according to 3.1.2 stability is guaranteed for all perturbations such that the ∞-norm of

$$\frac{(L - L_o)/L_o}{w} = \frac{\varepsilon_o - \varepsilon}{(\theta s + \varepsilon)(\theta s + 1 + \varepsilon_o)} \tag{98}$$

is less than or equal to 1. As a function of $s = j\omega$, $\omega \in \mathrm{R}$, the magnitude of the expression (98) has a peak value

$$\left| \frac{\varepsilon_o - \varepsilon}{\varepsilon(1 + \varepsilon_o)} \right| \tag{99}$$

at $\omega = 0$. It follows that the perturbed closed-loop system is stable if

$$\frac{|1-\frac{\varepsilon_o}{\varepsilon}|}{1+\varepsilon_o} < 1. \tag{100}$$

This condition is easily violated if ε is close to 0. Thus, the robustness criterion rules out values of ε close to zero, which according to (94) do not at all endanger stability. The reason that these values are ruled out is not that stability is endangered, but that the condition 3 of 3.1.1 under which the criterion holds is endangered. •

3.2 A modified stability robustness criterion

The weakness of the stability criterion indicated by the example may be corrected by a modification of the criterion that requires *more* information about the perturbations than the very simple bound of 3.1.2. To this end, we consider perturbations that are differently specified.

3.2.1 Assumption (Plant perturbations.) In the configuration of Fig. 3 the perturbed and nominal loop gain are given by

$$L = \frac{N}{D}h, \quad L_o = \frac{N_o}{D_o}, \tag{101}$$

respectively, where

1. the polynomials D and N have the same degrees as the nominal polynomials D_o and N_o, and L_o is proper,
2. h is a proper rational function that has all its poles in the open left-half plane, and nominally equals 1.

•

In this representation, D and N model the *structured* perturbations, that is, perturbations caused by parameter variations in the nominal model. No conditions are imposed on whether or not any roots of D move across the imaginary axis. The function h, on the other hand, represents *unstructured* perturbations caused by unmodeled dynamics. The uncertainty model (101) requires that any dynamics that may become unstable be explicitly included in the structured part of the model.

In the robustness analysis we again assume that the nominal closed-loop system is stable. Writing

$$h = \frac{n}{d}, \tag{102}$$

with n and d polynomials that are nominally equal to 1, we have for the perturbed and nominal closed-loop characteristic polynomials χ and χ_o, respectively,

$$\chi = Dd + Nn, \quad \chi_o = D_o + N_o. \tag{103}$$

Application of the principle of the argument to the biproper rational function

$$\sigma = \frac{\chi}{\chi_o d} = \frac{Dd + Nn}{(D_o + N_o)d} \tag{104}$$

leads easily to the conclusion that the perturbed closed-loop system is stable if and only if the image of σ does not encircle the origin as s traverses the imaginary axis. We rewrite σ in the form

$$\begin{aligned} \sigma = \frac{Dd + Nn}{(D_o + N_o)d} &= \frac{D}{D_o}\frac{D_o}{D_o + N_o} + h\frac{N}{N_o}\frac{N_o}{D_o + N_o} = \frac{D}{D_o}S_o + h\frac{N}{N_o}T_o, \\ &= 1 + (\frac{D}{D_o} - 1)S_o + (h\frac{N}{N_o} - 1)T_o, \end{aligned} \tag{105}$$

where S_o is the nominal sensitivity function and T_o the nominal complementary sensitivity function. A sufficient condition that the image of this function does not encircle the origin is that

$$\left|(\frac{D}{D_o} - 1)S_o + (h\frac{N}{N_o} - 1)T_o\right| < 1 \tag{106}$$

on the imaginary axis. We rewrite this in the form

$$\left|\frac{(D - D_o)/D_o}{w_1} \cdot w_1 S_o + \frac{(Nh - N_o)/N_o}{w_2} \cdot w_2 T_o\right| < 1, \tag{107}$$

where w_1 and w_2 are given rational functions whose significance will soon be explained. By the Cauchy-Schwarz inequality we have

$$\begin{aligned} &\left|\frac{(D - D_o)/D_o}{w_1} \cdot w_1 S_o + \frac{(Nh - N_o)/N_o}{w_2} \cdot w_2 T_o\right|^2 \\ &\leq \left(\left|\frac{(D - D_o)/D_o}{w_1}\right|^2 + \left|\frac{(Nh - N_o)/N_o}{w_2}\right|^2\right)\left(|w_1 S_o|^2 + |w_2 T_o|^2\right). \end{aligned} \tag{108}$$

We can now establish the following result.

3.2.2 Summary (Extended stability robustness criterion.) The closed-loop system remains stable under all perturbations such that

$$\left|\frac{(D - D_o)/D_o}{w_1}\right|^2 + \left|\frac{(Nh - N_o)/N_o}{w_2}\right|^2 < 1 \quad \text{on the imaginary axis} \tag{109}$$

if

$$|w_1 S_o|^2 + |w_2 T_o|^2 \leq 1 \quad \text{on the imaginary axis.} \tag{110}$$

•

It is conjectured that (110) is not only sufficient but also necessary. Figure 5 shows that for a fixed frequency ω the numbers $|w_1(j\omega)|$ and $|w_2(j\omega)|$ form the semi-axes of the quarter ellipse that bounds the relative perturbations $|(D(j\omega) - D_o(j\omega)|/|D_o(j\omega)|$ and $|(N(j\omega)h(j\omega) - N_o(j\omega)|/|N_o(j\omega)|$.

We illustrate the application of the criterion by a simple example.

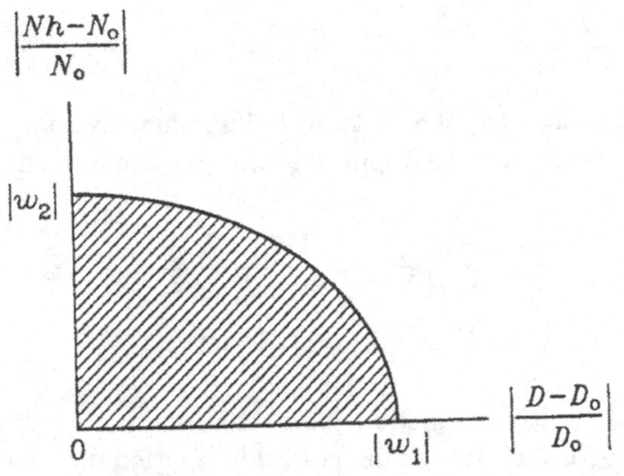

Figure 5: Allowable perturbations

3.2.3 Example (Extended robustness criterion.) By way of example we consider the same feedback system as in Example 3.1.3. For the nominal sensitivity and complementary sensitivity functions we have

$$S_o(s) = \frac{\theta s + \varepsilon_o}{\theta s + \varepsilon_o + 1}, \quad T_o(s) = \frac{1}{\theta s + \varepsilon_o + 1}. \tag{111}$$

To maximize the perturbations that according to the criterion leave the perturbed system stable we take

$$w_1(s) = \frac{1}{S_o(s)} = \frac{\theta s + \varepsilon_o + 1}{\theta s + \varepsilon_o}, \tag{112}$$

$$w_2(s) = \frac{1}{T_o(s)} = \theta s + \varepsilon_o + 1. \tag{113}$$

In Fig. 6 asymptotic Bode plots of the two functions are given. That of w_1 shows that the feedback system at low frequencies offers protection against structured perturbations of the plant denominator. The plot of w_2 shows that at high frequencies protection is available against unmodeled perturbations, such as parasitic effects.

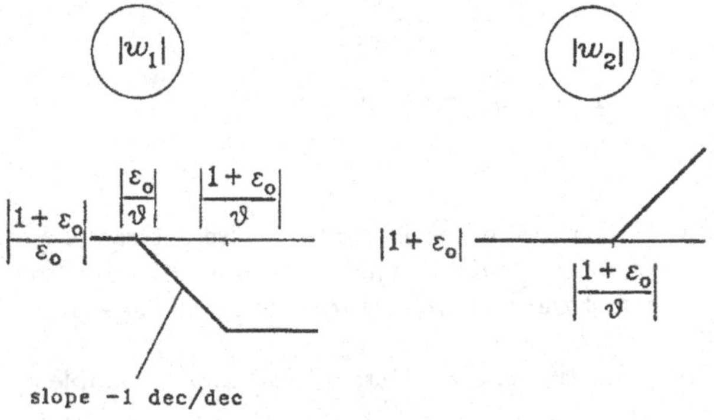

Figure 6: Asymptotic Bode plots of w_1 and w_2

Variations in the parameter ε affect the denominator

$$D(s) = \theta s + \varepsilon \tag{114}$$

of the plant. We have

$$\frac{D(s) - D_o(s)}{D_o(s)} = \frac{\varepsilon - \varepsilon_o}{\theta s + \varepsilon_o}, \qquad \frac{(D(s) - D_o(s))/D_o(s)}{w_1(s)} = \frac{\varepsilon - \varepsilon_o}{\theta s + \varepsilon_o + 1}. \tag{115}$$

The latter expression has a peak value at zero frequency. It follows that according to the criterion stability is guaranteed for all ε such that

$$\left|\frac{\varepsilon - \varepsilon_o}{\varepsilon_o + 1}\right| < 1, \tag{116}$$

or

$$-1 < \varepsilon < 1 + 2\varepsilon_o. \tag{117}$$

This result is much more realistic than that of the previous example. In particular, it does not exclude the case where $\varepsilon_o = 0$ (i.e., the nominal plant is an integrator), which is outside the scope of Doyle's criterion. ●

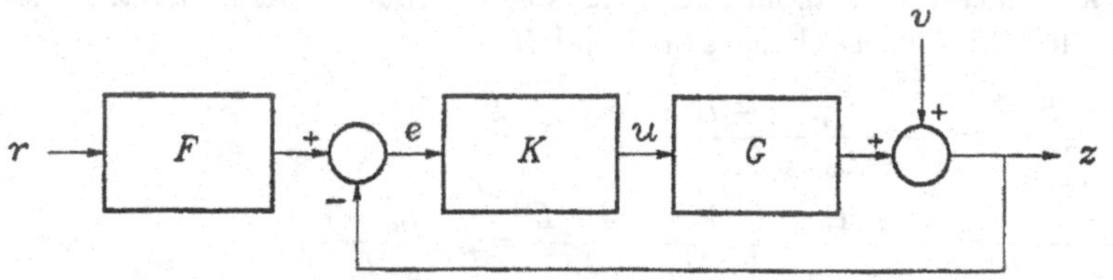

Figure 7: Two-degree-of-freedom configuration

3.3 Performance robustness

Although stability is a primary concern, feedback systems are designed so that they perform certain functions, such as attenuating disturbances and tracking command signals.

Consider the configuration of Fig. 7. The basic configuration of Fig. 1 has been expanded to a two-degree of freedom configuration including a prefilter F and a reference signal r. There are two important transfer functions that are decisive for the performance of the system.

1. The by now familiar *sensitivity function*

$$S = \frac{1}{1 + GK}, \tag{118}$$

which is the transfer function from the disturbance v to the control system output z.

2. The *closed-loop transfer function*

$$H = \frac{GK}{1+GK}F, \tag{119}$$

which is the transfer function from the reference input r to the control system output z.

Note that the closed-loop transfer function H may be written as

$$H = TF, \tag{120}$$

where

$$T = \frac{GK}{1+GK} \tag{121}$$

is the complementary sensitivity function.

We soon discuss the requirements that performance imposes on S and H, but at this point are interested in their *robustness*, that is, to what extent these two functions are affected by plant changes. Only plant changes are considered, because it is assumed that the compensator K and prefilter F are well enough under control of the designer to permit ignoring any uncertainty about them. It is easily found that under this assumption we may write for the *relative* changes of S and H

$$\frac{S - S_o}{S} = \frac{L_o}{1+L_o}\,\frac{L_o - L}{L_o} = T_o\,\frac{L_o - L}{L_o}, \tag{122}$$

$$\frac{H - H_o}{H} = \frac{T - T_o}{T} = \frac{1}{1+L_o}\,\frac{L_o - L}{L} = S_o\,\frac{L_o - L}{L}. \tag{123}$$

The first relation shows that the relative effect that changes in the loop gain have on the sensitivity is determined by the (nominal) *complementary sensitivity function*. Conversely, the relative effect of changes in the loop gain on the closed-loop transfer function is determined by the *sensitivity function*.

Hence, for performance robustness we need both S and T to be small. Of course, they only need be small in the frequency range where it matters. The sensitivity S, for instance, should be small at *low* frequencies, because this is where the behavior of the closed-loop transfer function is important. Because S and T add up to one, we can only aspire to make T really small at *high* frequencies.

3.4 The mixed sensitivity problem

As we have seen, the stability of the closed-loop system of Fig. 3 is robust against perturbations whose size is measured by the frequency dependent bounds w_1 and w_2 if

$$|S_o(j\omega)w_1(j\omega)|^2 + |T_o(j\omega)w_2(j\omega)|^2 \le 1 \quad \text{for all } \omega \in \mathbb{R}. \tag{124}$$

S_o and T_o are the sensitivity and complementary sensitivity functions of the nominal feedback system. Suppose that

$$\sup_{\omega\in\mathbb{R}} \left(|S_o(j\omega)w_1(j\omega)|^2 + |T_o(j\omega)w_2(j\omega)|^2\right) = \delta. \tag{125}$$

If $\delta \leq 1$, the stability robustness criterion is satisfied. Whatever value δ has, as long as it is nonzero we can rescale the weighting functions to w_1/δ and w_2/δ, and satisfy the criterion. This means that for given functions w_1 and w_2, the smaller δ is, the more robust the feedback system.

If δ is small, also the individual terms $S_o w_1$ and $T_o w_2$ are small, and hence, depending on the weighting functions w_1 and w_2, also S_o and T_o. As we have seen, this is what is needed for performance robustness.

For these reasons, *robustness optimization* may be defined as the problem of shaping the feedback loop such that δ as given by (125) is *minimal.* The freedom that is available is to choose the compensator transfer function K in a configuration as in Fig. 1. The problem of minimizing an expression of the form

$$\sup_{\omega \in \mathbb{R}} \left(|S(j\omega)w_1(j\omega)|^2 + |T(j\omega)w_2(j\omega)|^2 \right) \tag{126}$$

with respect to all compensators that stabilize the closed-loop system is known as the *mixed sensitivity problem.* It is an extension of the minimum sensitivity problem, and is called "mixed" because of the additional term involving complementary sensitivity. Note that we omit the subscript $_o$ because from now on we are concerned with the nominal system only. Moreover, we represent (126) by the notation

$$\|S^{\sim}Sw_1^{\sim}w_1 + T^{\sim}Tw_2^{\sim}w_2\|_\infty. \tag{127}$$

3.5 Performance design by the mixed sensitivity problem

The mixed sensitivity problem has been introduced as a robustness optimization problem. Although robustness surely is an important issue, the principal function of a feedback system is to *perform.* Performance requirements impose specifications on the sensitivity function and the closed-loop transfer function that can easily be accomodated in the mixed sensitivity problem.

It is convenient to replace the complementary sensitivity function T and closed-loop transfer function H as design targets with the *input sensitivity function*

$$U = \frac{K}{1+GK}. \tag{128}$$

The input sensitivity function is the transfer function from the disturbance v to the plant input u (Fig. 7). T and H may be expressed in U as

$$T = GU, \quad H = FGU. \tag{129}$$

We now redefine the mixed sensitivity problem as the problem of minimizing

$$\|V^{\sim}V(S^{\sim}SW_1^{\sim}W_1 + U^{\sim}UW_2^{\sim}W_2)\|_\infty \tag{130}$$

with respect to all stabilizing compensators. The purpose of the mixed sensitivity problem is to optimize the *feedback loop* of the configuration of Fig. 7. Determining the prefilter F is not part of the problem. V is a *disturbance shaping filter* as in Fig. 8, and W_1 and W_2 are frequency dependent weighting filters for the feedback system output and the plant input, respectively.

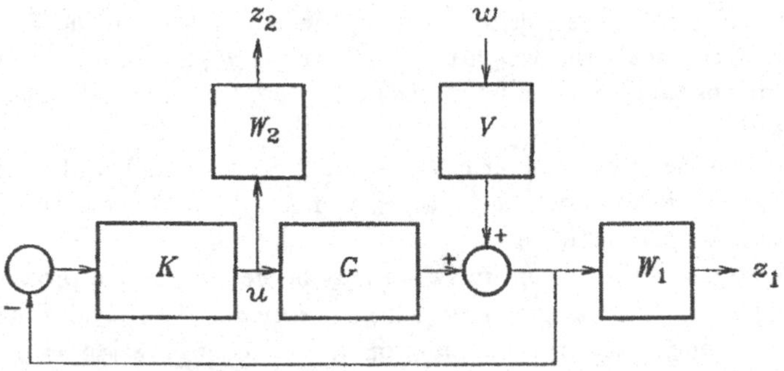

Figure 8: The mixed sensitivity problem

The mixed sensitivity criterion has the same form as the criterion for the robustness test, and, hence, may be used for robustness optimization. At the same time, by choosing V, W_1, and W_2 suitably, the feedback loop may be shaped for performance.

The first term $V^{\sim}VS^{\sim}SW_1^{\sim}W_1$ serves to shape the sensitivity function S. The shaping filter V should include as modes all the open-loop plant modes whose poles need be shifted to better locations in the complex plane. Assuming that the plant transfer function is

$$G = \frac{N}{D} \tag{131}$$

we choose the shaping filter V as

$$V = \frac{M}{D}. \tag{132}$$

The polynomial M has the same degree as D, so that V is biproper. Without loss of generality, M may be assumed to be Hurwitz. It will be seen that the optimal compensator shifts the open-loop poles, which are the roots of D, to the locations of the roots of M. If W_1 and W_2 are well-chosen, the roots of M are the *dominant* closed-loop poles. Hence, choosing M amounts to placing the dominant poles. The dominant poles determine important properties such as closed-loop bandwidth and closed-loop response.

The compensator that solves the mixed sensitivity problem, like that for the minimum sensitivity problem, is *equalizing*, that is, there exists a constant λ such that

$$\|V^{\sim}V(S^{\sim}SW_1^{\sim}W_1 + U^{\sim}UW_2^{\sim}W_2)\|_\infty = \lambda^2. \tag{133}$$

Equivalently,

$$|V(j\omega)S(j\omega)W_1(j\omega)|^2 + |V(j\omega)U(j\omega)W_2(j\omega)|^2 = \lambda^2, \qquad \omega \in \mathsf{R}. \tag{134}$$

It follows that

$$|V(j\omega)S(j\omega)W_1(j\omega)|^2 \leq \lambda^2, \qquad |V(j\omega)U(j\omega)W_2(j\omega)|^2 \leq \lambda^2 \tag{135}$$

for all $\omega \in \mathsf{R}$, and, hence,

$$|S(j\omega)| \;\leq\; \frac{|\lambda|}{|V(j\omega)W_1(j\omega)|}, \qquad \omega \in \mathsf{R}, \tag{136}$$

$$|U(j\omega)| \;\leq\; \frac{|\lambda|}{|V(j\omega)W_2(j\omega)|}, \qquad \omega \in \mathsf{R}. \tag{137}$$

If the weighting functions W_1 and W_2 are well-chosen, the first of the two terms in (134) dominates at *low* frequencies, and the second at *high* frequencies. Hence,

$$|S(j\omega)| \approx \frac{|\lambda|}{|V(j\omega)W_1(j\omega)|} \quad \text{at low frequencies,} \tag{138}$$

$$|U(j\omega)| \approx \frac{|\lambda|}{|V(j\omega)W_2(j\omega)|} \quad \text{at high frequencies.} \tag{139}$$

This means that the *shapes* of $|S|$ and $|U|$ at low and high frequencies, respectively, are predetermined by V, W_1 and W_2. For shaping S, mainly V is important. W_1, which need be chosen *proper* of the form

$$W_1 = \frac{A_1}{B_1}, \tag{140}$$

offers opportunity for refinement. Often, $W_1 = 1$ is an adequate choice.

The function of the weighting filter W_2 is to shape the high-frequency behavior of the input sensitivity function

$$U = \frac{K}{1+GK}. \tag{141}$$

Practical compensators require a proper or, even better, strictly proper transfer function K. Assume that the plant transfer function G is strictly proper. Then if K is proper, at high frequencies U behaves as K. Since by asssumption V is biproper, it is constant at high frequencies. Hence, by (139) we can force K to *decrease* at high frequencies, so that it is proper, if we let the weighting function W_2 *increase* at high frequencies. Therefore, we choose

$$W_2 = \frac{A_2}{B_2} \tag{142}$$

such that its *inverse* $1/W_2 = B_2/A_2$ is proper. This results in a compensator K whose pole excess equals the pole excess of $1/W_2$.

By controlling the pole excess and, more in general, the high frequency behavior of U and K, we also control the high frequency behavior of the complementary sensitivity function $T = GU$. The pole excess of T equals the sum of the pole excess of the plant G and that of U.

Later, when we know how to solve the mixed sensitivity problem, we illustrate these points by a simple design example.

3.5.1 Summary (The mixed sensitivity problem.) The mixed sensitivity problem is the problem to find a (rational) compensator transfer function K in the configuration of Fig. 8 such that

1. the closed-loop system is stable,

2. the ∞-norm

$$\|V^{\sim}V(S^{\sim}SW_1^{\sim}W_1 + U^{\sim}UW_2^{\sim}W_2)\|_{\infty} \tag{143}$$

is minimal.

$V = M/D$ is biproper, and $W_1 = A_1/B_1$ and $1/W_2 = B_2/A_2$ are proper rational functions. M is strictly Hurwitz, A_1 and B_1 are coprime and Hurwitz, and also A_2 and B_2 are coprime and Hurwitz. •

3.6 Solution of the mixed sensitivity problem

Like the minimum sensitivity problem, we solve the mixed sensitivity problem by the equalizer principle, which straightforwardedly may be generalized as follows.

3.6.1 Summary (Equalizer principle for the mixed sensitivity problem.) Suppose that the stabilizing compensator K_o is equalizing, that is,

$$|V(j\omega)S(j\omega)W_1(j\omega)|^2 + |V(j\omega)U(j\omega)W_2(j\omega)|^2 = \lambda^2, \quad \omega \in \mathbb{R}, \tag{144}$$

for some real constant λ_o. Then if there exists a nontrivial nonnegative real function $\Phi(j\omega)$, $\omega \in \mathbb{R}$, such that K_o minimizes

$$\int_{-\infty}^{\infty} \Phi(j\omega) \left(|V(j\omega)S(j\omega)W_1(j\omega)|^2 + |V(j\omega)U(j\omega)W_2(j\omega)|^2 \right) \, d\omega \tag{145}$$

with respect to all stabilizing compensators, the compensator K_o minimizes

$$\|V^{\sim}V(S^{\sim}SW_1^{\sim}W_1 + U^{\sim}UW_2^{\sim}W_2)\|_\infty \tag{146}$$

with respect to all stabilizing compensators. •

Like for the minimum sensitivity problem, we attempt to find equalizing compensators, that is, compensators such that

$$V^{\sim}V(S^{\sim}SW_1^{\sim}W_1 + U^{\sim}UW_2^{\sim}W_2) = \lambda^2 \tag{147}$$

for some real λ. Setting

$$K = \frac{Y}{X}, \tag{148}$$

and writing out all functions in terms of polynomials we find easily that we need

$$M^{\sim}M(X^{\sim}X\; A_1^{\sim}A_1\; B_2^{\sim}B_2 + Y^{\sim}Y\; A_2^{\sim}A_2\; B_1^{\sim}B_1) = \lambda^2\chi^{\sim}\chi\; B_1^{\sim}B_1\; B_2^{\sim}B_2, \tag{149}$$

where χ is the closed-loop polynomial

$$\chi = DX + NY. \tag{150}$$

Inspection of (149) shows that the first term on the left and the term on the right both have a factor $B_2^{\sim}B_2$. Hence, also the second term on the left has this factor. The only place where it can go is into $Y^{\sim}Y$. By a similar argument, $X^{\sim}X$ has a factor $B_1^{\sim}B_1$. It turns out, and later we explain why, that we need take

$$X = B_1 P, \qquad Y = B_2 Q, \tag{151}$$

with P and Q new polynomials that replace X and Y as unknowns. Finally, inspection of (149) reveals a factor $M^\sim M$ on the left that can only be accomodated on the right as a factor of $\chi^\sim\chi$. Because by assumption M is Hurwitz, and we are looking for compensators that make also χ Hurwitz, necessarily χ contains M as a factor. We take

$$\chi = \frac{1}{\lambda} MR, \tag{152}$$

with R another new polynomial, which replaces χ as unknown.

With these substitutions (149) and (150) reduce to

$$\frac{1}{\lambda} MR_\lambda = DB_1 P_\lambda + NB_2 Q_\lambda, \tag{153}$$

$$R_\lambda^\sim R_\lambda = A_1^\sim A_1 P_\lambda^\sim P_\lambda + A_2^\sim A_2 Q_\lambda^\sim Q_\lambda, \tag{154}$$

where the subscript λ explicitly indicates the dependence of the solution on λ.

If a solution P_λ, Q_λ, R_λ exists such that R_λ is Hurwitz, there exists a stabilizing equalizing compensator such that

$$\|V^\sim V(S^\sim S W_1^\sim W_1 + U^\sim U W_2^\sim W_2)\|_\infty = \lambda^2. \tag{155}$$

In fact, more can be said.

3.6.2 Assertion (Improvability of the solution of the mixed sensitivity problem.) For any given λ, there exists a stabilizing compensator such that

$$\|V^\sim V(S^\sim S W_1^\sim W_1 + U^\sim U W_2^\sim W_2)\|_\infty \le \lambda^2. \tag{156}$$

if and only if the equations

$$\frac{1}{\lambda} MR_\lambda = DB_1 P_\lambda + NB_2 Q_\lambda, \tag{157}$$

$$R_\lambda^\sim R_\lambda = A_1^\sim A_1 P_\lambda^\sim P_\lambda + A_2^\sim A_2 Q_\lambda^\sim Q_\lambda \tag{158}$$

have a solution P_λ, Q_λ, R_λ such that the compensator

$$K = \frac{B_2 Q_\lambda}{B_1 P_\lambda} \tag{159}$$

stabilizes the closed-loop system. •

The sufficiency of the condition that (157-158) have a stabilizing solution is clear. The proof of the necessity follows. Like for the minimum sensitivity problem, we consider *reduced solutions.*

The equations (157-158) generally cannot easily be solved, except for $\lambda = \infty$. Then the linear equation (157) reduces to

$$0 = DB_1 P_\infty + NB_2 Q_\infty, \tag{160}$$

which has the mimimum-degree solution

$$P_\infty = NB_2, \qquad Q_\infty = -DB_1. \tag{161}$$

Substitution of this solution into the quadratic equation (158) yields

$$R_\infty^\sim R_\infty = A_1^\sim A_1 B_2^\sim B_2 N^\sim N + A_2^\sim A_2 B_1^\sim B_1 D^\sim D. \tag{162}$$

From this, R_∞ may be determined by spectral factorization of the right-hand side.

At this point, concrete polynomials P_∞, Q_∞, and R_∞ have been obtained, whose degrees are known. It may be shown by application of the implicit function theorem (Kwakernaak, 1985) that under the stated assumptions on G, V, W_1, and W_2, for $|\lambda|$ sufficiently large the equations (157-158) have a solution such that the polynomials Q_λ, and R_λ have the same degrees as at ∞. In fact, there exists a lower bound $\lambda_o \geq 0$ such that for $|\lambda| \geq \lambda_o$ the equations (157-158) have a family of solutions P_λ, Q_λ, R_λ whose coefficients depend analytically on λ, such that for $|\lambda|$ sufficiently large R_λ has all its roots in the open left-half plane. The value of the lower bound λ_o will be established later.

As λ changes, say from ∞ down to λ_o, the coefficients of R_λ change, and the roots of R_λ describe a locus in the complex plane. As λ decreases to λ_o, one of two following situations arises:

1. All roots of R_λ stay in the open left-half plane until λ reaches the value λ_o. Then the equalizing compensator that corresponds to λ_o is stabilizing. As will be seen,

$$\|V^\sim V(S^\sim S W_1^\sim W_1 + U^\sim U W_2^\sim W_2)\|_\infty \geq \lambda_o^2 \tag{163}$$

   for *any* compensator, whether stabilizing or not. Since the stabilizing equalizing compensator that is obtained for λ_o achieves this bound, it is optimal.

2. One or several of the roots of R_λ cross over into the right-half plane (and stay there).

The former case is the exception rather than the rule. In the latter situation, a root of R_λ may cross over into the right-half plane through the imaginary axis, or through infinity. Inspection of the quadratic equation (158) shows that if R_λ has a root on the imaginary axis or at infinity (i.e., R_λ loses degree), both P_λ and Q_λ also have this root. Because the equations (157-158) are homogeneous, we may *cancel* the common factor of P_λ, Q_λ, and R_λ corresponding to this common root. The polynomials thus obtained still are a solution of (157-158), and are called a *reduced* solution, as in the minimum sensitivity problem.

In view of what we know from the minimum sensitivity problem we expect that the optimal solution is a reduced solution. Being reduced by itself it not enough, however.

3.6.3 Assertion (Optimality of reduced solutions for the mixed sensitivity problem.) Define the polynomial E_λ as

$$E_\lambda = A_1^\sim A_1 B_2 N P_\lambda^\sim - A_2^\sim A_2 B_1 D Q_\lambda^\sim, \tag{164}$$

and factor it as $E_\lambda = E_{-,\lambda} E_{+,\lambda}$, where $E_{-,\lambda}$ has all its roots in the closed-left half plane and $E_{+,\lambda}$ has all its roots in the open right-half plane. Then a reduced solution of (157-158) is optimal if and only if the following conditions are satisfied:

1. The corresponding compensator is stabilizing, and
2. $\deg(E_{+,\lambda}) > \deg(R_\lambda)$.

•

For completeness we conclude this subsection by stating the degrees of the solution polynomials. Denoting $\deg(D) = d$, $\deg(B_1) = b_1$, $\deg(B_2) = b_2$, and $\deg(A_2) = a_2$, it may be found that under the stated assumptions on the properness of G, V, W_1, and W_2 the degrees of the full-degree solution polynomials are $\deg(P_\lambda) = b_2 + d$, $\deg(Q_\lambda) = b_1 + d$, and $\deg(R_\lambda) = a_2 + b_1 + d$. As a result, the compensator transfer function K is always proper.

3.7 Optimality

We first present the proof of 3.6.3, and then describe how optimal reduced solutions may be computed.

3.7.1 Proof (Optimality of reduced solutions.) The proof of 3.6.3 is based on the YBJ parametrization of 2.6.2. As seen there, the sensitivity function may be expressed as $S = DU_{11} - DNF$, where the closed-loop system is stable if and only if the rational function F has all its poles in the left-half plane. Similarly, the input sensitivity function may be parametrized as $U = DU_{21} - DNF$. By a straightforward variational argument familiar from Wiener-Hopf optimization it follows that a necessary and sufficient condition for a stabilizing compensator to minimize

$$\int_{-\infty}^{\infty} \Phi(j\omega)\left(|V(j\omega)S(j\omega)W_1(j\omega)|^2 + |V(j\omega)U(j\omega)W_2(j\omega)|^2\right)\, d\omega \tag{165}$$

is that

$$\Phi V^\sim V D(W_1^\sim W_1 S^\sim N - W_2^\sim W_2 U^\sim D) \tag{166}$$

have all its poles in the left-half complex plane. Substituting $V = M/D$, $W_1 = A_1/B_1$, $W_2 = A_2/B_2$, $S = 1/(1+GK)$, $U = K/(1+GK)$, $G = N/D$, and $K = Y/X$, it follows that this is equivalent to the condition that

$$\Phi \frac{M^\sim M}{\chi^\sim \chi}\left(\frac{A_1^\sim A_1}{B_1^\sim B_1} X^\sim N - \frac{A_2^\sim A_2}{B_2^\sim B_2} Y^\sim D\right) \tag{167}$$

have all its poles in the left-half plane, where $\chi = DX + NY$ is the closed-loop characteristic polynomial. For equalizing solutions we have $X = B_1 P_\lambda$, $Y = B_2\, Q_\lambda$, $\chi = \frac{1}{\lambda} M R_\lambda$, so that for equalizing solutions to minimize (165) we need

$$\Phi \frac{\lambda^2}{R^\sim R}\left(\frac{A_1^\sim A_1 N P^\sim}{B_1} - \frac{A_2^\sim A_2 D Q^\sim}{B_2}\right) = \lambda^2 \Phi \frac{E_\lambda}{R^\sim R B_1 B_2} \tag{168}$$

to have all its poles in the left-half plane. To cancel the anti-Hurwitz factor $R^\sim$ in the denominator, the numerator of Φ must contain this factor. To make Φ real nonnegative on the imaginary axis, the numerator of Φ need be chosen as $R_\lambda^\sim R_\lambda$. Φ should be strictly proper, however, if equalizing solutions are to have any chance of minimizing (165). The degree of the denominator Φ is maximized while keeping all poles of (168) in the left-half plane by taking

$$\Phi = \frac{R_\lambda^\sim R_\lambda}{E_{+,\lambda}^\sim E_{+,\lambda}}. \tag{169}$$

Φ is strictly proper if and only the second condition of 3.6.3 holds. This completes a sketch of the proof of 3.6.3.

Note at this point that if at an earlier stage we had not chosen $X = B_1P$, $Y = B_2Q$ to account for the common factors in (149) we would have obtained additional non-Hurwitz factors in the denominator of (168) that cannot be accomodated. •

A crucial question, of course, is whether there actually exist reduced solutions satisfying the conditions of 3.6.2. We prove later that they do (in 3.10.2), but state the conclusion here.

3.7.2 Assertion (Existence of optimal reduced solutions.) Consider the set of solutions P_λ, Q_λ, R_λ, $\lambda \geq \lambda_o$, of (157-158) whose coefficients depend analytically on λ such that

$$P_\infty = NB_2, \quad Q_\infty = -DB_1, \tag{170}$$

$$R_\infty^\sim R_\infty = A_1^\sim A_1 B_2^\sim B_2 N^\sim N + A_2^\sim A_2 B_1^\sim B_1 D^\sim D, \tag{171}$$

with R_∞ Hurwitz. Then as λ decreases from ∞ to λ_o, the *first* reduced solution that is obtained (at $\lambda = \lambda_{\text{opt}}$)
is optimal. •

This result shows that optimal reduced solutions may be found by following the solution curve of the equations (157-158), starting at $\lambda = \infty$, until a reduced solution is encountered at $\lambda = \lambda_{\text{opt}}$. It may be proved that the solution is *unique*.

3.8 Numerical solution

The pair of polynomial equations (157-158) generally cannot be handled analytically. Numerically they can be solved quite conveniently by Newton iteration. Linearization yields the iterative set of linear polynomial equations

$$\frac{1}{\lambda} R_k = DB_1P_k + NB_2Q_k, \tag{172}$$

$$\begin{aligned} R_k^\sim R_{k-1} + R_{k-1}^\sim R_k - R_{k-1}^\sim R_{k-1} \\ = A_1^\sim A_1 P_k^\sim P_{k-1} + A_1^\sim A_1 P_{k-1}^\sim P_k - A_1^\sim A_1 P_{k-1}^\sim P_{k-1} \\ + A_2^\sim A_2 Q_k^\sim Q_{k-1} + A_2^\sim A_2 Q_{k-1}^\sim Q_k - A_2^\sim A_2 Q_{k-1}^\sim Q_{k-1}, \end{aligned} \tag{173}$$

$k = 1, 2, \cdots$, where the index k denotes the iteration number. Given a sufficiently good initial estimate P_0, Q_0, R_0, the solution sequence convergences quadratically to an accurate solution of (157-158).

To obtain optimal reduced solutions, the procedure is initialized by finding the solution (170-171) at infinity. This solution serves as starting solution for the Newton iteration for a new value of λ such that $1/\lambda$ is sufficiently small. Taking the resulting solution as a starting solution for a next value of λ, the solution curve as a function of λ is followed until one of the roots of R_λ crosses over into the right-half plane.

By a simple search procedure the optimal solution can be closed in arbitrarily tightly. Because at the reduced solution the Jacobian of the equations becomes singular, close to the reduced solution convergence slows down from quadratic to linear. If desired, the computation can be finalized by solving *directly* for the reduced solution (including λ as a variable in the Newton iteration), taking the result of the search procedure as a starting solution.

3.9 Example

We consider an example. In particular we show how by choosing the polynomial M in the weighting function $V = M/D$ the dominant closed-loop poles may be placed.

3.9.1 Example (Mixed sensitivity problem.) We apply the mixed sensitivity problem to the simple second-order plant

$$G(s) = \frac{1}{s^2}, \tag{174}$$

so that $N = 1$, $D = s^2$. The first step is to establish the shaping filter

$$V = \frac{M}{D}. \tag{175}$$

The optimal compensator shifts the open-loop poles (both at 0) to closed-loop locations that coincide with the locations of the roots of M. If these are the dominant closed-loop poles, they determine the closed-loop bandwidth. A closed-loop bandwidth of, say, 1 [rad/s] may be achieved by placing the dominant poles at $\frac{1}{2}\sqrt{2}(-1 \pm j)$. Thus, we let

$$M(s) = s^2 + s\sqrt{2} + 1. \tag{176}$$

It turns out that there is no need for fine tuning with the weighting function W_1, so we simply let

$$W_1(s) = 1. \tag{177}$$

W_2 controls the high-frequency behavior of the compensator, the input sensitivity function and the complementary sensitivity function. Suppose that we take

$$W_2(s) = c(1 + rs), \tag{178}$$

with c and r nonnegative constants to be determined. Since if $c \neq 0$ and $r \neq 0$ the function W_2 for high frequencies behaves as s, both the compensator transfer function K and the input sensitivity function U for high frequencies behave as $1/s$, and the complementary sensitivity function T as $1/s^2$. This provides a high-frequency roll-off that is desirable for robustness against nonstructured plant perturbations. If $c \neq 0$ but $r = 0$, K and U are constant at high frequencies, while T behaves as $1/s$.

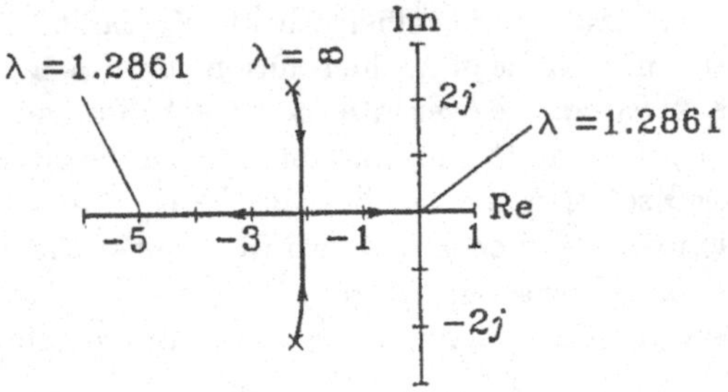

Figure 9: Loci of the roots of R_λ for $r = 0$

With this choice for the weighting functions, the polynomial equations (157-158) take the form

$$\begin{aligned} \frac{1}{\lambda}(s^2 + s\sqrt{2} + 1) &= s^2 P_\lambda(s) + Q_\lambda(s), & (179) \\ R_\lambda^\sim(s) R_\lambda(s) &= P_\lambda^\sim(s) P_\lambda(s) + c^2(1 - rs^2) Q_\lambda^\sim(s) Q_\lambda(s). & (180) \end{aligned}$$

Solution at $\lambda = \infty$ results in

$$P_\infty(s) = 1, \qquad Q_\infty(s) = -s^2, \tag{181}$$

while R_∞ follows by spectral factorization of

$$R_\infty^\sim(s) R_\infty(s) = 1 + c^2(1 - r^2 s^2) s^4. \tag{182}$$

If $r = 0$, this reduces to $R_\infty^\sim(s) R_\infty(s) = 1 + c^2 s^4$, which is easily seen to have the spectral factor

$$R_\infty(s) = cs^2 + s\sqrt{2c} + 1. \tag{183}$$

Inspection shows that for $r = 0$ the equations (179-180) may be solved with the polynomials P_λ, Q_λ and R_λ all having degree two.

Fig. 9 shows the loci of the roots of R_λ as a function of λ when we take the numerical value 1/10 for c. For $\lambda = \lambda_{\text{opt}} = 1.2861$ one of the roots passes into the right-half plane via the origin. The corresponding solution of the polynomial equations (179-180) is given by

$$\begin{aligned} P_{\lambda_o}(s) &= s^2 + 6.4256s, \\ Q_{\lambda_o}(s) &= 8.0872s^2 + 5.0114s, & (184) \\ R_{\lambda_o}(s) &= s^2 + 5.0114s. \end{aligned}$$

Cancelation of the common factor s corresponding to the common root 0 of the three polynomials results in the optimal compensator

$$K(s) = \frac{Q(s)}{P(s)} = \frac{8.0872s + 5.0114}{s + 6.4256}. \tag{185}$$

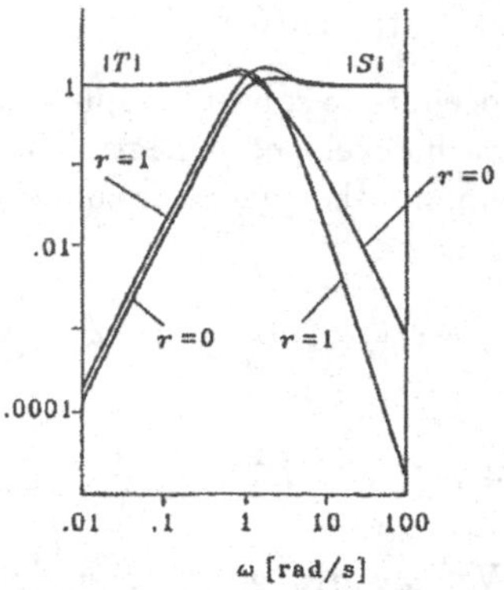

Figure 10: Bode plots of S and T for $r = 0$ and $r = 1$

As expected, K is proper but not strictly proper. The closed-loop poles of the optimal feedback system consist of the roots $\frac{1}{2}\sqrt{2}(-1 \pm j)$ of M, and the remaining root of R_{λ_o}, which is given by -5.0114. The former pair of roots clearly dominates the latter root. If c had been chosen larger, this might not have been the situation. Figure 10 shows Bode plots of the sensitivity function S and the complementary sensitivity function T. They confirm that the closed-loop bandwidth is 1 [rad/s].

The inverses of these functions are indicative for the robustness of the closed-loop system. The plot of the sensitivity functions shows good robustness against low-frequency structured perturbations.

The high-frequency robustness against unstructured perturbations may be improved by taking $r \neq 0$. To let the resulting extra roll-off set in at the break frequency 1 [rad/s] we choose $r = 1$. R_∞ now has three roots. Correspondingly, R_λ and P_λ have degree three, while Q_λ still has degree two. The locus of the roots of R_λ is shown in Fig. 11. A reduced solution is now obtained for $\lambda_{\text{opt}} = 1.8559$, with the remaining roots of $R_{\lambda_{\text{opt}}}$ located at $-2.0897 \pm j1.932$. The optimal compensator is

$$K(s) = \frac{15.6343s + 8.0998}{s^2 + 5.5937s + 15.0105}. \tag{186}$$

Fig. 10 shows the resulting changes in the plots of S and T. As expected, $|T|$ rolls of faster at high frequencies. The price is a slight increase of $|S|$ at low frequencies and a peak of $|S|$ at crossover frequency. Both effects can be diminished by taking a smaller value for r. This shifts the onset of the extra roll-off of $|T|$ to a higher frequency. •

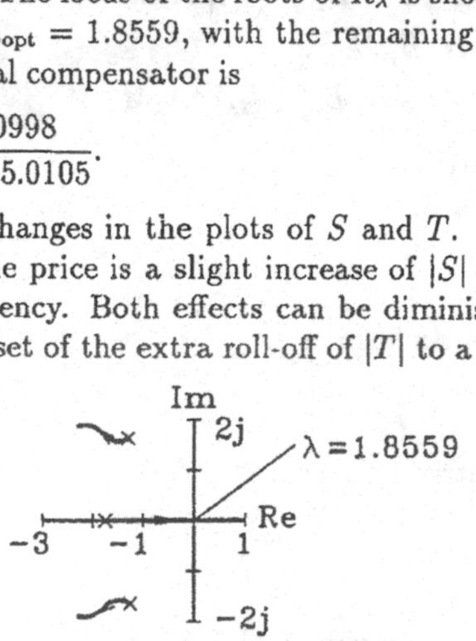

Figure 11: Locus of the roots of R_λ for $r = 1$

3.10 Solution by J-spectral factorization

In this subsection we pursue a procedure to solve the equations (157-158) that anticipates the J-spectral factorization approach developed in Section 3. Solution of R_λ from the linear equation (157) and substitution into the quadratic equation (158) yields the quadratic equation

$$P_\lambda^\sim(D^\sim D B_1^\sim B_1 - \frac{1}{\lambda^2}A_1^\sim A_1 M^\sim M)P_\lambda + P_\lambda^\sim D^\sim B_1^\sim B_2 N Q_\lambda$$

$$+ Q_\lambda^\sim N^\sim B_2^\sim B_1 D P_\lambda + Q_\lambda^\sim(N^\sim N B_2^\sim B_2 - \frac{1}{\lambda^2}A_2^\sim A_2 M^\sim M)Q_\lambda = 0 \tag{187}$$

in the polynomials P_λ and Q_λ. We may arrange this equation in matrix form as

$$[P_\lambda^\sim \quad Q_\lambda^\sim]\,\Pi_\lambda \begin{bmatrix} P_\lambda \\ Q_\lambda \end{bmatrix} = 0, \tag{188}$$

where Π_λ is the polynomial matrix

$$\Pi_\lambda = \begin{bmatrix} D^\sim D B_1^\sim B_1 - \frac{1}{\lambda^2}A_1^\sim A_1 M^\sim M & D^\sim B_1^\sim B_2 N \\ N^\sim B_2^\sim B_1 D & N^\sim N B_2^\sim B_2 - \frac{1}{\lambda^2}A_2^\sim A_2 M^\sim M \end{bmatrix}. \tag{189}$$

One way of solving the quadratic polynomial equation (187) is to represent it in indefinite quadratic polynomial form as

$$(\Gamma_{11}P + \Gamma_{12}Q)^\sim(\Gamma_{11}P + \Gamma_{12}Q) - (\Gamma_{21}P + \Gamma_{22}Q)^\sim(\Gamma_{21}P + \Gamma_{22}Q) = 0, \tag{190}$$

with Γ_{11}, Γ_{12}, Γ_{21}, and Γ_{22} polynomials to be determined. We temporarily suppress the subscript λ. Identification of the coefficients of P_λ and Q_λ in (187) and (190) shows that

$$\begin{aligned} \Gamma_{11}^\sim\Gamma_{11} - \Gamma_{21}^\sim\Gamma_{21} &= D^\sim D B_1^\sim B_1 - \tfrac{1}{\lambda^2}A_1^\sim A_1 M^\sim M, \\ \Gamma_{11}^\sim\Gamma_{12} - \Gamma_{21}^\sim\Gamma_{22} &= D^\sim B_1^\sim B_2 N, \\ \Gamma_{12}^\sim\Gamma_{12} - \Gamma_{22}^\sim\Gamma_{22} &= N^\sim N B_2^\sim B_2 - \tfrac{1}{\lambda^2}A_2^\sim A_2 M^\sim M. \end{aligned} \tag{191}$$

Defining the polynomial matrix

$$\Gamma_\lambda = \begin{bmatrix} \Gamma_{11} & \Gamma_{12} \\ \Gamma_{21} & \Gamma_{22} \end{bmatrix}, \tag{192}$$

we may arrange (191) in matrix form as

$$\Gamma_\lambda^\sim J \Gamma_\lambda = \Pi_\lambda, \tag{193}$$

where J is the signature matrix

$$J = \begin{bmatrix} 1 & 0 \\ 0 & -1 \end{bmatrix}. \tag{194}$$

Here if A is a polynomial or rational matrix, we define its adjoint $A^\sim$ by

$$A^{\sim}(s) = A^{\mathrm{T}}(-s), \tag{195}$$

with the superscipt T denoting the transpose. *If* we manage to find a matrix Γ satisfying (193), the equation (187) is solved if we let

$$\Gamma_{11}P + \Gamma_{12}Q = \Gamma_{21}P + \Gamma_{22}Q, \tag{196}$$

that is,

$$P_\lambda = \Gamma_{22,\lambda} - \Gamma_{12,\lambda}, \qquad Q_\lambda = \Gamma_{11,\lambda} - \Gamma_{21,\lambda}, \tag{197}$$

where the subscript λ has been reinstated.

Polynomial matrix factorizations of the form (193) are well-known (Jabubovič, 1970). They are a generalization of the spectral factorization of polynomial matrices. The factorization (193) exists as long as Π_λ has the signature matrix J on the imaginary axis. This is the case for $\lambda \geq \lambda_o$, where at this point we define λ_o as the first value of λ as λ decreases from ∞ for which $\det(\Pi_\lambda)$ has a root on the imaginary axis.

The factorization (193) is not at all unique. We shall prove that we need a factorization such that $\det(\Gamma_\lambda)$ is Hurwitz. In this case we refer to (193) as a (polynomial) *J-spectral factorization.* Also the J-spectral factor is not completely unique, because it may be multiplied on the left by any J-orthogonal constant matrix C (i.e., any real constant matrix such that $C^{\mathrm{T}}JC = J$).

3.10.1 Summary (Solution for equalizing compensators by J-spectral factorization.) The J-spectral factor Γ_λ determined by

$$\Gamma_\lambda^{\sim} J \Gamma_\lambda = \Pi_\lambda \tag{198}$$

exists for $|\lambda| \geq \lambda_o$, where λ_o is the first value of λ as it decreases from ∞ such that $\det(\Pi_\lambda)$ has a root on the imaginary axis. The polynomials

$$P_\lambda = \Gamma_{22,\lambda} - \Gamma_{12,\lambda}, \qquad Q_\lambda = \Gamma_{11,\lambda} - \Gamma_{21,\lambda}, \tag{199}$$

define an equalizing compensator that is stabilizing if and only if

$$|\lambda| \geq \lambda_{\mathrm{opt}}. \tag{200}$$

•

3.10.2 (Proof.) The starting point of the proof is the fact that for λ sufficiently large the equations (157-158) have a solution such that R_λ is Hurwitz, and, hence the corresponding equalizing compensator is stabilizing. Consider the function E_λ defined in 3.6.3. For $\lambda = \infty$ we have

$$E_\infty = A_1^{\sim} A_1 B_2^{\sim} B_2 N^{\sim} N + A_2^{\sim} A_2 B_1^{\sim} B_1 D^{\sim} D = R_\infty^{\sim} R_\infty, \tag{201}$$

which shows that $E_{-,\infty} = R_\infty$ and $E_{+,\infty} = R_\infty^{\sim}$. Hence, for λ sufficiently large, $E_{-,\lambda}$ and $E_{-,\lambda}$ have the same degrees as R_λ.

By direct calculation it is easily verified that

$$[P_\lambda^{\sim} \quad Q_\lambda^{\sim}]\,\Pi_\lambda \begin{bmatrix} -NB_2 \\ DB_1 \end{bmatrix} = \frac{1}{\lambda^2} M^{\sim} M E_\lambda. \tag{202}$$

Together with (188) this results in

$$[P_\lambda^\sim \quad Q_\lambda^\sim]\,\Pi_\lambda \begin{bmatrix} P_\lambda & -NB_2 \\ Q_\lambda & DB_1 \end{bmatrix} = \begin{bmatrix} 0 & \frac{1}{\lambda^2} M^\sim M E_\lambda \end{bmatrix}. \tag{203}$$

(This device originates from G. Meinsma.) The determinant of the matrix that appears as the final factor on the left-hand side is $DB_1P_\lambda + NB_2Q_\lambda$ and, hence, equals $\frac{1}{\lambda}MR_\lambda$. It follows that the left-hand side loses rank at the roots of R_λ. Because necessarily also the right-hand side loses rank at these same points in the complex plane, the polynomial E_λ has these same roots. Hence, at least for λ sufficiently large,

$$E_{-,\lambda} = R_\lambda. \tag{204}$$

To learn more about $E_{+,\lambda}$ we consider (202) in more detail, with Π_λ replaced with $\Gamma_\lambda^\sim J \Gamma_\lambda$. The resulting expression may be rewritten as

$$\begin{aligned} \begin{bmatrix} P_\lambda^\sim \Gamma_{11,\lambda}^\sim + Q_\lambda^\sim \Gamma_{21,\lambda}^\sim & P_\lambda^\sim \Gamma_{12,\lambda}^\sim + Q_\lambda^\sim \Gamma_{22,\lambda}^\sim \end{bmatrix} J \begin{bmatrix} DB_1\Gamma_{12,\lambda} - NB_2\Gamma_{11,\lambda} \\ DB_1\Gamma_{22,\lambda} - NB_2\Gamma_{21,\lambda} \end{bmatrix} \\ = \frac{1}{\lambda^2} M^\sim M E_\lambda. \end{aligned} \tag{205}$$

For P_λ, Q_λ satisfying (196) we define

$$\Gamma_{11,\lambda}P_\lambda + \Gamma_{12,\lambda}Q_\lambda = \Gamma_{21,\lambda}P_\lambda + \Gamma_{22,\lambda}Q_\lambda = \gamma_\lambda. \tag{206}$$

By rearrangement of (206) it follows that

$$\Gamma_\lambda \begin{bmatrix} P_\lambda \\ Q_\lambda \end{bmatrix} = \begin{bmatrix} 1 \\ 1 \end{bmatrix} \gamma_\lambda. \tag{207}$$

Because the left-hand side of (207) loses rank at the roots of $\det(\Gamma_\lambda)$, so does the right-hand side. As a result, γ_λ equals $\det(\Gamma_\lambda)$ within a nonzero constant factor. By suitable scaling of P_λ and Q_λ this factor may be made equal to 1.

From (207) we have

$$\begin{bmatrix} P_\lambda^\sim \Gamma_{11,\lambda}^\sim + Q_\lambda^\sim \Gamma_{21,\lambda}^\sim & P_\lambda^\sim \Gamma_{12,\lambda}^\sim + Q_\lambda^\sim \Gamma_{22,\lambda}^\sim \end{bmatrix} = \gamma_\lambda^\sim [1 \quad 1]. \tag{208}$$

Substitution into (205) yields

$$\gamma_\lambda^\sim [DB_1(\Gamma_{12,\lambda} - \Gamma_{22,\lambda}) + NB_2(\Gamma_{21,\lambda} - \Gamma_{11,\lambda})] = \frac{1}{\lambda^2} M^\sim M E_\lambda. \tag{209}$$

With (199) this reduces to

$$-\gamma_\lambda^\sim (DB_1P_\lambda + NB_2Q_\lambda) = \frac{1}{\lambda^2} M^\sim M E_\lambda. \tag{210}$$

Substituting $DB_1P_\lambda + NB_2Q_\lambda = \frac{1}{\lambda}MR_\lambda$ it finally follows that

$$-\gamma_\lambda^\sim R_\lambda = \frac{1}{\lambda} M^\sim E_\lambda. \tag{211}$$

Inspection shows that if we are to have $E_\lambda = E_{+,\lambda} R_\lambda$ for λ large enough, γ_λ and, hence, $\det(\Gamma_\lambda)$, should be Hurwitz.

This proves that taking $\det(\Gamma_\lambda)$ Hurwitz leads to the desired solution of the quadratic equation, which results in a stabilizing solution for λ large enough. Decreasing λ to λ_o results in a optimal reduced solution. Equalizing solutions with $|\lambda| < \lambda_o$ are necessarily destabilizing.

This completes the proof of 3.10.1. It also proves 3.7.2. •

3.11 Lower bound

We next turn to the derivation of the lower bound

$$\|V^\sim V(S^\sim S W_1^\sim W_1 + U^\sim U W_2^\sim W_2)\|_\infty \geq \lambda_o^2 \tag{212}$$

that holds for *any* compensator, whether stabilizing or not. We prove that λ_o is precisely the number defined in 3.10.1, namely, the first value of λ as λ decreases from ∞ for which

$$\det(\Pi_\lambda) = -\frac{1}{\lambda^2} M^\sim M(A_1^\sim A_1 B_2^\sim B_2 N^\sim N + A_2^\sim A_2 B_1^\sim B_1 D^\sim D - \frac{1}{\lambda^2} A_1^\sim A_1 A_2^\sim A_2 M^\sim M) \tag{213}$$

has a root on the imaginary axis.

Note that from the definitions of S and U we have $S + GU = 1$. Substitution of $S = 1 - GU$ and completion of the square in U result in

$$\begin{aligned} & V^\sim V(S^\sim S W_1^\sim W_1 + U^\sim U W_2^\sim W_2) \\ = \; & V^\sim V \frac{[(W_1^\sim W_1 G^\sim G + W_2^\sim W_2)U - W_1^\sim W_1 G]^\sim [(W_1^\sim W_1 G^\sim G + W_2^\sim W_2)U - W_1^\sim W_1 G]}{W_1^\sim W_1 G^\sim G + W_2^\sim W_2} \\ & + \frac{V^\sim V W_1^\sim W_1 W_2^\sim W_2}{W_1^\sim W_1 G^\sim G + W_2^\sim W_2}. \end{aligned} \tag{214}$$

Inspection shows that

$$\|V^\sim V(S^\sim S W_1^\sim W_1 + U^\sim U W_2^\sim W_2)\|_\infty \geq \left\|\frac{V^\sim V W_1^\sim W_1 W_2^\sim W_2}{W_1^\sim W_1 G^\sim G + W_2^\sim W_2}\right\|_\infty. \tag{215}$$

Equality is assumed if the first term on the right of of (214) is zero, which is normally achieved by a non-stabilizing compensator. The ∞-norm of the right-hand side of (215) may be determined as the first value of λ^2 as λ decreases from infinity for which

$$\lambda^2 - \frac{V^\sim V W_1^\sim W_1 W_2^\sim W_2}{W_1^\sim W_1 G^\sim G + W_2^\sim W_2} \tag{216}$$

assumes the value zero somewhere on the imaginary axis. Since

$$\frac{V^\sim V W_1^\sim W_1 W_2^\sim W_2}{W_1^\sim W_1 G^\sim G + W_2^\sim W_2} = \frac{A_1^\sim A_1 A_2^\sim A_2 M^\sim M}{A_1^\sim A_1 B_2^\sim B_2 N^\sim N + A_2^\sim A_2 B_1^\sim B_1 D^\sim D}, \tag{217}$$

this value of λ^2 is precisely λ_o^2. This proves (212).

3.11.1 Summary (Lower bound.) Define the nonnegative number λ_o as the first value of λ as λ decreases from infinity for which $\det(\Pi_\lambda)$ has a root on the imaginary axis or at infinity. Then

$$\|V^\sim V(S^\sim S W_1^\sim W_1 + U^\sim U W_2^\sim W_2)\|_\infty \geq \lambda_o^2 \tag{218}$$

for any compensator, whether stabilizing or not. •

3.12 Example of solution by J-spectral factorization

To illustrate the solution procedure by J-spectral factorization we take a simpler example than the problem we studied before.

3.12.1 Example (Solution of a mixed sensitivity problem by J-spectral factorization.) Consider the mixed sensitivity problem for the plant

$$G(s) = \frac{1}{s}, \tag{219}$$

with the weighting functions

$$V(s) = \frac{s+1}{s}, \qquad W_1(s) = 1, \qquad W_2(s) = c. \tag{220}$$

Here c is a positive constant. It is easy to find that the polynomial matrix Π_λ is given by

$$\Pi_\lambda(s) = \begin{bmatrix} -\frac{1}{\lambda^2} - (1 - \frac{1}{\lambda^2})s^2 & -s \\ s & (1 - \frac{c^2}{\lambda^2}) + \frac{c^2}{\lambda^2}s^2 \end{bmatrix}. \tag{221}$$

By using (213) we find for the determinant

$$\det(\Pi_\lambda(s)) = -\frac{1}{\lambda^2}(1 - s^2)\left((1 - \frac{c^2}{\lambda^2}) - c^2(1 - \frac{1}{\lambda^2})s^2\right). \tag{222}$$

Inspection shows that a root of this determinant reaches 0 for $\lambda = c$, and infinity for $\lambda = 1$. Hence, the lower bound λ_o is given by

$$\lambda_o = \max(1,\ c). \tag{223}$$

We discuss J-spectral factorization in much more detail later. At this point we simply state (as may be verified by back substitution) that for $\lambda > \sqrt{1 + c^2}$ the matrix Π_λ may be J-spectrally factored as $\Pi_\lambda = \Gamma_\lambda^\sim J \Gamma_\lambda$, where

$$\Gamma_\lambda(s) = \frac{1}{\sqrt{1 - \frac{1+c^2}{\lambda^2}}} \begin{bmatrix} -\frac{1}{\lambda^2} + (1 - \frac{1}{\lambda^2})s & (1 - \frac{c^2}{\lambda^2}) - \frac{c^2}{\lambda^2}s \\ -\frac{c}{\lambda}(s + \zeta_\lambda)\sqrt{1 - \frac{1}{\lambda^2}} & \frac{c}{\lambda}(s + \zeta_\lambda)\sqrt{1 - \frac{1}{\lambda^2}} \end{bmatrix}. \tag{224}$$

The constant ζ_λ is given by

$$\zeta_\lambda = \frac{1}{c}\sqrt{\frac{1-\frac{c^2}{\lambda^2}}{1-\frac{1}{\lambda^2}}}. \tag{225}$$

For $\lambda_o \leq \lambda < \sqrt{1+c^2}$ a different but similar expression may be obtained for Γ_λ. The determinant of Γ_λ may be checked to have the factor $(s+1)(s+\zeta_\lambda)$, so that it is Hurwitz as needed.

Given the elements $\Gamma_{ij,\,\lambda}$, i, $j = 1, 2$ of Γ_λ, the equalizing compensator follows from

$$\begin{aligned} K_\lambda(s) &= \frac{Q_\lambda(s)}{P_\lambda(s)} = \frac{\Gamma_{11,\,\lambda}(s) - \Gamma_{21,\,\lambda}(s)}{\Gamma_{22,\,\lambda}(s) - \Gamma_{12,\,\lambda}(s)} \\ &= \frac{(1-\frac{1}{\lambda^2}+\frac{c}{\lambda}\sqrt{1-\frac{1}{\lambda^2}})s + (-\frac{1}{\lambda^2}+\frac{1}{\lambda}\sqrt{1-\frac{c^2}{\lambda^2}})}{(\frac{c^2}{\lambda^2}+\frac{c}{\lambda}\sqrt{1-\frac{c^2}{\lambda^2}})s + (-1+\frac{c^2}{\lambda^2}+\frac{1}{\lambda}\sqrt{1-\frac{1}{\lambda^2}})}. \end{aligned} \tag{226}$$

It may be found that this expression is also valid for $\lambda_o \leq \lambda < \sqrt{1+c^2}$. Inspection shows that for $\lambda > \sqrt{1+c^2}$ the (single) pole of K is positive, and, hence, lies in the right-half complex plane. As λ approaches $\sqrt{1+c^2}$, the pole approaches 0, but at the same time the (single) zero of K also approaches 0. As a result, at $\lambda = \lambda_{\text{opt}}$, where

$$\lambda_{\text{opt}} = \sqrt{1+c^2}, \tag{227}$$

a pole and zero of K at 0 cancel, and

$$K_{\lambda_{\text{opt}}}(s) = 1. \tag{228}$$

This is the desired optimal solution. We note, though, that as λ approaches λ_{opt}, the coefficients of the J-spectral factor Γ_λ approach ∞. Hence, the optimal solution forms a ***singular point*** for the J-spectral factorization. This turns out to be a characteristic feature of $\mathcal{H}_\infty$ optimal solutions. In fact, *at* λ_{opt} a J-spectral factor *does* exist. It has the form

$$\Gamma_{\lambda_{\text{opt}}}(s) = \frac{\frac{1}{2}\sqrt{2}}{\sqrt{1+c^2}}\begin{bmatrix} \frac{1}{2}(cs^2+1)(s+1)+1 & \frac{1}{2}(cs^2+1)(s+1)-1 \\ \frac{1}{2}(cs^2+1)(s+1)-1 & \frac{1}{2}(cs^2+1)(s+1)+1 \end{bmatrix}, \tag{229}$$

and, hence, has a completely different degree structure than the J-spectral factor for $\lambda \neq \lambda_{\text{opt}}$. This factorization also results in $K_{\lambda_{\text{opt}}}(s) = 1$. •

4 The Standard $\mathcal{H}_\infty$-Optimal Regulation Problem

The "standard" $\mathcal{H}_\infty$-optimal regulation problem applies to a wide class of multivariable control problems. It involves the minimization of the ∞-norm of the closed-loop transfer function of a general linear feedback configuration.

The ∞-norm of a rational transfer matrix H is defined as

$$\|H\|_\infty = \sup_{\omega \in \mathbb{R}} \lambda_{\max}^{1/2}\left(H^\sim(j\omega)H(j\omega)\right), \tag{230}$$

where $\lambda_{\max}(A)$ denotes the largest eigenvalue of the (constant) matrix A. The number $\lambda^{1/2}_{\max}(A^{\mathrm{H}}A)$ is the largest of the *singular values*

$$\sigma_i(A) = \lambda_i^{1/2}(A^{\mathrm{H}}A), \quad i = 1,\, 2,\, \cdots, n, \tag{231}$$

of the $m \times n$ matrix A. The superscript H denotes the Hermitian, and λ_i is the ith eigenvalue. Note that

$$\lambda_{\max}(A^{\mathrm{H}}A) = \lambda_{\max}(AA^{\mathrm{H}}), \tag{232}$$

so that $\|H\|_\infty = \|H^{\mathrm{T}}\|_\infty = \|H^\sim\|_\infty$. It is useful to remember the following facts:

4.0.1 Assertion (∞-Norm of a rational matrix function.) Let λ be a nonnegative real number. Then the following statements are equivalent:

1. $\|H\|_\infty \leq \lambda$.
2. $H^\sim(j\omega)H(j\omega) \leq \lambda^2 I$ for all $\omega \in \mathrm{R}$.
3. The rational function $\det(\mu^2 I - H^\sim H)$ has no zeros on the imaginary axis or at ∞ for $|\mu| > \lambda$.

•

The latter result allows computing $\|H\|_\infty$ as the *first* value of μ as μ decreases from ∞ for which $\det(\mu^2 I - H^\sim H)$ has a zero on the imaginary axis or at infinity.

4.0.2 Example (Computation of the ∞-norm.) By way of example we consider determining the ∞-norm of the rational matrix function H given by

$$H(s) = \begin{bmatrix} 1 \\ \frac{1}{s+1} \end{bmatrix}. \tag{233}$$

It is easily found that

$$\det[\mu^2 I - H^\sim(s)H(s)] = \mu^2 - 1 - \frac{1}{1-s^2} = \frac{(\mu^2-2)-(\mu^2-1)s^2}{1-s^2}. \tag{234}$$

This rational function has a zero at the origin for $\mu^2 = 2$, and one at ∞ for $\mu^2 = 1$. The former situation is reached first as μ decreases from ∞. It follows that $\|H\|_\infty = \sqrt{2}$. •

The standard problem is defined by the multivariable configuration of Fig. 12. The block marked "G" is the "plant," that is, the system to be controlled. The signal w represents *external*, uncontrollable inputs such as disturbances, measurement noise, and reference inputs. The signal u is the *control* input. The output z has the meaning of *control error*, which ideally should be zero. The signal y, finally, is the *measured* output, available for feedback via the compensator, which is the block marked "K."

The plant is represented by the rational transfer matrix G such that

$$\begin{bmatrix} z \\ y \end{bmatrix} = \begin{bmatrix} G_{11} & G_{12} \\ G_{21} & G_{22} \end{bmatrix} \begin{bmatrix} w \\ u \end{bmatrix}. \tag{235}$$

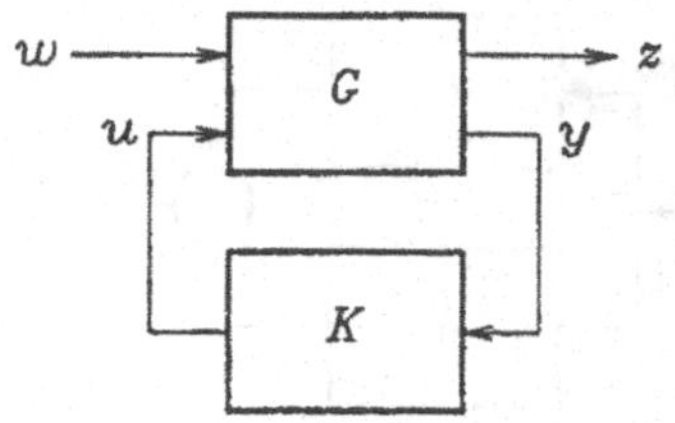

Figure 12: The standard $\mathcal{H}_\infty$ problem

The dimensions of the subblocks of G follows from

$$\begin{array}{cc} & \begin{array}{cc} k_1 & k_2 \end{array} \\ \begin{array}{c} m_1 \\ m_2 \end{array} & \left[\begin{array}{cc} G_{11} & G_{12} \\ G_{21} & G_{22} \end{array}\right]. \end{array} \tag{236}$$

The dimensions of all rational and polynomial matrices encountered in this section are listed in the Appendix.

Without loss of generality we introduce the following assumptions:

4.0.3 Assumptions (Standard problem.)

1. The dimension of the external input w is at least as great as that of the observed output y, that is, $k_1 \geq m_2$, and G_{21} has full normal row rank.

2. The dimension of the control error z is at least as great as that of the control input u, that is, $m_1 \geq k_2$, and G_{12} has full normal column rank.

•

The assumptions are met for many problems of interest. If they are not satisfied, the problem may be transformed to an equivalent problem for which the assumptions hold.

In the configuration of Fig. 12, the closed-loop transfer function H from the external input w to the control error z is easily found to be given by

$$H = G_{11} + G_{12}K(I - G_{22}K)^{-1}G_{21}, \tag{237}$$

The standard $\mathcal{H}_\infty$-optimal regulation problem may now be defined as follows:

4.0.4 Definition (Standard $\mathcal{H}_\infty$-optimal regulation problem.). Determine the compensator K such that

1. the closed-loop system is stable,

2. the ∞-norm $\|H\|_\infty$ of the closed-loop transfer matrix H is minimal.

•

What "stable" exactly means is discussed in Subsection 4.2.

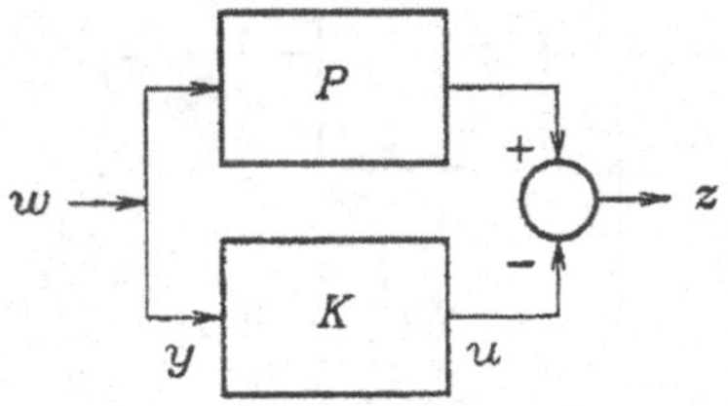

Figure 13: The model matching problem

4.1 Special cases

The standard problem may be specialized to several problems of particular interest.

4.1.1 Example (The model matching problem .) The model matching problem (Fig. 13) is the problem to find a stable system K that approximates a given stable or unstable single-input-single-output or multi-input-multi-output system P as closely as possible in the sense that $\|P-K\|_\infty$ is minimal. Denoting the output of the block "K" as u and its input as y, we have $z = Pw - u$ and $y = w$. Hence, this is a standard problem with plant

$$G = \begin{bmatrix} P & -I \\ I & 0 \end{bmatrix}. \tag{238}$$

•

4.1.2 Example (The MIMO minimum sensitivity problem.) The block diagram of Fig. 2 of the minimum sensitivity problem may be rearranged and relabeled as in Fig. 14. Note that we drop the negative feedback convention and write P for the plant transfer function. Since $z = y = Vw + Pu$, this is a standard problem with transfer matrix

$$G = \begin{bmatrix} V & P \\ V & P \end{bmatrix}. \tag{239}$$

P, V and K may be scalar or matrix-valued. •

4.1.3 Example (The MIMO mixed sensitivity problem.) Again abandoning the negative feedback convention and renaming the plant to P, Fig. 15 represents a MIMO version of the mixed sensitivity problem. Like the SISO problem, it may be used for frequency response shaping and robustness protection. From $z_1 = W_1Vw + W_1Pu$, $z_2 = W_2u$, and $y = Vw + Pu$, it follows that this is a standard problem with transfer matrix

$$G = \left[\begin{array}{c|c} W_1V & W_1P \\ 0 & W_2 \\ \hline V & P \end{array}\right]. \tag{240}$$

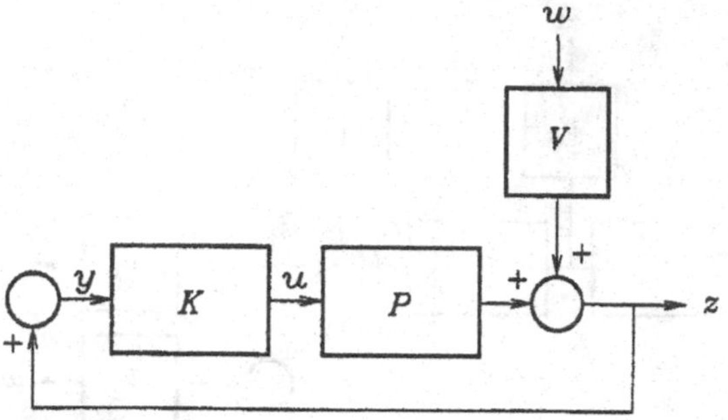

Figure 14: The (MIMO) minimum sensitivity problem

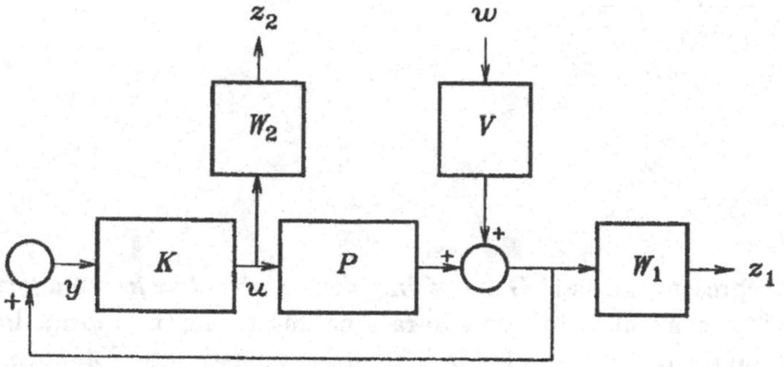

Figure 15: The MIMO mixed sensitivity problem

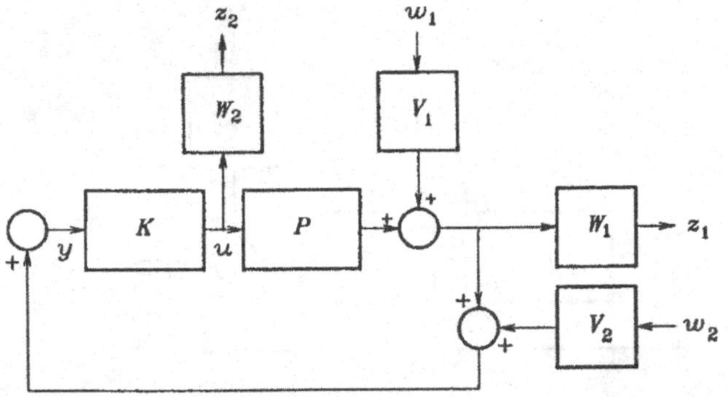

Figure 16: The mixed sensitivity problem with measurement noise

•

4.1.4 Example (The mixed sensitivity problem with measurement noise.) Including measurement noise in the configuration of Fig. 15 leads to the configuration of Fig. 16. From $z_1 = W_1V_1w_1 + W_1Pu$, $z_2 = W_2u$, $y = V_1w_1 + V_2w_2 + Pu$ it follows that this is a standard problem with transfer matrix

$$G = \left[\begin{array}{cc|c} W_1V_1 & 0 & W_1P \\ 0 & 0 & W_2 \\ \hline V_1 & V_2 & P \end{array}\right]. \tag{241}$$

•

4.2 Stability

Throughout, we represent the plant G and K in *polynomial matrix fraction form*, although on one occasion it is convenient to resort to rational matrix fraction form. In particular, we assume that the plant is represented in left coprime matrix fraction form

$$G = D^{-1}N, \tag{242}$$

where D is an $(m_1 + m_2) \times (m_1 + m_2)$ nonsingular square polynomial matrix, and N an $(m_1+m_2)\times(k_1+k_2)$ polynomial matrix. This representation is equivalent to characterizing the plant by the set of differential equations

$$D\begin{bmatrix} z \\ y \end{bmatrix} = N\begin{bmatrix} w \\ u \end{bmatrix}. \tag{243}$$

Partitioning

$$D = \overset{\begin{matrix} m_1 & m_2 \end{matrix}}{[D_1 \quad D_2]}, \qquad N = \overset{\begin{matrix} k_1 & k_2 \end{matrix}}{[N_1 \quad N_2]}, \tag{244}$$

we have

$$D_1 z + D_2 y = N_1 w + N_2 u. \tag{245}$$

We represent the compensator K in *right* matrix fraction form as

$$K = YX^{-1}, \tag{246}$$

with X an $m_2 \times m_2$ nonsingular square polynomial matrix, and Y a $k_2 \times m_2$ polynomial matrix. Writing

$$u = YX^{-1}y = Yq, \tag{247}$$

with q the signal defined by $q = X^{-1}y$, we may describe the compensator by the set of differential equations

$$u = Yq, \qquad y = Xq. \tag{248}$$

Substitution of the latter equations into (245) shows that the closed-loop system is described by the set of differential equations

$$[D_1 \quad D_2X - N_2Y] \begin{bmatrix} z \\ q \end{bmatrix} = N_1 w, \tag{249}$$

with (248) as output equations for u and y. *Closed-loop stability* is determined by the roots of

$$\det\left([D_1 \quad D_2X - N_2Y]\right). \tag{250}$$

These roots are precisely the closed-loop poles.

We look more closely at the closed-loop poles by assuming, temporarily but without loss of generality, that by elementary row operations the polynomial matrices D and N are brought into the form

$$D = [D_1 \quad D_2] = \begin{bmatrix} D_{11} & D_{12} \\ 0 & D_{22} \end{bmatrix}, \qquad N = [N_1 \quad N_2] = \begin{bmatrix} N_{11} & N_{12} \\ N_{21} & N_{22} \end{bmatrix}, \tag{251}$$

with D_{11} square of dimensions $m_1 \times m_1$, and the dimensions of the other submatrices consistent. With this, (250) reduces to

$$\det\left(\begin{bmatrix} D_{11} & D_{12}X - N_{12}Y \\ 0 & D_{22}X - N_{22}Y \end{bmatrix}\right) = \det(D_{11}) \cdot \det(D_{22}X - N_{22}Y). \tag{252}$$

Writing D_{22} and N_{22} as

$$D_{22} = A_o D_o, \qquad N_{22} = A_o N_o, \tag{253}$$

with D_o and N_o left coprime and the square polynomial matrix A_o a left common divisor of D_{22} and N_{22}, it follows that the closed-loop poles are the roots of

$$\det(D_{11}) \cdot \det(A_o) \cdot \det(D_oX - N_oY). \tag{254}$$

The roots of $\det(D_{11})\det(A_o)$ are independent of the compensator, and are called the *fixed* closed-loop poles. The roots of $\det(D_oX - N_oY)$, on the other hand, are competely under the control of the compensator, and, hence, are called the *assignable* closed-loop poles.

Here we use the well-known fact that if D_o and N_o are left coprime, the matrix polynomial equation $F = D_oX - N_oY$ has a solution X, Y for any square F. Note that with the partitioning (251) we have

$$G_{22} = D_{22}^{-1}N_{22} = D_o^{-1}N_o. \tag{255}$$

This confirms the well-known fact that the closed-loop system is stabilizable if and only if G_{22} is stabilizable (Francis and Doyle, 1987).

We do not impose full stabilizability, however, but call a compensator *stabilizing* if the assignable poles are all in the open left-half plane. For the solution of the standard problem this is all we require. The fixed poles may or may not be in the left-half plane.

4.2.1 Summary (Closed-loop stability.) Let

$$G = D^{-1}N = [D_1 \quad D_2]^{-1}[N_1 \quad N_2] \tag{256}$$

be a left coprime polynomial matrix fraction representation of the plant, and $K = YX^{-1}$ a right coprime representation of the compensator. Then the closed-loop poles are the roots of

$$\det([D_1 \quad D_2X - N_2Y]). \tag{257}$$

The assignable closed-loop poles are the roots of

$$\det(D_oX - N_oY), \tag{258}$$

with $G_{22} = D_o^{-1}N_o$ a left coprime polynomial matrix fraction representation of G_{22}. •

The example that follows clarifies why we do not require the fixed poles to be in the left-half complex plane.

4.2.2 Example (Model matching problem: Stability.) Consider the model matching problem of 4.1.1. Writing P in left coprime polynomial matrix fraction form as $P = d^{-1}n$ we have for the plant transfer function

$$G(s) = \begin{bmatrix} d^{-1}n & -I \\ I & 0 \end{bmatrix} = \begin{bmatrix} d & 0 \\ 0 & I \end{bmatrix}^{-1} \begin{bmatrix} n & -d \\ I & 0 \end{bmatrix}. \tag{259}$$

Inspection shows that $D_{11} = d$, $D_{22} = I$, and $N_{22} = 0$. It follows that $A_o = I$, $D_o = I$, and $N_o = 0$. As a result, the fixed poles are the roots of d, that is, the poles of the "plant" P. Some or all of these may lie in the right-half complex plane. The assignable poles are the roots of $D_oX + N_oY = X$. The condition that K "stabilizes" the closed-loop system in this case requires K to be a stable transfer matrix. It does not restrain the "plant" P to be stable. •

4.3 Generalization of the equalizer principle

Our first step in solving the standard $\mathcal{H}_\infty$ optimal control problem is to generalize the equalizer principle from the scalar versions previously considered.

The generalization that first comes to mind, and indeed is correct, is that if a stabilizing compensator has the property

$$H^\sim(j\omega)H(j\omega) = \lambda^2 I \qquad \text{for all } \omega \in \mathsf{R}, \tag{260}$$

and at the same time minimizes

$$\operatorname{tr} \int_{-\infty}^{\infty} \Phi(j\omega)H^\sim(j\omega)H(j\omega)\, d\omega \tag{261}$$

for some Φ that is nonnegative-definite on the imaginary axis, then this compensator also minimizes $\|H\|_\infty$.

This idea has to be slightly revised, however, because often (but not always) the matrix $H^\sim H$ structurally has less than full rank, so that there is no chance that equalization in the sense of (260) is possible. The correct generalization may be stated as follows. To make the dependence of the closed-loop transfer matrix H on K explicit we write it as H_K.

4.3.1 Summary (Equalizer principle.) Suppose that the stabilizing compensator K_o is equalizing, that is,

$$H_{K_o}^\sim H_{K_o} = \lambda_o^2 I - L_o^\sim L_o, \tag{262}$$

for some nonnegative constant λ_o, where the (rational) matrix L_o is rank deficient. Then if there exists a para-Hermitian matrix Φ that is nonnegative-definite and nonzero on the imaginary axis such that the pair K_o, L_o minimizes

$$\operatorname{tr} \int_{-\infty}^{\infty} \Phi(j\omega)[H_K^\sim(j\omega)H_K(j\omega) + L^\sim(j\omega)L(j\omega)]\, d\omega \tag{263}$$

with respect to all stabilizing compensators K and all rank deficient L, the compensator K_o minimizes $\|H_K\|_\infty$ with respect to all stabilizing compensators, and $\|H_{K_o}\|_\infty = \lambda_o$. •

The proof is not difficult.

4.3.2 Proof (Equalizer principle.) Because by assumption L_o is rank deficient, for every $\omega \in \mathsf{R}$ the largest eigenvalue of

$$H_{K_o}^\sim(j\omega)H_{K_o}(j\omega) = \lambda_o^2 I - L_o^\sim(j\omega)L_o(j\omega) \tag{264}$$

equals λ_o^2, that is, the largest singular value of H_{K_o} is constant and equal to λ_o. This is why the compensator is called equalizing. It follows that $\|H_{K_o}\|_\infty = \lambda_o$. Suppose that there exists another stabilizing compensator K_* such that $\|H_{K_*}\|_\infty < \|H_{K_o}\|_\infty = \lambda_o$. Then $H_{K_*}^\sim(j\omega)H_{K_*}(j\omega) \leq \lambda_o^2 I$ for all $\omega \in \mathsf{R}$ and, hence,

$$\operatorname{tr} \int_{-\infty}^{\infty} \Phi(j\omega)H_{K_*}^\sim(j\omega)H_{K_*}(j\omega)\, d\omega \;<\; \operatorname{tr} \int_{-\infty}^{\infty} \Phi(j\omega)\lambda_o^2\, d\omega \tag{265}$$

$$= \operatorname{tr} \int_{-\infty}^{\infty} \Phi(j\omega)[H_{K_o}^{\sim}(j\omega)H_{K_o}(j\omega) + L_o^{\sim}(j\omega)L_o(j\omega)]\, d\omega. \tag{266}$$

This would mean that the pair K_*, 0 achieves a smaller value for (263) than K_o, L_o, which contradicts the assumption that the pair K_o, L_o minimizes (263). Hence, the hypothesis that there exists a compensator K_* such that $\|H_*\|_\infty < \|H_o\|_\infty$ is false, and K_o minimizes $\|H_K\|_\infty$. •

4.4 Rational inequality

We now reach a crucial point, where we derive a rational inequality that is satisfied by compensators K for which $\|H_K\|_\infty \leq \lambda$. For equalizing compensators the inequality reduces to *equality.* First we introduce an additional assumption:

4.4.1 Assumption (No zeros on imaginary axis.) Let

$$G = [D_1 \quad D_2]^{-1}[N_1 \quad N_2] = \begin{bmatrix} \bar{N}_1 \\ \bar{N}_2 \end{bmatrix} \begin{bmatrix} \bar{D}_1 \\ \bar{D}_2 \end{bmatrix}^{-1} \tag{267}$$

be left- and right-coprime polynomial matrix fraction representations of G, respectively, with D_1 and D_2 having m_1 and m_2 columns, N_1 and N_2 having k_1 and k_2 columns, $\bar{N}_1$ and $\bar{N}_2$ having m_1 and m_2 rows, and $\bar{D}_1$ and $\bar{D}_2$ having k_1 and k_2 rows, respectively. Then we assume that

$$[N_1 \quad D_1] \quad \text{and} \quad \begin{bmatrix} \bar{N}_1 \\ \bar{D}_1 \end{bmatrix} \tag{268}$$

both have full rank everywhere on the imaginary axis. •

The assumption can probably be relaxed or eliminated but at this point is convenient. The main result of this subsection is first stated and then proved. It has been found in cooperation with G. Meinsma.

4.4.2 Summary (Rational inequality.) For λ nonnegative, let Π_λ be the rational matrix

$$\Pi_\lambda = \begin{bmatrix} D_2^{\sim} \\ -N_2^{\sim} \end{bmatrix} (N_1 N_1^{\sim} - \lambda^2 D_1 D_1^{\sim})^{-1} [D_2 \quad -N_2]. \tag{269}$$

Then $\|H_K\|_\infty \leq \lambda$ for the compensator $K = YX^{-1}$ if and only if

$$[X^{\sim} \quad Y^{\sim}]\, \Pi_\lambda \begin{bmatrix} X \\ Y \end{bmatrix} \geq 0 \quad \text{on the imaginary axis.} \tag{270}$$

•

4.4.3 (Proof.) The starting point of the proof is the fact that according to (4.0.1) $\|H_K\|_\infty \leq \lambda$ if and only if

$$H_K(j\omega)H_K^{\sim}(j\omega) \leq \lambda^2 I \quad \text{for all } \omega \in \mathbb{R}. \tag{271}$$

(Note that we reversed H_K and $H_K^{\sim}$.) From (249) we see that

$$\begin{bmatrix} z \\ q \end{bmatrix} = [D_1 \quad D_2X - N_2Y]^{-1}N_1 w, \tag{272}$$

so that the closed-loop transfer function is given by

$$H_K = [I \quad 0]\,[D_1 \quad D_2X - N_2Y]^{-1}N_1. \tag{273}$$

Defining the polynomial matrices P and R by the right-to-left matrix fraction conversion

$$[I \quad 0]\,[D_1 \quad D_2X - N_2Y]^{-1} = R^{-1}P \tag{274}$$

we have $H_K = R^{-1}PN_1$, so that the inequality (271) is equivalent to

$$\lambda^2 RR^{\sim} - PN_1N_1^{\sim}P^{\sim} \geq 0 \quad \text{on the imaginary axis.} \tag{275}$$

By clearing fractions we have from (274) $R[I \quad 0] = P[D_1 \quad D_2X - N_2Y]$, so that

$$R = PD_1, \qquad 0 = P(D_2X - N_2Y). \tag{276}$$

As a result, we need

$$P(\lambda^2 D_1D_1^{\sim} - N_1N_1^{\sim})P^{\sim} \geq 0 \quad \text{on the imaginary axis.} \tag{277}$$

Define λ_1 as the first value of λ as λ decreases from ∞ for which $\det(\lambda^2 D_1D_1^{\sim} - N_1N_1^{\sim})$ has a root on the imaginary axis. Then for $|\lambda| \geq \lambda_1$ there exists a *J-spectral cofactorization*

$$N_1N_1^{\sim} - \lambda^2 D_1D_1^{\sim} = Q_\lambda J' Q_\lambda^{\sim}, \tag{278}$$

such that Q_λ is square with $\det(Q_\lambda)$ Hurwitz, and J' a matrix of the form

$$J' = \begin{bmatrix} I & 0 \\ 0 & -I \end{bmatrix}. \tag{279}$$

Define the polynomial matrix

$$T_\lambda = [T_{1,\lambda} \quad T_{2,\lambda}] = PQ_\lambda, \tag{280}$$

with $T_{2,\lambda}$ square. Then since $P(N_1N_1^{\sim} - \lambda^2 D_1D_1^{\sim})P^{\sim} = PQ_\lambda J'Q_\lambda^{\sim}P^{\sim} = T_\lambda J' T_\lambda^{\sim} = T_{1,\lambda}T_{1,\lambda}^{\sim} - T_{2,\lambda}T_{2,\lambda}^{\sim}$, the inequality (277) is equivalent to

$$\|T_2^{-1}T_1\|_\infty \leq 1. \tag{281}$$

Next we introduce the polynomial matrix

$$\bar{T}_\lambda = \begin{bmatrix} \bar{T}_{1,\lambda} \\ \bar{T}_{2,\lambda} \end{bmatrix}, \tag{282}$$

with $\bar{T}_{2,\lambda}$ square, such that $T_\lambda \bar{T}_\lambda = 0$. Then since $T_2^{-1}T_1 = -\bar{T}_2\bar{T}_1^{-1}$ we have

$$\|\bar{T}_2\bar{T}_1^{-1}\|_\infty \leq 1, \tag{283}$$

which in turn is equivalent to

$$\bar{T}^{\sim}J'\bar{T} \geq 0. \tag{284}$$

Next, from $0 = P(D_2X - N_2Y) = T_\lambda Q_\lambda^{-1}(D_2X - N_2Y)$ it follows that $Q_\lambda^{-1}(D_2X - N_2Y) = \bar{T}_\lambda M_\lambda$, with M_λ some square rational matrix. The final step of the proof follows by noting that on the imaginary axis

$$\begin{aligned} 0 &\leq M^{\sim}\bar{T}_\lambda^{\sim}J'\bar{T}_\lambda M_\lambda = (D_2X - N_2Y)^{\sim}(Q_\lambda^{\sim})^{-1}J'Q_\lambda^{-1}(D_2X - N_2Y) \\ &= (D_2X - N_2Y)^{\sim}(N_1N_1^{\sim} - \lambda^2 D_1D_1^{\sim})^{-1}(D_2X - N_2Y) \\ &= [X^{\sim} \quad Y^{\sim}]\,\Pi_\lambda \begin{bmatrix} X \\ Y \end{bmatrix}. \end{aligned} \tag{285}$$

•

Before explaining how the rational inequality may be used to parametrize all compensators K such that $\|H_K\|_\infty \leq \lambda$ we show that if the equality reduces to equality then the compensator is equalizing.

4.4.4 Summary (Equalizing compensators.) Compensators $K = YX^{-1}$ such that

$$[X^{\sim} \quad Y^{\sim}]\,\Pi_\lambda \begin{bmatrix} X \\ Y \end{bmatrix} = 0 \tag{286}$$

are equalizing. •

4.4.5 Proof (Equalizing compensators.) It follows from the end of the proof 4.4.3 that (286) implies $\bar{T}^{\sim}J'\bar{T} = 0$. Equivalently, $\bar{T}_2\bar{T}_1^{-1} = I$, so that also $T_2^{-1}T_1 = I$. As a result,

$$\begin{aligned} \det(TJ'T^{\sim}) &= \det(T_1T_1^{\sim} - T_2T_2^{\sim}) = \frac{\det(I - T_2^{-1}T_1T_1^{\sim}(T_2^{-1})^{\sim})}{\det(T_2)\det(-T_2^{\sim})} \\ &= \frac{\det(I - T_1^{\sim}(T_2^{-1})^{\sim}T_2^{-1}T_1)}{\det(T_2)\det(-T_2^{\sim})} = 0, \end{aligned} \tag{287}$$

so that $TJ'T^{\sim}$ is rank deficient. This implies that also $H_KH_K^{\sim} - \lambda^2 I$ is rank deficient, because

$$H_KH_K^{\sim} - \lambda^2 I = R^{-1}P(N_1N_1^{\sim} - \lambda^2 D_1D_1^{\sim})P^{\sim}(R^{-1})^{\sim} = R^{-1}TJ'T^{\sim}(R^{-1})^{\sim}. \tag{288}$$

It follows that the largest singular value of $H_K(j\omega)$ is constant and equal to λ for $\omega \in \mathbb{R}$. Hence, K is equalizing. •

4.5 All compensators such that $\|H\|_\infty \leq \lambda$

The first step in establishing explicitly which compensators achieve $\|H\|_\infty \leq \lambda$ is to determine a *rational* J-spectral factorization of the rational matrix function Π_λ of the form

$$\Pi_\lambda = Z_\lambda^{\sim} J Z_\lambda, \tag{289}$$

where Z_λ is a square rational matrix such that both Z_λ and Z_λ^{-1} have all their poles in the open left-half complex plane, and

$$J = \begin{bmatrix} I & 0 \\ 0 & -I \end{bmatrix}. \tag{290}$$

We shall see that such a factorization exists for $|\lambda| > \lambda_o$, where λ_o is a lower bound such that $\|H_K\|_\infty \geq \lambda_o$ for any compensator (whether stabilizing or not). We shall also see that this rational J-spectral factorization (289) may be reduced to *two* polynomial J-spectral factorizations–one for the denominator, the other for the numerator.

The rational J-spectral factorization (289) is a generalization of the well-known ordinary spectral factorization for para-Hermitian rational matrices. It applies to para-Hermitian matrices that are not definite on the imaginary axis, such as Π_λ.

The second step is to observe that the inequality

$$[X^{\sim} \quad Y^{\sim}]\, \Pi_\lambda \begin{bmatrix} X \\ Y \end{bmatrix} \geq 0 \quad \text{on the imaginary axis} \tag{291}$$

is equivalent to

$$[\tilde{X}^{\sim} \quad \tilde{Y}^{\sim}]\, \Pi_\lambda \begin{bmatrix} \tilde{X} \\ \tilde{Y} \end{bmatrix} \geq 0 \quad \text{on the imaginary axis,} \tag{292}$$

where $\tilde{X}$ and $\tilde{Y}$ given by

$$\tilde{X} = XM^{-1}, \qquad \tilde{Y} = YM^{-1}, \tag{293}$$

with M any square polynomial matrix whose determinant is strictly Hurwitz. For brevity we call such $\tilde{X}$ and $\tilde{Y}$ *stable*. Given any stable $\tilde{X}$ and $\tilde{Y}$ satisfying (292) we can always determine the corresponding X and Y by simplifying $\tilde{Y}\tilde{X}^{-1}$ to YX^{-1} with X and Y polynomial. Thus, if we can find all stable rational $\tilde{X}$ and $\tilde{Y}$ satisfying (292) we can also find all polynomial X and Y satisfying (291).

On the other hand, any stable rational $\tilde{X}$ and $\tilde{Y}$ can be expressed as

$$\begin{bmatrix} \tilde{X} \\ \tilde{Y} \end{bmatrix} = Z_\lambda^{-1} \begin{bmatrix} A \\ B \end{bmatrix}, \tag{294}$$

where A and B are suitable stable rational matrices (of the correct dimensions such that A is square). Because both Z_λ and its inverse are stable, for any stable rational $\tilde{X}$ and $\tilde{Y}$ there exist stable rational A and B, and vice-versa.

With the transformation (294), the inequality (292) takes the simple form

$$A^{\sim}A - B^{\sim}B \geq 0 \quad \text{on the imaginary axis,} \tag{295}$$

or

$$\|BA^{-1}\|_\infty \leq 1. \tag{296}$$

We summarize our findings as follows.

4.5.1 Summary (All compensators such that $\|H_K\|_\infty \leq \lambda$.) Given the rational J-spectral factorization

$$\Pi_\lambda = Z_\lambda^\sim J Z_\lambda, \tag{297}$$

any compensator K such that $\|H_K\|_\infty \leq \lambda$ may be represented as

$$K = \tilde{Y}\tilde{X}^{-1}, \tag{298}$$

where

$$\begin{bmatrix} \tilde{X} \\ \tilde{Y} \end{bmatrix} = Z_\lambda^{-1} \begin{bmatrix} A \\ B \end{bmatrix}, \tag{299}$$

with A and B rational stable and A square such that

$$\|BA^{-1}\|_\infty \leq 1. \tag{300}$$

•

There are two obvious special choices for compensators K such that $\|H_K\|_\infty \leq \lambda$.

1. $A = I$, $B = 0$. This is the so-called *central* solution, because it represents the center of the "ball" $\|BA^{-1}\|_\infty \leq 1$.
2. $A = B = I$. This solution is only possible if A and B are both square, which occurs when $k_2 = m_2$. It is *equalizing,* and results in $\|H_K\|_\infty = \lambda$.

4.6 All stabilizing compensators such that $\|H\|_\infty \leq \lambda$

Since we are interested in *stabilizing* optimal solutions, we consider the question which of the compensators K that achieve $\|H_K\|_\infty \leq \lambda$ are stabilizing. The answer is surprisingly simple. We first state and then prove it.

4.6.1 Summary (All stabilizing compensators such that $\|H\|_\infty \leq \lambda$.)

1. For given λ, all stabilizing compensators $K = \tilde{Y}\tilde{X}^{-1}$ such that $\|H_K\|_\infty \leq \lambda$, if any exist, are given by

$$\begin{bmatrix} \tilde{X} \\ \tilde{Y} \end{bmatrix} = Z_\lambda^{-1} \begin{bmatrix} A \\ B \end{bmatrix}, \tag{301}$$

with A and B stable rational such that

$$\|BA^{-1}\|_\infty \leq 1 \tag{302}$$

and the numerator of $\det(A)$ is Hurwitz.

2. There exists a stabilizing compensator $K = \tilde{X}\tilde{Y}^{-1}$ such that $\|H_K\|_\infty \leq \lambda$ if and only if *any* of the compensators (301) (for instance the central solution $A = I$, $B = 0$) is stabilizing.

•

This result allows searching for *suboptimal* stabilizing solutions by choosing a value of λ, and testing whether any of the compensators given in (301) is stabilizing (for instance the central or the equalizing solution). If it is not, λ need be increased. If the compensator is stabilizing, we have a formula for *all* suboptimal stabilizing compensators. Note that for λ sufficiently large a stabilizing compensator always exists (take any stabilizing compensator K, and choose λ larger than $\|H_K\|_\infty$.)

Later we discuss how $\mathcal{H}_\infty$-*optimal* solution are obtained.

4.6.2 Proof (All stabilizing compensators such that $\|H\|_\infty \leq \lambda$.) Given a compensator $K = YX^{-1}$, with X and Y polynomial, we know from 4.2.1 that the closed-loop poles are the roots of the determinant of

$$[D_1 \quad [D_2 \quad -N_2]\begin{bmatrix} X \\ Y \end{bmatrix}]. \tag{303}$$

When X and Y are replaced with rational stable $\tilde{X}$ and $\tilde{Y}$, respectively, this is still true, except that we need consider the roots of the *numerator* of the determinant. To see which of the compensators K that achieve $\|H_K\|_\infty \leq \lambda$ stabilize the closed-loop system we consider the zeros of the determinant of

$$[\rho D_1 \quad [D_2 \quad -N_2]Z_\lambda^{-1}\begin{bmatrix} A \\ \varepsilon B \end{bmatrix}]. \tag{304}$$

The nonzero real number ρ will be chosen later. It does not affect the location of the zeros. The real parameter ε varies between 0 and 1. For $\varepsilon = 1$ we obtain the situation for which we want to determine stability. Defining

$$[D_2 \quad -N_2]Z_\lambda^{-1} = [L_{1,\lambda} \quad L_{2,\lambda}] = L_\lambda, \tag{305}$$

we hence consider the determinant of

$$[\rho D_1 \quad L_{1,\lambda}A + \varepsilon L_{2,\lambda}B]. \tag{306}$$

For $\varepsilon = 0$ this expression reduces to

$$[\rho D_1 \quad L_{1,\lambda}A] = [\rho D_1 \quad L_{1,\lambda}]\begin{bmatrix} I & 0 \\ 0 & A \end{bmatrix}. \tag{307}$$

This shows that the compensator corresponding to $\varepsilon = 0$ is stabilizing if and only if the central solution is stabilizing (corresponding to $A = I$) and the numerator of the determinant of A is Hurwitz.

We now prove that if ε changes continuously from 0 to 1, none of the roots of the determinant of (306) crosses over the imaginary axis. This means that each root remains in that half of the complex plane where it starts for $\varepsilon = 0$, and, hence, the compensator determined by A and B is stabilizing if and only if the central solution is stabilizing and the numerator of the determinant of A is Hurwitz.

To prove that no root of the numerator of the determinant of (306) crosses over the imaginary axis, we rewrite (306) in the form

$$[\rho D_1 \quad L_{1,\lambda}]\left(I+\varepsilon[\rho D_1 \quad L_{1,\lambda}]^{-1}[0 \quad L_{2,\lambda}]\begin{bmatrix} I & 0 \\ 0 & BA^{-1}\end{bmatrix}\right)\begin{bmatrix} I & 0 \\ 0 & A\end{bmatrix}. \tag{308}$$

Inspection shows that as ε changes from 0 to 1, none of the zeros of the determinant crosses over the imaginary axis if

$$\| [\rho D_1 \quad L_{1,\lambda}]^{-1}[0 \quad L_{2,\lambda}]\begin{bmatrix} I & 0 \\ 0 & BA^{-1}\end{bmatrix} \|_\infty < 1. \tag{309}$$

Because $\|BA^{-1}\| \leq 1$, it is sufficient to prove that

$$\| [\rho D_1 \quad L_{1,\lambda}]^{-1}[0 \quad L_{2,\lambda}] \|_\infty < 1. \tag{310}$$

This is equivalent to

$$[\rho D_1 \quad L_{1,\lambda}]^{-1}[0 \quad L_{2,\lambda}][0 \quad L_{2,\lambda}]^{\sim}([\rho D_1 \quad L_{1,\lambda}]^{-1})^{\sim} \leq I \tag{311}$$

on the imaginary axis, which in turn is implied by

$$L_\lambda J L_\lambda^{\sim} + \rho^2 D_1 D_1^{\sim} > 0 \quad \text{on the imaginary axis,} \tag{312}$$

or

$$[D_2 \quad -N_2]\,\Pi_\lambda^{-1}\begin{bmatrix} D_2^{\sim} \\ -N_2^{\sim}\end{bmatrix} + \rho^2 D_1 D_1^{\sim} > 0 \quad \text{on the imaginary axis.} \tag{313}$$

Since it may be proved that

$$[D_2 \quad -N_2]\,\Pi_\lambda^{-1}\begin{bmatrix} D_2^{\sim} \\ -N_2^{\sim}\end{bmatrix} \geq N_1 N_1^{\sim} - \lambda^2 D_1 D_1^{\sim} \quad \text{on the imaginary axis,} \tag{314}$$

it follows that no root of the numerator of the determinant of (306) crosses over the imaginary axis if

$$N_1 N_1^{\sim} + (\rho^2 - \lambda^2) D_1 D_1^{\sim} > 0 \quad \text{on the imaginary axis.} \tag{315}$$

Taking $\rho > \lambda$, the proof is complete. •

4.7 Reduction to two polynomial J-spectral factorizations and lower bound

Determining suboptimal solutions of the $\mathcal{H}_\infty$ optimal control problem requires the J-spectral factorization

$$\Pi_\lambda = Z_\lambda^\sim J Z_\lambda \tag{316}$$

of the rational matrix

$$\Pi_\lambda = \begin{bmatrix} D_2^\sim \\ -N_2^\sim \end{bmatrix} (N_1 N_1^\sim - \lambda^2 D_1 D_1^\sim)^{-1} [D_2 \quad -N_2]. \tag{317}$$

We show how to reduce this factorization to two polynomial J-spectral factorizations, one for the denominator, the other for the numerator. The former is the polynomial J-spectral cofactorization

$$N_1 N_1^\sim - \lambda^2 D_1 D_1^\sim = Q_\lambda J' Q_\lambda^\sim, \tag{318}$$

with Q_λ square such that its determinant is Hurwitz. The existence of this spectral factorization is discussed later in this section. Once Q_λ has been determined, we may obtain polynomial matrices Δ_λ and Λ_λ by the left-to-right fraction conversion

$$Q_\lambda^{-1} [D_2 \quad -N_2] = \Delta_\lambda \Lambda_\lambda^{-1}. \tag{319}$$

With this we have

$$\Pi_\lambda = \begin{bmatrix} D_2^\sim \\ -N_2^\sim \end{bmatrix} (Q_\lambda^{-1})^\sim J' Q_\lambda^{-1} [D_2 \quad -N_2] = (\Lambda_\lambda^{-1})^\sim \Delta_\lambda^\sim J' \Delta_\lambda \Lambda_\lambda^{-1}. \tag{320}$$

By the second polynomial J-spectral factorization

$$\Delta_\lambda^\sim J' \Delta_\lambda = \Gamma_\lambda^\sim J \Gamma_\lambda, \tag{321}$$

with Γ_λ square such that its determinant is Hurwitz, we obtain the rational J-spectral factor Z_λ as

$$Z_\lambda = \Gamma_\lambda \Lambda_\lambda^{-1}. \tag{322}$$

4.7.1 Summary (Reduction to two J-spectral factorizations.) The rational J-spectral factor Z_λ defined by $\Pi_\lambda = Z_\lambda^\sim J Z_\lambda$ is given by

$$Z_\lambda = \Gamma_\lambda \Lambda_\lambda^{-1}, \tag{323}$$

where the polynomial matrices Γ_λ and Λ_λ are obtained as follows:

1. Determine the J-spectral cofactor Q_λ from

$$N_1 N_1^\sim - \lambda^2 D_1 D_1^\sim = Q_\lambda J' Q_\lambda^\sim. \tag{324}$$

2. Find the polynomial matrices Δ_λ and Λ_λ by solving the linear matrix polynomial equation

$$Q_\lambda \Delta_\lambda = [D_2 \quad -N_2]\Lambda_\lambda. \tag{325}$$

3. Determine the J-spectral factor Γ_λ from

$$\Delta_\lambda^\sim J' \Delta_\lambda = \Gamma_\lambda^\sim J \Gamma_\lambda. \tag{326}$$

•

We next discuss the existence of the two polynomial J-spectral factorizations. The factorization $N_1 N_1^\sim - \lambda^2 D_1 D_1^\sim = Q_\lambda J' Q_\lambda^\sim$ is well-defined for $|\lambda| \geq \lambda_1$, where λ_1 is the first value of λ as λ decreases from ∞ for which a root of $\det(N_1 N_1^\sim - \lambda^2 D_1 D_1^\sim)$ reaches the imaginary axis or infinity. Similarly, the J-spectral factorization $\Delta_\lambda^\sim J' \Delta_\lambda = \Gamma_\lambda^\sim J \Gamma_\lambda$ exists for $|\lambda| \geq \lambda_2$, where λ_2 is the first value of λ as λ decreases from ∞ for which $\det(\Delta_\lambda^\sim J' \Delta_\lambda)$ has a root on the imaginary axis or at infinity.

The numbers λ_1 and λ_2 have a further meaning as *lower bounds* for $\|H\|_\infty$. Consideration of the proof 4.4.3 of the rational inequality leads to the conclusion that if $|\lambda| < \lambda_1$ there exists *no* compensator K such that $\|H_K\|_\infty \leq \lambda$. Hence,

$$\|H_K\|_\infty \geq \lambda_1 \tag{327}$$

for any compensator, whether stabilizing or not.

A second bound may be found by studying the *dual* problem, which originates by replacing the plant G with its transpose $\bar{G} = G^{\mathrm{T}}$. It is easy to see that if any compensator $\bar{K}$ achieves $\|\bar{H}_{\bar{K}}\|_\infty \leq \lambda$ for the dual problem, then the compensator $K = \bar{K}^{\mathrm{T}}$ accomplishes the same for the original problem. To solve the dual problem, we need a *right* polynomial matrix fraction representation

$$G = \bar{N}\bar{D}^{-1} \tag{328}$$

of the original problem, because this yields the left coprime factorization

$$\bar{G} = (\bar{D}^{\mathrm{T}})^{-1} \bar{N}^{\mathrm{T}} \tag{329}$$

needed for the dual problem. Partitioning

$$\bar{D} = \begin{matrix} k_1 \\ k_2 \end{matrix} \begin{bmatrix} \bar{D}_1 \\ \bar{D}_2 \end{bmatrix}, \qquad \bar{N} = \begin{matrix} m_1 \\ m_2 \end{matrix} \begin{bmatrix} \bar{N}_1 \\ \bar{N}_2 \end{bmatrix}, \tag{330}$$

we obtain the lower bound

$$\|H_K\|_\infty = \|H_{K^{\mathrm{T}}}^{\mathrm{T}}\|_\infty \geq \lambda_2 \tag{331}$$

where λ_2 is the first value of λ as λ decreases from ∞ such that a root of

$$\det(\bar{N}_1^{\mathrm{T}} \bar{N}_1 - \lambda^2 \bar{D}_1^{\mathrm{T}} \bar{D}_1) \tag{332}$$

reaches the imaginary axis or ∞.

It is no coincidence that the same notation λ_2 is used as for the number that bounds the existence of the second polynomial J-spectral factorization, because it may be proved that the two numbers are equal. This leads to the satisfactory conclusion that the two polynomial J-spectral factorizations are well-defined for those values of λ that are at all of interest. Moreover, we have the lower bound

$$\|H_K\|_\infty \geq \max(\lambda_1,\ \lambda_2). \tag{333}$$

4.7.2 Summary (Existence of spectral factorizations and lower bound.) The two polynomial J-spectral factorizations of 4.7.1 exist for all $|\lambda| \geq \max(\lambda_1,\ \lambda_2)$, where λ_1 is the first value of λ as λ decreases for which $\det(N_1 N_1^\sim - \lambda^2 D_1 D_1^\sim)$ has a root on the imaginary axis or at infinity, and λ_2 is the first value of λ as λ decreases from infinity for which $\det(\bar{N}_1^\sim \bar{N}_1 - \lambda^2 \bar{D}_1^\sim \bar{D}_1)$ has a root on the imaginary axis or at infinity. Moreover,

$$\|H_K\|_\infty \geq \max(\lambda_1,\ \lambda_2) \tag{334}$$

for any compensator K, whether stabilizing or not. •

4.8 Example

By way of example we solve the mixed sensitivity problem of Example 3.12.1 as a standard problem.

4.8.1 Example (Solution of a mixed sensitivity problem as standard problem.) The plant transfer function for the standard problem defined by the mixed sensitivity problem of Example 3.12.1 is given by

$$G(s) = \left[\begin{array}{c|c} \frac{s+1}{s} & \frac{1}{s} \\ 0 & c \\ \hline \frac{s+1}{s} & \frac{1}{s} \end{array}\right] = \left[\begin{array}{cc|c} 1 & 0 & -1 \\ 0 & 1 & 0 \\ 0 & 0 & s \end{array}\right]^{-1} \left[\begin{array}{c|c} 0 & 0 \\ 0 & c \\ s+1 & 1 \end{array}\right]. \tag{335}$$

It is straightforward to find that

$$N_1(s)N_1^\sim(s) - \lambda^2 D_1(s) D_1^\sim(s) = \begin{bmatrix} -\lambda^2 & 0 & 0 \\ 0 & -\lambda^2 & 0 \\ 0 & 0 & 1-s^2 \end{bmatrix}. \tag{336}$$

Since the determinant of this matrix is $\lambda^4(1-s^2)$ we have $\lambda_1 = 0$, so that the J-spectral factorization is defined for all λ. It is easy to see that

$$Q_\lambda(s) = \begin{bmatrix} 0 & \lambda & 0 \\ 0 & 0 & \lambda \\ s+1 & 0 & 0 \end{bmatrix}, \qquad J' = \begin{bmatrix} 1 & 0 & 0 \\ 0 & -1 & 0 \\ 0 & 0 & -1 \end{bmatrix}. \tag{337}$$

The next step is the conversion

$$Q_\lambda^{-1}(s)[D_2(s) \quad -N_2(s)] = \begin{bmatrix} \frac{s}{s+1} & -\frac{1}{s+1} \\ -\frac{1}{\lambda} & 0 \\ 0 & -\frac{c}{\lambda} \end{bmatrix}$$

$$= \begin{bmatrix} 1 & -1 \\ -\frac{1}{\lambda} & 0 \\ \frac{c}{\lambda} & -\frac{c}{\lambda}(s+1) \end{bmatrix} \begin{bmatrix} 1 & 0 \\ -1 & s+1 \end{bmatrix}^{-1} = \Delta_\lambda(s)\Lambda_\lambda^{-1}(s). \tag{338}$$

The second J-spectral factorization is defined by

$$\Delta_\lambda^\sim(s)J'\Delta_\lambda(s) = \begin{bmatrix} 1-\frac{1+c^2}{\lambda^2} & -1+\frac{c^2}{\lambda^2}(s+1) \\ -1+\frac{c^2}{\lambda^2}(-s+1) & 1-\frac{c^2}{\lambda^2}(1-s^2) \end{bmatrix} = \Gamma_\lambda^\sim(s)J\Gamma_\lambda(s). \tag{339}$$

The determinant of this matrix equals $\frac{c^2}{\lambda^2}(1-\frac{1}{\lambda^2})s^2 - \frac{1}{\lambda^2}(1-\frac{c^2}{\lambda^2})$. Hence, the spectral factorization exists for $|\lambda| \geq \lambda_2 = \max(1,\ c)$. Using a method that will be discussed in Subsection 4.10, it may be found that for $\lambda > \sqrt{1+c^2}$ a J-spectral factor Γ_λ is given by

$$\Gamma_\lambda(s) = \begin{bmatrix} \sqrt{1-\frac{1+c^2}{\lambda^2}} & \frac{(-1+\frac{c^2}{\lambda^2})+\frac{c^2}{\lambda^2}s}{\sqrt{1-\frac{1+c^2}{\lambda^2}}} \\ 0 & \frac{\frac{c}{\lambda}\sqrt{1-\frac{1}{\lambda^2}}(s+\zeta_\lambda)}{\sqrt{1-\frac{1+c^2}{\lambda^2}}} \end{bmatrix}, \tag{340}$$

where

$$\zeta_\lambda = \frac{1}{c}\sqrt{\frac{1-\frac{c^2}{\lambda^2}}{1-\frac{1}{\lambda^2}}}. \tag{341}$$

For $\lambda_2 \leq \lambda < \sqrt{1+c^2}$ a similar expression may be developed. It follows that

$$Z_\lambda^{-1}(s) = \Lambda_\lambda(s)\Gamma_\lambda^{-1}(s) = \frac{1}{\sqrt{1-\frac{1+c^2}{\lambda^2}}} \begin{bmatrix} 1 & -\frac{(-1+\frac{c^2}{\lambda^2})+\frac{c^2}{\lambda^2}s}{\frac{c}{\lambda}\sqrt{1-\frac{1}{\lambda^2}}(s+\zeta_\lambda)} \\ -1 & \frac{-\frac{1}{\lambda^2}+(1-\frac{1}{\lambda^2})s}{\frac{c}{\lambda}\sqrt{1-\frac{1}{\lambda^2}}(s+\zeta_\lambda)} \end{bmatrix}. \tag{342}$$

As a result, the transfer function of all stabilizing compensators K such that $\|H_K\|_\infty \leq \lambda$ may be expressed as

$$K(s) = -\frac{A - \frac{\lambda}{c}\frac{(1-\frac{1}{\lambda^2})s-\frac{1}{\lambda^2}}{s\sqrt{1-\frac{1}{\lambda^2}}+\frac{1}{c}\sqrt{1-\frac{c^2}{\lambda^2}}}B}{A - \frac{\lambda}{c}\frac{\frac{c^2}{\lambda^2}s+(1-\frac{c^2}{\lambda^2})}{s\sqrt{1-\frac{1}{\lambda^2}}+\frac{1}{c}\sqrt{1-\frac{c^2}{\lambda^2}}}B}, \tag{343}$$

where without loss of generality A and B may be chosen as polynomials such that $A^\sim A \geq B^\sim B$ on the imaginary axis, and A is Hurwitz.

It is easily checked that if we take $A = B$, the compensator is precisely the equalizing compensator of Example 3.12.1 (within a minus sign). By cancelation, for $\lambda = \lambda_{\text{opt}} = \sqrt{1+c^2}$ the transfer function of the equalizing compensator reduces to

$$K_{\lambda_{\text{opt}}}(s) = -1, \tag{344}$$

which agrees with that we found in 3.12.1. This reduction indeed takes place for *any* choice of the polynomials A and B.

We note that although the compensator remains well-defined for all λ, the coeffficients of the spectral factor Γ_λ approach infinity as λ gets closer to $\sqrt{1+c^2}$. As in 3.12.1, the optimal value of λ forms a *singular point* for the J-spectral factorization. •

4.9 Reduced and optimal solutions of the $\mathcal{H}_\infty$ optimal regulator problem

Given any value of λ larger than the lower bound $\max(\lambda_1, \lambda_2)$, the results we obtained allow determining a simple parametrization for all compensators that perform at least as well as λ. All that is needed for this parametrization is a rational J-spectral factorization that can be reduced to two polynomial J-spectral factorizations.

If any of these compensators is stabilizing (for instance the "central" solution), all stabilizing compensators that achieve better than λ can easily be singled out. If the central solution is not stabilizing, no stabilizing compensator exists that achieves $\|H_K\|_\infty \leq \lambda$.

These results allow searching for stabilizing compensators whose performance is arbitrarily close to the optimal value λ_{opt}. Example 4.8.1 illustrates that as the optimal solution is approached two phenomena occur:

1. The coefficients of the J-spectral factor Γ_λ grow without bound[1].

2. The numerator X and denominator Y of the compensator acquire a common factor that may be canceled.

The *reduced* solutions that result after cancelation solve the $\mathcal{H}_\infty$ optimal regulation problem. It may be proved that a suitable spectral density matrix Φ may be constructed such that these solutions satisfy the equalizer principle. The present solution of the standard $\mathcal{H}_\infty$-optimal regulator problem, however, follows by J-spectral factorization and does not rely on the equalizer principle.

In what follows we describe two further results that are needed to complete the solution of the $\mathcal{H}_\infty$ optimal regulator problem successfully:

1. Polynomial J-spectral factorization by symmetric factor extraction.

2. A direct way (based on symmetric factor extraction) to compute reduced solutions.

4.10 Polynomial J-spectral factorization by symmetric factor extraction

Polynomial J-spectral factorization is an essential ingredient of the present solution of the $\mathcal{H}_\infty$ optimal regulation problem. We sketch an algorithm that is suitable for numerical solution. It is based on Callier's method for ordinary polynomial spectral factorization by symmetric factor extraction (Callier, 1985), and has been implemented in the form of an experimental MATLAB function.

[1]In exceptional cases the coefficients of the spectral cofactor Q_λ become unbounded. This complication may be avoided by solving the dual problem.

Given is an $n \times n$ para-Hermitian square polynomial matrix P with real coefficients. We assume it to be *diagonally reduced* (Callier, 1985). We refer to the roots of $\det(P)$ as the roots of P. Because P is para-Hermitian, if ζ is a root, so is $-\zeta$. Besides being diagonally reduced P is assumed to have no roots on the imaginary axis. Because P has real coefficients, if ζ is a complex root, so is its complex conjugate ζ^*.

The basic idea of symmetric factor extraction is to find corresponding to a root ζ of P a polynomial *elementary factor* T such that

$$P = T^{\sim} P'' T, \tag{345}$$

where also P'' is polynomial. Depending on whether ζ is real or complex we distinguish three different cases for the elementary factor T.

1. *Case 1.* If the root ζ is *real* the elementary factor is of the form

$$T(s) = \begin{bmatrix} 1 & 0 & \cdots & 0 & 0 & -\alpha_1 & 0 & 0 & \cdots & 0 & 0 \\ 0 & 1 & \cdots & 0 & 0 & -\alpha_2 & 0 & 0 & \cdots & 0 & 0 \\ \cdots & \cdots & \cdots & \cdots & \cdots & \cdots & \cdots & \cdots & \cdots & \cdots & \cdots \\ 0 & 0 & \cdots & 0 & 1 & -\alpha_{k-1} & 0 & 0 & \cdots & 0 & 0 \\ 0 & 0 & \cdots & 0 & 0 & s-\zeta & 0 & 0 & \cdots & 0 & 0 \\ 0 & 0 & \cdots & 0 & 0 & -\alpha_{k+1} & 1 & 0 & \cdots & 0 & 0 \\ \cdots & \cdots & \cdots & \cdots & \cdots & \cdots & \cdots & \cdots & \cdots & \cdots & \cdots \\ 0 & 0 & \cdots & 0 & 0 & -\alpha_n & 0 & 0 & \cdots & 0 & 1 \end{bmatrix}, \tag{346}$$

with $\alpha_1, \alpha_2, \cdots, \alpha_n$ real constants to be determined. The nonunit column of T is the kth column, with k to be determined. The factor T may be seen as a variant of the Hermite standard form of a polynomial matrix factor of degree 1. It has a single real root ζ. In more compact form we write T as

$$T(s) = \begin{bmatrix} I & -a_1 & 0 \\ 0 & s-\zeta & 0 \\ 0 & -a_2 & I \end{bmatrix}. \tag{347}$$

The constant vectors a_1 and a_2 may be found as follows. If ζ is a root of P, clearly $P(\zeta)$ is singular, so that there exists a real constant *null vector* e corresponding to ζ such that

$$P(\zeta)e = 0. \tag{348}$$

Because $T^{\sim}(\zeta)P''(\zeta)T(\zeta)e = 0$ and ζ is a root of T, we may determine a_1 and a_2 by letting $T(\zeta)e = 0$. Writing out this identity component-by-component it is easily found that a_1 and a_2 follow from

$$a = e/e_k, \tag{349}$$

where e_k is the kth component of e, and the constant vector a is defined by

$$a = \begin{bmatrix} a_1 \\ 1 \\ a_2 \end{bmatrix}. \tag{350}$$

It remains to determine the "remaining factor" P''. To this end, we first extract the factor T "on the right" and write

$$P = P'T, \tag{351}$$

with the square polynomial matrix P' to be determined. Multiplying the equality $P = P'T$ out element-by-element it is easy to see that all entries of P and P' are equal except those in their kth columns. Denoting the kth column of P' as p'_k it follows that

$$P(s)a = p'_k(s)(s-\zeta). \tag{352}$$

From this, p'_k may easily be computed. The polynomial matrix P'' now may be obtained by the left extraction

$$P' = T^{\sim} P'', \tag{353}$$

which we rewrite as the right extraction

$$(P')^{\sim} = (P'')^{\sim} T. \tag{354}$$

This extraction follows by the same procedure as before. Because P'' is para-Hermitian, it is sufficient to compute the kth diagonal entry p''_{kk} of P''. It may be solved from the equation

$$p'^{\sim}_k(s)a = p''_{kk}(s)(s-\zeta). \tag{355}$$

The nondiagonal elements of the kth column of P'' equal the corresponding entries of p'_k, the nondiagonal elements of the kth row of P'' follow by adjugation, while the remaining elements of P'' equal the corresponding elements of P. This defines P''.

2. *Case 21.* If the root ζ is *complex* there are two possibilities for the elementary factor T. The first (case 21) is that T has a single nonunit kth column of degree two, so that T is of the form

$$T(s) = \begin{bmatrix} I & -a_1 - b_1 s & 0 \\ 0 & (s-\zeta)(s-\zeta^*) & 0 \\ 0 & -a_2 - b_2 s & I \end{bmatrix}, \tag{356}$$

with a_1, a_2, b_1 and b_2 real constant coefficient vectors to be determined. Extraction of this factor takes care of both the root ζ and its complex conjugate ζ^*. Extraction on the right in the form

$$P = P'T \tag{357}$$

results in a matrix P' whose elements equal those of P except those in the kth column. Writing the complex root ζ and the corresponding (complex) null vector e in Cartesian form as

$$\zeta = \sigma + j\omega, \qquad e = p + jq, \tag{358}$$

denoting the kth column of P' as p'_k, and defining the two constant vectors a and b by

$$a = \begin{bmatrix} a_1 \\ 1 \\ a_2 \end{bmatrix}, \qquad b = \begin{bmatrix} b_1 \\ 0 \\ b_2 \end{bmatrix}, \tag{359}$$

it may be found that

$$[a \quad b]\begin{bmatrix} p_k & q_k \\ \sigma p_k - \omega q_k & \sigma q_k + \omega p_k \end{bmatrix} = [p \quad q], \tag{360}$$

which may be solved for a and b. The kth column p'_k of P' may be obtained from the equality

$$P(s)(a + bs) = p'_k(s)(s - \zeta)(s - \zeta^*). \tag{361}$$

The nondiagonal elements of the kth column of the remaining factor P'' in the extraction $P = T^{\sim}P''T$ equal the corresponding elements of p'_k, while the nondiagonal elements of the kth row of P'' follow by adjugation. The kth diagonal element p''_{kk} of P'' may be found from the equality

$$p'^{\sim}_k(s)(a + bs) = p''_{kk}(s)(s - \zeta)(s - \zeta^*). \tag{362}$$

3. *Case 22.* The second situation that occurs when the root ζ is complex (case 22) is that the elementary factor T has two nonunit columns (columns k and l) of degree one, so that it has the form

$$T(s) = \begin{bmatrix} I & -a_1 & 0 & -b_1 & 0 \\ 0 & s-\alpha & 0 & -\beta & 0 \\ 0 & -a_2 & I & -b_2 & 0 \\ 0 & -\gamma & 0 & s-\delta & 0 \\ 0 & -a_3 & 0 & -b_3 & I \end{bmatrix}, \tag{363}$$

with a_1, a_2, a_3, b_1, b_2, and b_3 real constant vectors to be determined, and α, β, γ, and δ real constants to be found. Extraction on the right in the form $P = P'T$ now results in a polynomial matrix P' whose elements equal the corresponding elements of P except those in the kth and lth columns. Writing the complex root ζ and the corresponding complex null vector e in Cartesian form as

$$\zeta = \sigma + j\omega, \qquad e = p + jq, \tag{364}$$

and denoting the two constant vectors a and b by

$$a = \begin{bmatrix} a_1 \\ 1 \\ a_2 \\ 0 \\ a_3 \end{bmatrix}, \qquad b = \begin{bmatrix} b_1 \\ 0 \\ b_2 \\ 1 \\ b_3 \end{bmatrix}, \tag{365}$$

it may be found that

$$[a \quad b] \begin{bmatrix} p_k & q_k \\ p_l & q_l \end{bmatrix} = [p \quad q], \tag{366}$$

with p_k and p_l the kth and lth elements of p, and a similar notation for q. This expression allows solving for a and b.

The constants α, β, γ, and δ follow from the identity

$$\begin{bmatrix} \alpha & \beta \\ \gamma & \delta \end{bmatrix} = \sigma I - \omega \begin{bmatrix} p_k & q_k \\ p_l & q_l \end{bmatrix} \begin{bmatrix} 0 & -1 \\ 1 & 0 \end{bmatrix} \begin{bmatrix} p_k & q_k \\ p_l & q_l \end{bmatrix}^{-1}, \tag{367}$$

with I the 2×2 unit matrix. In the right extraction $P = P'T$, the elements of P' equal the corresponding elements of P except those of the kth and lth columns. The kth and lth columns p'_k and p'_l of P' may be solved from the identity

$$P(s)[a \quad b] = [p'_k(s) \quad q'_k(s)] \begin{bmatrix} s-\alpha & -\beta \\ -\gamma & s-\delta \end{bmatrix}. \tag{368}$$

The elements of the factor P'' in the extraction $P = T^{\sim}P''T$ equal the elements of P, except all elements of the kth and lth columns and rows. The elements of the kth column of P'', except those in the kth and lth positions, equal the corresponding elements of p'_k. Similarly, the elements of the lth column of P'', except those in the kth and lth positions, equal the corresponding elements of p'_l. The elements of the kth and lth rows of P'', except those in the kth and lth positions, follow by adjugation. The elements p''_{kk}, p''_{kl}, p''_{lk}, and p''_{ll} of P'', finally, may be obtained from the identity

$$\begin{bmatrix} p'^{\sim}_k(s) \\ p'^{\sim}_l(s) \end{bmatrix} [a \quad b] = \begin{bmatrix} p''_{kk}(s) & p''_{kl}(s) \\ p''_{lk}(s) & p''_{ll}(s) \end{bmatrix} \begin{bmatrix} s-\alpha & -\beta \\ -\gamma & s-\delta \end{bmatrix}. \tag{369}$$

The following rules determine which case applies and which column k or columns k and l are to be selected:

1. If the root ζ is real, case 1 applies.

2. If the root ζ is complex, define $\mathcal{M}$ as the set of indices i of the diagonal elements p_{ii} of P whose degree equals the degree of the diagonal element of highest degree. Furthermore, writing the null vector e corresponding to the zero ζ in Cartesian form as $e = p + jq$, define the $n \times 2$ matrix C as

$$C = [p_{\mathcal{M}} \quad q_{\mathcal{M}}], \tag{370}$$

   where $p_{\mathcal{M}}$ is a column vector whose entries consists of those elements p_i of p such that $i \in \mathcal{M}$, and $q_{\mathcal{M}}$ is similarly defined. Then if rank$(C) = 1$, case 21 applies, while if rank$(C) = 2$, case 22 applies. In particular, if $\mathcal{M}$ contains one element only (i.e., P has a unique highest degree diagonal element), case 21 applies.

3. In case 1, the column index k is chosen so that the kth diagonal element of P is the diagonal element of highest degree. If there are several such elements, for numerical reasons it is recommended to choose the one such that the magnitude $|e_k|$ of the kth entry of the null vector e is maximal.

4. In case 21, k is chosen as any of the elements of the set $\mathcal{M}$. If the choice of k is not unique, for numerical reasons it is best to choose k such that the magnitude $|e_k|$ of the kth element of the null vector e is maximal.

5. In case 22, k and l are chosen as any two different elements of the set $\mathcal{M}$. If the choice is not unique, again it is best to choose k and l so as to maximize $|e_k|$ and $|e_l|$.

For spectral factorization, each root ζ that is extracted on the right is chosen to have negative real part. Then the factor that is extracted on the left has $-\zeta$ as its root. The extraction procedure is repeated until the supply of roots is exhausted. The order in which factors corresponding to the successive roots are extracted is not important. Eventually, P is reduced to the form

$$P = (T_m T_{m-1} \cdots T_1)^{\sim} P_0 (T_m T_{m-1} \cdots T_1), \tag{371}$$

with $T_1, T_2, \cdots, T_m$ the elementary factors that successively have been extracted, and P_0 a symmetric real constant matrix. P_0 may be factored as

$$P_0 = T_0^{\mathrm{T}} J T_0 \tag{372}$$

by successively using each diagonal element of P_0 to clear the corresponding row and column, and normalizing this diagonal element to ± 1 depending on its sign. For ordinary spectral factorization this amounts to Choleski factorization. For J-spectral factorization the columns of T_0 need be permuted such that J is in standard form.

In the case of *ordinary* spectral factorization, according to Callier (1985) at each stage of the procedure an extraction of type 1, 21 or 22 is always possible. In the case of J-spectral factorization the algorithm may fail at some intermediate stage. The reason for the failure is a *singularity* that manifests itself in one of the following forms:

1. In case 1, the kth element e_k of the null vector e turns out to be zero, so that the division (349) breaks down.

2. In case 21, the magnitude $|e_k|$ of the kth element of e turns out to be zero. This makes the coefficient matrix

$$\begin{bmatrix} p_k & q_k \\ \sigma p_k - \omega q_k & \sigma q_k - \omega p_k \end{bmatrix} \tag{373}$$

singular, so that inversion needed to solve (360) fails.

3. In case 22, the coefficient matrix

$$\begin{bmatrix} p_k & q_k \\ p_l & q_l \end{bmatrix} \tag{374}$$

turns out to be singular, with the result that the inversion required for the solution of (366) is impossible.

In the solution of the $\mathcal{H}_\infty$-optimal regulator problem this singularity occurs at the optimal solution, and is responsible for the cancelation that results in reduced solutions. As we shall see in Subsection 4.11, the phenomenon may be exploited to solve directly for reduced solutions.

4.10.1 Example (J-spectral factorization.) By way of example we consider the J-spectral factorization of the polynomial matrix

$$P_\lambda(s) = \begin{bmatrix} -\frac{1}{\lambda^2} - (1-\frac{1}{\lambda^2})s^2 & -s \\ s & (1-\frac{c^2}{\lambda^2}) + \frac{c^2}{\lambda^2}s^2 \end{bmatrix} \tag{375}$$

encountered in Example 3.12.1. From

$$\det(P_\lambda(s)) = -\frac{1}{\lambda^2}(1-s^2)\left((1-\frac{c^2}{\lambda^2}) - c^2(1-\frac{1}{\lambda^2})s^2\right) \tag{376}$$

it follows that P_λ has the roots ± 1 and $\pm\zeta_\lambda$, where

$$\zeta_\lambda = \frac{1}{c}\sqrt{\frac{1-\frac{c^2}{\lambda^2}}{1-\frac{1}{\lambda^2}}}. \tag{377}$$

We first extract a factor corresponding to the root -1. Because this root is real, case 1 applies. Inspection shows that

$$P_\lambda(-1) = \begin{bmatrix} -1 & 1 \\ -1 & 1 \end{bmatrix}, \tag{378}$$

so that the null vector corresponding to the root -1 is

$$e = \begin{bmatrix} 1 \\ 1 \end{bmatrix}. \tag{379}$$

We choose the active column index, arbitarily, as $k = 1$. This makes the factor that is extracted equal to

$$T_1(s) = \begin{bmatrix} s+1 & 0 \\ -1 & 1 \end{bmatrix}. \tag{380}$$

From

$$P_\lambda(s)e = (s+1)\begin{bmatrix} -\frac{1}{\lambda^2} - (1-\frac{1}{\lambda^2})s \\ 1 - \frac{c^2}{\lambda^2} + \frac{c^2}{\lambda^2}s \end{bmatrix} = (s+1)p_1'(s) \tag{381}$$

and

$$e^{\mathrm{T}}P_\lambda(s)e = (-s^2+1)(1 - \frac{1+c^2}{\lambda^2}) = (-s^2+1)p_{11}''(s) \tag{382}$$

it follows that the remaining factor P'' is given by

$$P_\lambda''(s) = \begin{bmatrix} 1 - \frac{1+c^2}{\lambda^2} & (1-\frac{c^2}{\lambda^2}) - \frac{c^2}{\lambda^2}s \\ (1-\frac{c^2}{\lambda^2}) + \frac{c^2}{\lambda^2}s & (1-\frac{c^2}{\lambda^2}) + \frac{c^2}{\lambda^2}s^2 \end{bmatrix}. \tag{383}$$

At this point we could extract a second first-order factor corresponding to the root ζ_λ, but it is simpler to use the constant 11-element to clear the first row and column. This is possible as long as $\lambda^2 \neq 1 + c^2$, and results in

$$P_\lambda''(s) = \begin{bmatrix} \delta_\lambda & 0 \\ 0 & -\frac{\frac{c^2}{\lambda^2}(1-\frac{1}{\lambda^2})}{1-\frac{1+c^2}{\lambda^2}}(-s^2+\zeta_\lambda^2) \end{bmatrix}, \tag{384}$$

where $\delta_\lambda = \mathrm{sign}(1 - \frac{1+c^2}{\lambda^2})$. Correspondingly, a unimodular factor U_1 may be extracted that is given by

$$U_1(s) = \begin{bmatrix} \sqrt{\left|1 - \frac{1+c^2}{\lambda^2}\right|} & \frac{(1-\frac{c^2}{\lambda^2}) - \frac{c^2}{\lambda^2}s}{\delta_\lambda\sqrt{\left|1-\frac{1+c^2}{\lambda^2}\right|}} \\ 0 & 1 \end{bmatrix}. \tag{385}$$

Inspection of P_λ'' as given by (384) shows that the final factor to be extracted is

$$T_2(s) = \begin{bmatrix} 1 & 0 \\ 0 & \frac{\frac{c}{\lambda}\sqrt{1-\frac{1}{\lambda^2}}}{\sqrt{\left|1-\frac{1+c^2}{\lambda^2}\right|}}(s+\zeta_\lambda) \end{bmatrix}, \tag{386}$$

with the remaining factor

$$P_\lambda'''(s) = \begin{bmatrix} \delta_\lambda & 0 \\ 0 & -\delta_\lambda \end{bmatrix}. \tag{387}$$

The J-spectral factorization we are looking for is $P_\lambda = T_\lambda^\sim J T_\lambda$, with

$$T_\lambda = T_2U_1T_1 = \frac{1}{\sqrt{\left|1 - \frac{1+c^2}{\lambda^2}\right|}}\begin{bmatrix} -\frac{1}{\lambda^2} + (1-\frac{1}{\lambda^2})s & (1-\frac{c^2}{\lambda^2}) - \frac{c^2}{\lambda^2}s \\ -\frac{c}{\lambda}(s+\zeta_\lambda)\sqrt{1-\frac{1}{\lambda^2}} & \frac{c}{\lambda}(s+\zeta_\lambda)\sqrt{1-\frac{1}{\lambda^2}} \end{bmatrix}. \tag{388}$$

provided $\delta_\lambda = 1$. If $\delta_\lambda = -1$ then the rows of T_λ need be interchanged. •

4.11 Determination of reduced solutions by symmetric factor extraction

We next discuss how to obtain reduced optimal solutions *directly*, without performing the cancelation explicitly.

Given the rational J-spectral factorization $\Pi_\lambda = Z_\lambda^\sim J Z_\lambda$, with $Z_\lambda = \Gamma_\lambda \Lambda_\lambda^{-1}$, the compensators we are looking for are given by $K = YX^{-1}$,

$$\begin{bmatrix} X \\ Y \end{bmatrix} = Z_\lambda^{-1} \begin{bmatrix} A \\ B \end{bmatrix} = \Lambda_\lambda \Gamma_\lambda^{-1} \begin{bmatrix} A \\ B \end{bmatrix} = \Lambda_\lambda W_\lambda, \tag{389}$$

where

$$W_\lambda = \Gamma_\lambda^{-1} \begin{bmatrix} A \\ B \end{bmatrix}, \tag{390}$$

with $A^\sim A \geq B^\sim B$ on the imaginary axis, and the numerator of $\det(A)$ Hurwitz. The matrix Γ_λ follows from the polynomial J-spectral factorization

$$\Gamma_\lambda^\sim J \Gamma_\lambda = P_\lambda, \tag{391}$$

where $P_\lambda = \Delta_\lambda^\sim J' \Delta_\lambda$. $A^\sim A \geq B^\sim B$ on the imaginary axis is equivalent to $W_\lambda^\sim P_\lambda W_\lambda \geq 0$ on the imaginary axis.

When λ approaches the optimal value λ_{opt}, a cancelation occurs in YX^{-1}. This cancelation may be traced to a singularity in the J-spectral factorization of P_λ. For $\lambda \neq \lambda_{\text{opt}}$ the J-spectral factorization is *regular* in the sense that at each stage symmetric factor extraction is possible. For $\lambda = \lambda_{\text{opt}}$, however, the extraction fails at some intermediate step. Usually this happens at the *final* extraction step. It is conjectured that if it happens earlier, there is a multiple cancelation in YX^{-1}. We do not investigate this nongeneric situation. The singularity that causes the failure has been described in the preceding section.

To explain what happens to the parametrization of the compensators we consider the three different possibilities for the final extraction. Just prior to the final extraction, P_λ takes the form

$$P_\lambda = \hat{\Gamma}_\lambda^\sim \hat{P}_\lambda \hat{\Gamma}_\lambda, \tag{392}$$

where $\hat{\Gamma}_\lambda$ includes all factors that have been extracted so far, and $\hat{P}_\lambda$ is the remaining factor.

1. *Case 1:* The last extraction corresponds to a *real* root ζ. In this case, all diagonal entries of $\hat{P}_\lambda$ are constants, except the kth diagonal entry, which has degree two. All nondiagonal elements of the kth column and row of $\hat{P}_\lambda$ have degree one, while all other elements of $\hat{P}_\lambda$ are constants. The last extraction is singular if the kth element of the null vector e is zero.

   We may use each nonzero constant diagonal element of $\hat{P}_\lambda$ to clear the corresponding row and column, and to normalize the diagonal element to ± 1. This is equivalent to extraction of a unimodular factor, that is, a factor of degree zero. If the final extraction is singular, during this process process of clearing one of the diagonal elements of $\hat{P}_\lambda$ becomes zero, which means that the corresponding row and column cannot be cleared. If no zero diagonal element appears, the final result admits a nonsingular extraction, which contradicts the assumption of singularity. By suitable row and column permutations the result of this procedure can be arranged in the form

$$\hat{P}_\lambda = \begin{bmatrix} 1 & & & & & & & & & \\ & 1 & & & & & & & & \\ \cdots & \cdots & \cdots & \cdots & & & & & & \\ & & & 1 & & & & & & \\ & & & & 0 & \hat{p}_{\lambda,kl} & & & & \\ & & & & \hat{p}_{\lambda,lk} & \hat{p}_{\lambda,ll} & & & & \\ & & & & & & -1 & & & \\ & & & & & & & -1 & & \\ & & & & & & & \cdots & \cdots & \cdots \\ & & & & & & & & & -1 \end{bmatrix}, \tag{393}$$

where $\hat{p}_{\lambda,kl}$ and $\hat{p}_{\lambda,lk}$ are polynomials of degree one, $\hat{p}_{\lambda,ll}$ is a polynomial of degree two, $l = k+1$, and all blank entries are zeros. The null vector corresponding to a singular extraction is given by

$$e_\lambda = \begin{bmatrix} 0 \\ \cdots \\ 0 \\ 1 \\ 0 \\ \cdots \\ 0 \end{bmatrix}, \tag{394}$$

with the one in the kth position. We note that for the real constant vector e_λ we have

$$e_\lambda^{\mathrm{T}} \hat{P}_\lambda e_\lambda = 0. \tag{395}$$

Inspection shows that we may achieve

$$W_\lambda^\sim P_\lambda W_\lambda = W_\lambda^\sim \hat{\Gamma}_\lambda^\sim \hat{P}_\lambda \hat{\Gamma}_\lambda W_\lambda \geq 0 \tag{396}$$

on the imaginary axis by letting

$$W_\lambda = \hat{\Gamma}_\lambda^{-1} \begin{bmatrix} \hat{A} & 0 \\ 0 & \hat{a} \\ 0 & 0 \\ \hat{B} & 0 \end{bmatrix}, \tag{397}$$

with $\hat{A}$ and $\hat{B}$ rational or polynomial such that $\hat{A}^\sim \hat{A} \geq \hat{B}^\sim \hat{B}$ on the imaginary axis, and $\hat{a}$ any real polynomial or rational function. The dimensions of the block

$$\begin{bmatrix} \hat{A} & 0 \\ 0 & \hat{a} \end{bmatrix} \tag{398}$$

equal those of A, with A as in 4.5.1 and 4.6.1. The dimensions of the block

$$\begin{bmatrix} 0 & 0 \\ 0 & \hat{B} \end{bmatrix} \tag{399}$$

equal those of B, and the top block consists of a single row of zeros.

The relation (397) generates the reduced solutions we are looking for. The compensators are stabilizing if the numerators of $\hat{a}$ and $\det(\hat{A})$ are Hurwitz. Equalizing solutions follow by letting $\hat{a} = 1$, $\hat{A} = \hat{B} = I$ (only possible if $k_2 = m_2$), central solutions by letting $\hat{a} = 1$, $\hat{A} = I$ and $\hat{B} = 0$.

2. *Case 21.* In the case of a singular final extraction of type 21 the factor $\hat{P}_\lambda$ may again be brought into the form (393), where $\hat{p}_{\lambda,kl}$ and $\hat{p}_{\lambda,lk}$ now have degree two, and $\hat{p}_{\lambda,ll}$ has degree four. The null vector has the form

$$e_\lambda = \begin{bmatrix} 0 \\ \cdots \\ 0 \\ e_{\lambda,k} \\ e_{\lambda,l} \\ 0 \\ \cdots \\ 0 \end{bmatrix}, \tag{400}$$

and is in general complex-valued. In the singular situation $e_{\lambda,l} = 0$, and e_λ may be normalized as in (394). Again, $e_\lambda^{\mathrm{T}} \hat{P}_\lambda e_\lambda = 0$, and all W_λ that generate reduced solutions are given by (397).

3. *Case 22.* If the final extraction corresponds to a complex root with an elementary factor of type 22, the factor $\hat{P}_\lambda$ may be arranged to have the form

$$\hat{P}_\lambda = \begin{bmatrix} 1 & & & & & & & & & \\ & 1 & & & & & & & & \\ \cdots & \cdots & \cdots & \cdots & & & & & & \\ & & & 1 & & & & & & \\ & & & & \hat{p}_{\lambda,kk} & \hat{p}_{\lambda,kl} & & & & \\ & & & & \hat{p}_{\lambda,lk} & \hat{p}_{\lambda,ll} & & & & \\ & & & & & & -1 & & & \\ & & & & & & & -1 & & \\ & & & & & & & \cdots & \cdots & \cdots \\ & & & & & & & & & -1 \end{bmatrix}, \tag{401}$$

where $\hat{p}_{\lambda,kk}$, $\hat{p}_{\lambda,kl}$, $\hat{p}_{\lambda,lk}$, and $\hat{p}_{\lambda,ll}$ are polynomials of degree two. The null vector e_λ is of the form (400), and is generally complex-valued. It may easily be established that if the extraction is singular, the null vector is *real*, and $e_\lambda^{\mathrm{T}} P_\lambda e_\lambda = 0$. Correspondingly, reduced solutions are now generated by

$$W_\lambda = \hat{\Gamma}_\lambda^{-1} \begin{bmatrix} \hat{A} & 0 \\ 0 & e_{\lambda,k}\hat{a} \\ 0 & e_{\lambda,l}\hat{a} \\ \hat{B} & 0 \end{bmatrix}. \tag{402}$$

We summarize the results as follows.

4.11.1 Summary (Optimal reduced solutions.) Optimal solutions of the $\mathcal{H}_\infty$-optimal regulator problem are determined by the phenomenon that at $\lambda = \lambda_{\text{opt}}$ the final extraction in the J-spectral factorization that yields Γ_λ becomes singular. Let $P_\lambda = \hat{\Gamma}_\lambda^{\sim}\hat{P}_\lambda\hat{\Gamma}_\lambda$ represent the factorization of $P_\lambda = \Delta_\lambda^{\sim} J \Delta_\lambda$ just prior to the extraction of the final factor. Suppose that $\hat{P}_\lambda$ is arranged as in (393) for the cases 1 and 21 or as in (401) for case 22. Then if the final extraction is singular, the corresponding null vector

$$e_{\lambda_{\text{opt}}} = \begin{bmatrix} 0 \\ \cdots \\ 0 \\ e_{\lambda_{\text{opt}},k} \\ e_{\lambda_{\text{opt}},l} \\ 0 \\ \cdots \\ 0 \end{bmatrix} \tag{403}$$

is real, and satisfies

$$e_{\lambda_{\text{opt}}}^{\mathrm{T}} \hat{P}_{\lambda_{\text{opt}}} e_{\lambda_{\text{opt}}} = 0. \tag{404}$$

Correspondingly, all optimal compensators $K = YX^{-1}$ are given by

$$\begin{bmatrix} X \\ Y \end{bmatrix} = \Lambda_{\lambda_{\text{opt}}} \hat{\Gamma}_{\lambda_{\text{opt}}}^{-1} \begin{bmatrix} \hat{A} & 0 \\ 0 & e_{\lambda_{\text{opt}},k}\hat{a} \\ 0 & e_{\lambda_{\text{opt}},l}\hat{a} \\ \hat{B} & 0 \end{bmatrix}, \tag{405}$$

where $\hat{a}$, $\hat{A}$, and $\hat{B}$ are rational stable such that $\hat{a}$ and $\det(\hat{A})$ are Hurwitz, and $\hat{A}^{\sim}\hat{A} \geq \hat{B}^{\sim}\hat{B}$ on the imaginary axis. •

It is easy to see that the factor $\hat{a}$ always cancels in the compensator transfer function K, so that without loss of generality we may choose it equal to 1. Central solutions follow by setting $\hat{A} = I$ and $\hat{B} = 0$, and equalizing solutions by taking $\hat{A} = \hat{B} = I$ (only if $k_2 = m_2$).

A numerical procedure for determining optimal solutions may be outlined as follows:

4.11.2 Summary (Numerical procedure for computing optimal solutions.)

1. First employ a simple search procedure on λ, with as indicator the stability of the closed-loop system, to delimit a suitably small interval in which λ_{opt} is located.

2. Define ε_λ as a scalar indicator for determining λ_{opt}. If the final extraction in the spectral factorization for Γ_λ is of type 1, $\varepsilon_\lambda = e_{\lambda,k}$. If the final extraction is of type 21,

$$\varepsilon_\lambda = \det(\begin{bmatrix} p_{\lambda,k} & q_{\lambda,k} \\ \sigma_\lambda p_{\lambda,k} - \omega_\lambda q_{\lambda,k} & \sigma_\lambda q_{\lambda,k} + \omega_\lambda p_{\lambda,k} \end{bmatrix}). \tag{406}$$

If the final extraction is of type 22,

$$\varepsilon_\lambda = \det(\begin{bmatrix} p_{\lambda,k} & q_{\lambda,k} \\ p_{\lambda,l} & q_{\lambda,l} \end{bmatrix}). \tag{407}$$

Then find λ_{opt} by a root finding procedure that makes $\varepsilon_{\lambda_{\text{opt}}} = 0$.

3. Once λ_{opt} has been established, all optimal solutions follow from 4.11.1.

•

The indicator ε_λ may have several zero crossings. This is why step 1 of the precedure is needed to delimit a sufficiently small interval. The procedure has been realized in the form of a set of MATLAB macros and functions.

4.12 Example

Our solution of the $\mathcal{H}_\infty$ optimal regulator problem is now quite complete. The following example illustrates it.

4.12.1 Example (Reduced solutions by symmetric factor extraction.) We consider a simplified version of the mixed sensitivity problem of Example 3.10.1. In terms of the standard problem, the plant G is

$$G(s) = \left[\begin{array}{c|c} \frac{s^2+s\sqrt{2}+1}{s^2} & \frac{1}{s^2} \\ 0 & 1 \\ \hline \frac{s^2+s\sqrt{2}+1}{s^2} & \frac{1}{s^2} \end{array}\right] = \left[\begin{array}{cc|c} 1 & 0 & -1 \\ 0 & 1 & 0 \\ 0 & 0 & s^2 \end{array}\right]^{-1} \left[\begin{array}{c|c} 0 & 0 \\ 0 & 1 \\ s^2+s\sqrt{2}+1 & 1 \end{array}\right]. \tag{408}$$

This means that we have chosen $c = 1$ and $r = 0$. It is simple to find that

$$N_1(s)N_1^\sim(s) - \lambda^2 D_1(s)D_1^\sim(s) = \begin{bmatrix} -\lambda^2 & 0 & 0 \\ 0 & -\lambda^2 & 0 \\ 0 & 0 & 1+s^4 \end{bmatrix}. \tag{409}$$

Since the determinant of this matrix is $\lambda^4(1+s^4)$ its J-spectral cofactorization exists for all λ. We have

$$Q_\lambda(s) = \begin{bmatrix} 0 & \lambda & 0 \\ 0 & 0 & \lambda \\ s^2+s\sqrt{2}+1 & 0 & 0 \end{bmatrix}, \quad J' = \begin{bmatrix} 1 & 0 & 0 \\ 0 & -1 & 0 \\ 0 & 0 & -1 \end{bmatrix}. \tag{410}$$

From the left-to-right conversion $Q_\lambda^{-1}[D_2 \quad -N_2] = \Delta_\lambda \Lambda_\lambda^{-1}$ we obtain

$$\Delta_\lambda(s) = \begin{bmatrix} -1+s\sqrt{2} & 1 \\ -\frac{1}{\lambda}(1+s\sqrt{2}) & -\frac{1}{\lambda} \\ -\frac{1}{\lambda} & \frac{1}{\lambda}(1+s\sqrt{2}) \end{bmatrix}, \qquad \Lambda_\lambda(s) = \begin{bmatrix} 1+s\sqrt{2} & 1 \\ -1 & 1+s\sqrt{2} \end{bmatrix}. \tag{411}$$

We have taken care to make Δ_λ column reduced, so that

$$\Delta_\lambda^{\sim}(s)J'\Delta_\lambda(s) = \begin{bmatrix} (1-\frac{2}{\lambda^2}) - 2(1-\frac{1}{\lambda^2})s^2 & -1+(-1+\frac{2}{\lambda^2})s\sqrt{2} \\ -1-(-1+\frac{2}{\lambda^2})s\sqrt{2} & (1-\frac{2}{\lambda^2})+\frac{2}{\lambda^2}s^2 \end{bmatrix} = P_\lambda(s) \tag{412}$$

is diagonally reduced. The determinant of $P_\lambda(s)$ is $-\frac{4}{\lambda^2}(1-\frac{1}{\lambda^2})(1+s^4)$, so that the J-spectral factorization is well-defined for $|\lambda| \geq 1$. It moreover follows that P_λ has the four roots $\frac{1}{2}\sqrt{2}(\pm 1 \pm j)$. Exceptionally, the roots are independent of λ.

We extract a factor corresponding to the root $\zeta = -\frac{1}{2}\sqrt{2}(1+j)$ and its complex conjugate. The null vector for this root may be found to be given by

$$e_\lambda = \begin{bmatrix} 1 \\ \alpha_\lambda + j\beta_\lambda \end{bmatrix}, \tag{413}$$

where

$$\alpha_\lambda = \frac{2(1-\frac{2}{\lambda^2})}{1-\frac{4}{\lambda^2}+\frac{8}{\lambda^4}}, \qquad \beta_\lambda = \frac{1-\frac{8}{\lambda^2}+\frac{8}{\lambda^4}}{1-\frac{4}{\lambda^2}+\frac{8}{\lambda^4}}. \tag{414}$$

Hence, in Cartesian form $e_\lambda = p_\lambda + jq_\lambda$, where

$$p_\lambda = \begin{bmatrix} 1 \\ \alpha_\lambda \end{bmatrix}, \qquad q_\lambda = \begin{bmatrix} 0 \\ \beta_\lambda \end{bmatrix}. \tag{415}$$

Inspection of P_λ shows that the index set $\mathcal{M}$ is given by $\mathcal{M} = \{1,\ 2\}$. It follows that

$$C = [p_\mathcal{M} \quad q_\mathcal{M}] = \begin{bmatrix} 1 & 0 \\ \alpha_\lambda & \beta_\lambda \end{bmatrix}. \tag{416}$$

Generically, $\det(C) = \beta_\lambda \neq 0$ so that the extraction is of type 22. The extraction is singular if and only if λ is a root of

$$1 - \frac{8}{\lambda^2} + \frac{8}{\lambda^4} = 0. \tag{417}$$

The largest positive real root of this equation is

$$\lambda_{\text{opt}} = \sqrt{4+2\sqrt{2}} = 2.6131. \tag{418}$$

This is the first value of λ as λ decreases from ∞ for which the extraction becomes singular, and, hence, corresponds to the optimal solution. Accordingly,

$$\alpha_{\lambda_{\text{opt}}} = \frac{2(1-\frac{2}{\lambda_{\text{opt}}^2})}{1-\frac{4}{\lambda_{\text{opt}}^2}+\frac{8}{\lambda_{\text{opt}}^4}} = 1+\sqrt{2}. \tag{419}$$

The optimal compensator may now be found as follows. Since $\hat{\Gamma}_\lambda = I$ and

$$e_{\lambda_{opt}} = \begin{bmatrix} 1 \\ \alpha_{\lambda_{opt}} \end{bmatrix}, \tag{420}$$

we have

$$\begin{aligned} \begin{bmatrix} X(s) \\ Y(s) \end{bmatrix} &= \Lambda_{\lambda_{opt}}(s) \begin{bmatrix} 1 \\ \alpha_{\lambda_{opt}} \end{bmatrix} = \begin{bmatrix} 1+s\sqrt{2} & 1 \\ -1 & 1+s\sqrt{2} \end{bmatrix} \begin{bmatrix} 1 \\ 1+\sqrt{2} \end{bmatrix} \\ &= \begin{bmatrix} (2+\sqrt{2})+s\sqrt{2} \\ \sqrt{2}+(1+\sqrt{2})s \end{bmatrix}. \end{aligned} \tag{421}$$

Here we have taken $\hat{a} = 1$. $\hat{A}$ and $\hat{B}$ do not appear for reasons of dimension. The resulting optimal compensator is unique, and has the transfer function

$$K(s) = \frac{Y(s)}{X(s)} = \frac{\sqrt{2}+(1+\sqrt{2})s}{(2+\sqrt{2})+s\sqrt{2}}. \tag{422}$$

•

5 Conclusions

In these notes, the "polynomial" approach to the solution of $\mathcal{H}_\infty$-optimal regulation problems has been explored. In particular, a complete solution of the standard problem has been presented, including a formula and implementable algorithm to find all optimal (as opposed to suboptimal) optimal compensators.[2]

The use of $\mathcal{H}_\infty$-optimization for practical control system design has been discussed in the context of the mixed sensitivity problem. Both robustness and performance design may be dealt with. Further work need be done to further insight in the selection of frequency dependent weighting functions to achieve various design goals. The principles and intuition that have been developed over the last twenty years for LQG optimization and related methods, which are essentially time domain oriented, do not help much when it comes to $\mathcal{H}_\infty$-optimal regulation.

On the theoretical side by no means all problems have been solved. The question whether state space methods are better than the polynomial approach appears to be open. Although state space computational methods are currently better developed than polynomial methods there is no reason why this should not change. A definite advantage of the polynomial method is that it better matches the frequency domain slant of the $\mathcal{H}_\infty$-optimal regulation problem formulation. An illustration of this is the fact that the polynomial approach does not require the plant transfer function G in the standard problem to be proper, an assumption that currently cannot be dispensed with in the state space approach. Nonproper plant transfer functions arise naturally in mixed sensitivity problems (where often the weighting function W_2 need be taken nonproper).

An intrinsic difficulty with the $\mathcal{H}_\infty$-optimal regulation problem is that its solution–except in the single-input-single-output case–is not at all unique. Two special solutions

[2] A documented set of MATLAB macros and functions that implement the algorithm may be obtained by writing to the author or sending an e-mail message to twhuib@math.utwente.nl.

are the central and equalizing solutions. It is not clear at all whether either of these two solutions is to be preferred over the other. *A fortiori*, it is not clear if any of the many other solutions is preferable. One way of obtaining unique solutions is to consider what has become known as the ***super-optimization*** problem. In this problem, not only the ***largest*** singular value implicit in the definition of the ∞-norm is minimized, but, successively, also as many of the smaller singular values as possible. In addition to work of Young (1986), Tsai *et al.* (1988), Postlethwaite *et al.* (1989), and Limebeer *et al.* (1989), preliminary results for the mixed-sensitivity problem have been reported by Kwakernaak (1986) and Kwakernaak and Nyman (1989). Work in progress indicates that the present polynomial solution of the standard $\mathcal{H}_\infty$-optimal regulation problem permits the solution of the superoptimization problem for the standard problem.

Appendix: Matrix Dimensions for Section 4

G	$(m_1+m_2)\times(k_1+k_2)$		
G_{11}	$m_1\times k_1$	G_{12}	$m_1\times k_2$
G_{21}	$m_2\times k_1$	G_{22}	$m_2\times k_2$
K	$k_2\times m_2$	H	$m_1\times k_1$

Subsection 4.2

N	$(m_1+m_2)\times(k_1+k_2)$	D	$(m_1+m_2)\times(m_1+m_2)$
D_1	$(m_1+m_2)\times m_1$	D_2	$(m_1+m_2)\times m_2$
N_1	$(m_1+m_2)\times k_1$	N_2	$(m_1+m_2)\times k_2$
Y	$k_2\times m_2$	X	$m_2\times m_2$
N_{11}	$m_1\times k_1$	N_{12}	$m_1\times k_2$
N_{21}	$m_2\times k_1$	N_{22}	$m_2\times k_2$
D_{11}	$m_1\times m_1$	D_{12}	$m_1\times m_2$
D_{22}	$m_2\times m_2$	A_o	$m_2\times m_2$
D_o	$m_2\times m_2$	N_o	$m_2\times k_2$

Subsection 4.3

Φ	$k_1\times k_1$	L	$k_1\times k_1$

Subsection 4.4

$\bar{N}_1$	$m_1\times(k_1+k_2)$	$\bar{N}_2$	$m_2\times(k_1+k_2)$
$\bar{D}_1$	$k_1\times(k_1+k_2)$	$\bar{D}_2$	$k_2\times(k_1+k_2)$

Π_λ	$(k_2+m_2)\times(k_2+m_2)$		
P	$m_1\times(m_1+m_2)$	R	$m_1\times m_1$
Q_λ	$(m_1+m_2)\times(m_1+m_2)$	J'	$\mathrm{diag}(m_2,\ m_1)$
T_λ	$m_1\times(m_1+m_2)$		
$T_{1,\lambda}$	$m_1\times m_2$	$T_{2,\lambda}$	$m_1\times m_1$
$\bar{T}_\lambda$	$(m_1+m_2)\times m_2$		
$\bar{T}_{1,\lambda}$	$m_1\times m_2$	$\bar{T}_{2,\lambda}$	$m_2\times m_2$
M_λ	$m_2\times m_2$		

Subsection 4.5

Z_λ	$(k_2+m_2)\times(k_2+m_2)$	J	$\mathrm{diag}(m_2,\ k_2)$
$\tilde{Y}$	$k_2\times m_2$	$\tilde{X}$	$m_2\times m_2$
M	$m_2\times m_2$		
A	$m_2\times m_2$	B	$k_2\times m_2$

Subsection 4.6

L_λ	$(m_1+m_2)\times(k_2+m_2)$		
$L_{1,\lambda}$	$(m_1+m_2)\times m_2$	$L_{2,\lambda}$	$(m_1+m_2)\times k_2$
Δ_λ	$(m_1+m_2)\times(k_2+m_2)$	Λ_λ	$(k_2+m_2)\times(k_2+m_2)$
Γ_λ	$(k_2+m_2)\times(k_2+m_2)$		
$\bar{G}$	$(k_1+k_2)\times(m_1+m_2)$	$\bar{K}$	$m_2\times k_2$
$\bar{N}$	$(m_1+m_2)\times(k_1+k_2)$	$\bar{D}$	$(k_1+k_2)\times(k_1+k_2)$

Subsection 4.11

W_λ	$(k_2+m_2)\times m_2$	P_λ	$(k_2+m_2)\times(k_2+m_2)$
$\hat{P}_\lambda$	$(k_2+m_2)\times(k_2+m_2)$	$\hat{\Gamma}_\lambda$	$(k_2+m_2)\times(k_2+m_2)$
$\hat{A}$	$(m_2-1)\times(m_2-1)$	$\bar{B}$	$(k_2-1)\times(m_2-1)$

References

P. Boekhoudt (1988), *The $\mathcal{H}_\infty$ Control Design Method: A Polynomial Approach.* Ph. D. Dissertation, University of Twente.

F. M. Callier (1985), "On polynomial matrix spectral factorization by symmetric extraction." *IEEE Trans. Auto. Con.*, Vol. AC-30, pp. 453-464.

F. M. Callier and C. A. Desoer (1982), *Multivariable Feedback Systems.* New York, etc.: Springer-Verlag.

J. C. Doyle (1979), "Robustness of multiloop linear feedback systems." *Proc. 17th IEEE Conf. Decision Control*, pp. 12-18.

J. C. Doyle, K. Glover, P. P. Khargonekar, and B. A. Francis (1989), "State space solutions to standard $\mathcal{H}_2$ and $\mathcal{H}_\infty$ control problems." *IEEE Trans. Aut. Cont.*, Vol. 34, pp. 831-847.

B. A. Francis (1987), *A Course in H_∞ Control Theory.* Lecture Notes in Control and Information Sciences. Corrected first printing. Berlin etc.: Springer-Verlag.

B. A. Francis and J. C. Doyle (1987). "Linear control theory with an H_∞ optimality criterion." *SIAM J. Control Opt.*, Vol. 25, pp. 815-844.

K. Glover and J. C. Doyle (1989), "A state space approach to H_∞ optimal control." In: H. Nijmeijer and J. M. Schumacher (Eds.), *Three Decades of Mathematical System Theory.* Lecture Notes in Control and Information Sciences, Vol. 135. Berlin, etc.: Springer-Verlag.

M. Green (1989), "H_∞ controller synthesis by J-lossless coprime factorization." Preprint, Department of Electrical Engineering, Imperial College, London.

V. A. Jacubovič (1970), "Factorization of symmetric matrix polynomials." *Dokl. Akad. Nauk. SSSR,* Tom 194, No. 3.

T. Kailath (1980), *Linear Systems.* Englewood Cliffs, N. J.: Prentice-Hall.

V. Kučera (1979), *Discrete Linear Control: The Polynomial Equation Approach.* Chichester, etc.: John Wiley.

H. Kwakernaak (1983), "Robustness optimization of linear feedback systems." *Preprints, 22nd IEEE Conference on Decision and Control*, San Antonio, Texas, U.S.A.

H. Kwakernaak (1985), "Minimax frequency domain performance and robustness optimization of linear feedback systems." *IEEE Trans. Auto. Cont.*, Vol. AC-30, pp. 994-1004.

H. Kwakernaak (1986), "A polynomial approach to minimax frequency domain optimization of multivariable systems." *Int. J. Control,* Vol. 44, pp. 117-156.

H. Kwakernaak (1987), "A polynomial approach to $\mathcal{H}^\infty$-optimization of control systems." In: R. F. Curtain (Ed.), "*Modelling, Robustness and Sensitivity Reduction in Control Systems.* Berlin, etc.: Springer-Verlag.

H. Kwakernaak (1990), "Progress in the polynomial solution of the standard $\mathcal{H}_\infty$ optimal control problem." In: *Preprints of the 1990 IFAC Congress,* Tallinn, USSR.

H. Kwakernaak and P.-O. Nyman (1989), "An equalizing approach to superoptimization." *Proc. 1989 IEEE Int. Conf. Control and Applications,* Jerusalem.

H. Kwakernaak and H. Westdijk (1985), "Regulability of a multiple inverted pendulum system." *Control–Theory and Advanced Technology,* Vol. 1, pp. 1-9.

D. J. N. Limebeer, G. D. Halikias, and K. Glover (1989), "State-space algorithms for the computation of superoptimal matrix interpolation functions." *Int. J. Cont.*, Vol. 50, pp. 2431-2466.

I. Postlethwaite, M. C. Tsai, and D.-W. Gu (1989), "A state-space approach to discrete-time super-optimal $\mathcal{H}^\infty$ control problems." *Int. J. Cont.*, Vol. 49, pp 247-268.

M. C. Tsai, D.-W. Gu, and I. Postlethwaite (1988), "A state-space approach to super-optimal $\mathcal{H}^\infty$ control problems." *IEEE Trans. Aut. Cont.*, Vol. AC-33, pp. 833-843.

M. Verma and E. Jonckheere (1984), "L_∞-compensation with mixed sensitity as a broad-band matching problem." *Systems and Control Letters*, Vol. 4, pp. 125-130.

D. C. Youla, J. J. Bongiorno, and H. A. Jabr (1976), "Modern Wiener-Hopf Design of Optimal Controllers. Part I: The Single-Input-Output Case." *IEEE Trans. Auto. Cont.*, Vol. AC-21, pp. 3-13.

N. J. Young (1986), "The Nevanlinna-Pick problem for matrix-valued functions." *J. Op. Th.*, Vol. 15, pp. 239-265.

G. Zames (1981), "Feedback and optimal sensitivity: Model reference transformations, multiplicative seminorms, and aproximate inverses." *IEEE Trans. Auto. Cont.*, Vol. AC-26, pp. 301-320.

Notes on l^1-Optimal Control

by

J.B. Pearson

Department of Electrical and Computer Engineering

Rice University, Houston, TX 77251-1892

1. Introduction

These notes are intended to be an introduction to the design of control systems in which the objective is to minimize the maximum errors resulting from bounded inputs and norm bounded uncertainties.

The system that we will study is shown in Figure 1.

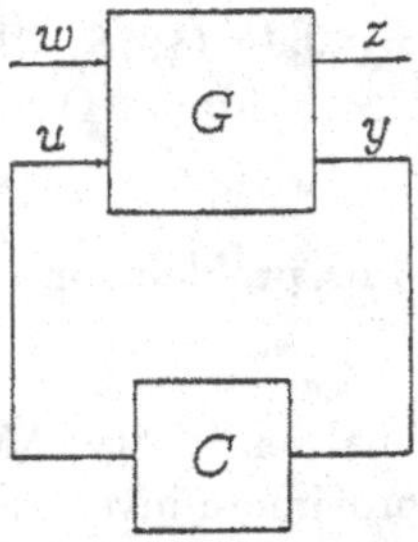

Figure 1

We will call G the plant and C the controller, or compensator. The system inputs are w and u, the exogenous and control inputs respectively, and the outputs are z and y, the regulated and measured outputs.

The object of control system design is to determine C so that z is "small" in the face of uncertainties in the exogenous inputs w and our knowledge of G. We will be more precise in defining these objectives as we proceed.

We will begin by assuming that the only uncertainty is in our knowledge of the exogenous inputs, and that our system is linear and time invariant. In this case, we write

$$\hat{z} = \hat{\Phi}\hat{w} \tag{1.1}$$

where $\hat{\Phi}$ is the transfer function of the closed loop system from input $\hat{w}$ to output $\hat{z}$.

We know that this transfer function can be written in terms of the YJBK parameter $\hat{Q}$ [1, 2] as

$$\hat{\Phi} = \hat{H} - \hat{U}\hat{Q}\hat{V} \tag{1.2}$$

where $\hat{H}, \hat{U}, \hat{V}$ are determined by G only and all stabilizing controllers C are parametrized by $\hat{Q}$.

In these notes, we consider (1.1) to represent a mapping from bounded sequences $\{w_0, w_1, \cdots\}$ to bounded sequences $\{z_0, z_1, \cdots\}$ so that

$$\begin{aligned} \hat{\Phi} &: l^\infty \to l^\infty \\ \hat{\Phi}\hat{w} &= \phi * w \end{aligned} \tag{1.3}$$

where $*$ denotes convolution.

When the sequences are bounded, the system is said to be Bounded-Input-Bounded-Output (BIBO) stable and the transfer function

$$\hat{\Phi}(\lambda) = \sum_{i=0}^{\infty} \phi_i \lambda^i \tag{1.5}$$

satisfies

$$\sum_{i=0}^{\infty} |\phi_i| < \infty, \quad \text{i.e. } \{\phi_i\} \in l^1$$

Lower case letters will represent scalar or vector sequences. Capitals represent matrix sequences. The hat is used to represent the λ-transform of a sequence and hatted capitals represent both scalar and matrix transfer functions. A *transfer function* is the λ-transform of the unit pulse response of a system.

When $\{\phi_i\} \in l^1$, we know that $\hat{\Phi}(\lambda)$ as given by (1.5) represents an analytic function in the disk $|\lambda| < 1$. The functions that are of primary interest to us are the rational functions that are BIBO stable. We will denote by A, the class of all BIBO stable functions, i.e. transforms of l^1 sequences, and by RA, the subclass of all rational functions in A.

We will define the system gain to be

$$\|\hat{\Phi}\|_A := \sup_{\|w\|_\infty \le 1} \|\phi * w\|_\infty \tag{1.6}$$

$$= \sum_{i=0}^{\infty} |\phi_i| =: \|\phi\|_1$$

where

$$\|w\|_\infty := \sup_i |w_i|$$

and our control problem will be to choose the controller $\hat{C}$ so that the system gain is as small as possible.

When $\hat{G}$ is linear, time-invariant and is admissible, i.e. can be stabilized by some $\hat{C}$, then it is known that in (1.2) $\hat{H}$, $\hat{U}$ and $\hat{V}$ are all stable rational functions and a stable rational $\hat{Q}$ parametrizes all stabilizing $\hat{C}$'s. The utility of (1.2) is that $\hat{\Phi}$ is linear-affine in $\hat{Q}$ whereas it is nonlinear in $\hat{C}$. Our control problem can now be stated as

(OPT)

$$\mu_0 = \inf_{\hat{Q} \in RA} \|\hat{H} - \hat{U}\hat{Q}\hat{V}\|_A \tag{1.7}$$

Our motivation for obtaining $\hat{Q} \in RA$ is that the resulting $\hat{C}$ will then be rational and easy to realize as a physical device.

The next section will briefly summarize the necessary mathematics to solve our problem *(OPT)*.

2. Mathematical Details [3]

A *linear functional* is a mapping from a vector space X to the real (complex) numbers $\mathbb{R}$ ($\mathbb{C}$) such that

$$f(\alpha x + \beta y) = \alpha f(x) + \beta f(y)$$

for every $x, y \in X$ and $\alpha, \beta \in \mathbb{R}$ ($\mathbb{C}$).

A linear functional on a normed linear space is *bounded* if there exists a constant M such that

$$|f(x)| \le M\|x\|$$

for every $x \in X$. The smallest such constant M is called the norm of f, i.e.

$$\|f\| = \inf\{M : |f(x)| \le M\|x\| \ \forall x \in X\}$$

It is well known that a linear functional on a normed linear space is bounded if and only if it is continuous.

The linear functionals on a vector space can be regarded as a vector space by defining addition and multiplication by a scalar as

$$f_1(x) + f_2(x) = (f_1 + f_2)(x)$$
$$\alpha f(x) = (\alpha f)(x)$$

The bounded linear functionals on X become a normed space by defining

$$\|f\| = \inf\{M : |f(x)| \le M\|x\| \ \forall x \in X\}$$

$$= \sup_{x \neq 0} \frac{|f(x)|}{\|x\|} = \sup_{\|x\| \leq 1} |f(x)|$$

$$= \sup_{\|x\|=1} |f(x)|$$

This normed space is designated as the *dual space* X^*, and is known to be a Banach space (a complete normed linear space).

The linear functionals on X will be denoted by $f, g, h, \ldots$, or $x_1^*, x_2^*, \ldots$ or by $<x,x^*>$ where the bracket denotes the value of the functional x^* at x, i.e.

$$x^*(x) = <x,x^*>$$

As usual l^p denotes the Banach space of sequences $\{f\}$ with norm

$$\|f\|_p = \left[\sum_{i=0}^{\infty} |f_i|^p \right]^{1/p} < \infty$$

The following theorem characterizes bounded linear functionals on l^p.

Theorem 2.1 [3]

Every bounded linear functional on l^p $1 \leq p < \infty$ is represented uniquely in the form

$$f(x) = \sum_{i=0}^{\infty} \eta_i \xi_i \quad x = \{\xi_i\} \in l^p$$

where

$$y = \{\eta_i\} \in l^q, \text{ and } \frac{1}{p} + \frac{1}{q} = 1.$$

Furthermore, every element of l^q defines a member of $(l^p)^*$ in this way and we have

$$\|f\| = \|y\|_q = \begin{cases} (\sum_{i=0}^{\infty} |\eta_i|^q)^{1/q} & 1 < p < \infty \\ \sup_k |\eta_k| & p = 1 \end{cases}$$

In general

$$<x,x^*> \leq \|x\| \, \|x^*\|$$

A vector $x^* \in X^*$ is *aligned* with a vector $x \in X$ if

$$<x,x^*> = \|x\| \, \|x^*\|$$

The vectors $x \in X$ and $x^* \in X^*$ are *orthogonal* if $<x,x^*> = 0$.

Let $S \subset X$, (a normed linear space). The *orthogonal compliment (annihilator)* of S denoted by $S^\perp$ consists of all $x^* \in X^*$ such that $<x,x^*> = 0 \ \forall \ x \in S$.

The basic theorems that will enable us to solve the control problem are the following:

Theorem 2.2 [3]

Let $x \in X$ (a real normed linear space) and d denote its distance from the subspace $M \subset X$. Then

$$d = \inf_{m \in M} \|x - m\| = \max_{\substack{x^* \in M^\perp \\ \|x^*\| \leq 1}} <x,x^*>$$

where the maximum on the right is achieved for some $x_0^* \in M^\perp$.

If the infimum on the left is achieved by some $m_0 \in M$, then x_0^* is aligned with $x - m_0$.

Theorem 2.3 [3]

Let M be a subspace of a real normed space X. Let $x^* \in X^*$ be a distance d from $M^\perp$. Then

$$d = \min_{m^* \in M^\perp} \|x^* - m^*\| = \sup_{\substack{x \in M \\ \|x\| \le 1}} <x,x^*>$$

where the minimum on the left is achieved for $m_0^* \in M^\perp$. If the supremum on the right is achieved for some $x_0 \in M$, then $x^* - m_0^*$. is aligned with x_0.

The main point of this theorem is to show that the existence of solutions to minimum norm problems is guaranteed if the problem is formulated in a dual space.

In the following, we will need the Smith-McMillan form of a rational matrix. This is defined as follows:

Assume $\hat{H}(\lambda) \in RA_{mn}$ is an mxn matrix with rank r. Then it is known [4] that

$$\hat{H}(\lambda) = \hat{U}_1 \hat{M} \hat{U}_2$$

where $\hat{U}_1$ and $\hat{U}_2$ are unimodular polynomial matrices and

$$\hat{M} = \begin{bmatrix} \varepsilon_1/\psi_1 & & 0 \\ & \varepsilon_r/\psi_r & \\ 0 & & 0 \end{bmatrix}$$

where

$$\text{where } \varepsilon_1 | \varepsilon_2 | \varepsilon_3 \ldots | \varepsilon_r$$

$$\psi_r | \psi_{r-1} | \cdots | \psi_1$$

and (ε_i,ψ_i) are coprime i=1,--,r.

The system poles are the zeros of the ψ_i and the system zeros are the zeros of the ε_i.

Suppose $\hat{H}$ has row rank and no zeros on the unit circle. Then we can write

$$\hat{M} = \begin{bmatrix} \varepsilon_1^+ & & 0 \\ & & \\ 0 & & \varepsilon_m^+ \end{bmatrix} \begin{bmatrix} \varepsilon_1^-/\psi_1 & & & 0 \\ & & & \\ 0 & & \varepsilon_m^-/\psi_m & 0 \end{bmatrix}$$

$$= \hat{M}_1 \hat{M}_2$$

where ε_i^+ has its zeros inside the unit disk and ε_i^- its zeros outside. Since the poles of $\hat{M}$ are outside the disk $\hat{M}_2$ has a right inverse in RA_{nm}, one such being

$$\hat{M}_2^R = \begin{bmatrix} \psi_1/\varepsilon_1^- & 0 \\ & \psi_m/\varepsilon_m^- \\ 0 & 0 \end{bmatrix}$$

If $\hat{H}$ has column rank, then the same factorization can be made where

$$\hat{M} = \hat{M}_3 \hat{M}_4$$

with $\hat{M}_4$ an nxn polynomial matrix with zeros inside the unit disk and $\hat{M}_3$ an mxn matrix with a left inverse in RA_{nm}.

A special case of the Smith-McMillan form is when $\hat{H}(\lambda)$ is a polynomial matrix. In this case $\psi_i = 1 \ \forall i$ and

$$\hat{M} = \begin{bmatrix} \varepsilon_1 & & 0 & \\ & \varepsilon_r & & \\ 0 & & 0 & \end{bmatrix}$$

is called the Smith form of $\hat{H}$.

A final result concerns the adjoint of a linear operator. Let X and Y be normed spaces and T be a bounded linear operator

$$T:X \to Y$$

The *adjoint* of T, devoted by T^* is defined as

$$T^*:Y^* \to X^*$$

Such that

$$<Tx,y^*> = <x,T^*y^*>$$

T^* is a bounded linear operator and $||T|| = ||T^*||$.

Our use of the adjoint will involve the equivalence [5]

$$Ker(T^*) = R(T)^\perp$$

where Ker is kernel (null space) and R is range (image).

3. SISO Problems

Recall that our problem (OPT) was

$$(OPT) \qquad \mu_0 = \inf_{\hat{Q} \in RA} ||\hat{H} - \hat{U}\hat{Q}\hat{V}||_A$$

In order to use our previous results, we first define

$$\hat{K} := \hat{U}\hat{Q}\hat{V}$$

and

$$S := \{ \hat{K} \in RA : \exists\, \hat{Q} \in RA \text{ satisfying } \hat{K} = \hat{U}\hat{Q}\hat{V} \}$$

Anticipating problems involving the existence of optimal solutions, we next define

$$S_A := \{ \hat{K} \in A : \exists\, \hat{Q} \in A \text{ satisfying } \hat{K} = \hat{U}\hat{Q}\hat{V} \}$$

and consider the problem

$$(OPTA) \qquad \mu_A = \inf_{\hat{K} \in S_A} ||\hat{H} - \hat{K}||_A$$

We will show that solutions to $(OPTA)$ in the SISO case are rational so they are also solutions to (OPT).

First, consider the set S_A. In the SISO case, $\hat{U}$, $\hat{Q}$, and $\hat{V}$ are scalars so we can combine $\hat{U}$ and $\hat{V}$ and consider only the equation

$$\hat{K} = \hat{U}\hat{Q} \tag{3.1}$$

In our problems, $\hat{U}$ will always belong to RA and we want to know how to characterize all $\hat{K} \in A$ that produce solutions to (3.1) with $\hat{Q} \in A$.

We will assume for simplicity that $\hat{U}$ has only real simple zeros in the unit disk with no zeros on the unit circle. Then (3,1) will have a solution $\hat{Q} \in A$ if and only if $\hat{K}(a_i) = 0$ where $\hat{U}(a_i) = 0$ for i = 1,2,...,n. The condition $\hat{K}(a_i) = 0$ can be described as follows:

$$\hat{K}(a_i) = \sum_{j=0}^{\infty} k_j (a_i)^j = 0$$

and from Theorem 2.1, we have

$$<k,\underline{a}_i> = 0 \quad i=1,...,n$$

where $k \in l^1$, $\underline{a}_i \in c^0$

$$\underline{a}_i = \{1, a_i, a_i^2, ...\}$$

and c^0 is the subspace of l^∞ consisting of all bounded sequences that approach zero as the index goes to infinity.

Since $(c^0)^* = l^1$, we can formulate our problem (*OPTA*) in a dual space and use Theorem 2.3 to establish existence of an optimal solution as follows.

Define

$$M = \{r \in c^0 \mid r = \sum_{i=1}^{n} \alpha_i \underline{a}_i\}$$

Then

$$M^\perp = \{k \in l^1 \mid <r,k> = 0 \; \forall \; r \in M\}$$

and our problem is

$$(OPTA) \qquad \mu_A = \min_{k \in M^\perp} \|h-k\|_1$$

which has a solution by Theorem 2.3.

Now, how do we find the solution?

Recall that from Theorem 2.3, the minimum μ_A can be found from

$$\mu_A = \max_{\substack{r \in M \\ \|r\| \leq 1}} <r,h>$$

where max replaces "sup" because M is finite dimensional.

Consider

$$\begin{aligned} <r,h> &= <\sum_{i=1}^{n} \alpha_i \underline{a}_i, h> \\ &= \sum_{i=1}^{n} \alpha_i <\underline{a}_i, h> \\ &= \sum_{i=1}^{n} \alpha_i \hat{H}(a_i) \end{aligned}$$

and the constraint $\|r\|_\infty \leq 1$ can be written as

$$\|\sum_{i=1}^{n} \alpha_i \underline{a}_i\|_\infty \leq 1$$

or

$$|\sum_{i=1}^{n} \alpha_i (a_i)^j| \leq 1 \qquad j=0,1,2,...$$

So (*OPTA*) has become a linear programming (LP) problem

$$(OPTA) \qquad \mu_A = \max_{\alpha_i} \sum_{i=1}^{n} \alpha_i \hat{H}(a_i) \tag{3.2}$$

Subject to

$$|\sum_{i=1}^{n} \alpha_i (a_i)^j| \leq 1 \qquad j=0,1,2,...$$

This LP problem has a finite number of variables and an infinite number of constraints, is called semi-infinite, and initially it seems this might cause trouble. This is only apparent however and the trouble disappears due to the fact that $|a_i| < 1, i=1,..,n$ so clearly there will be a value of $j=N$ say, such that if the constraints are satisfied for all $j \leq N$, they will be satisfied for all $j > N$. Therefore our problem is really very simple and we can find the optimal r_0 by solving a finite LP problem.

The next step is to use r_0 to construct the optimal k_0. Recall from Theorem 2.2 that when the minimum is achieved, $h - k_0$ is aligned with r_0, i.e.

$$\mu_A = \langle r_0, h - k_0 \rangle = \langle r_0, h \rangle$$

Define

$$\phi_0 := h - k_0$$

Then

$$\mu_A = \langle r_0, \phi_0 \rangle = \|r_0\|_\infty \|\phi_0\|_1$$
$$= \sum_{i=0}^{\infty} (r_0)_i (\phi_0)_i = \sum_{i=1}^{\infty} |(\phi_0)_i|$$

since $\|r_0\|_\infty = 1$
and alignment implies that the following conditions are satisfied

$$a) \ (\phi_0)_i = 0 \ \text{ when } |(r_0)_i| < 1$$

$$b) \ (\phi_0)_i (r_0)_i \geq 0 \ \ \forall i \tag{3.3}$$

$$c) \ \sum_{i=0}^{\infty} |(\phi_0)_i| = \mu_A$$

In addition, since $\phi_0 = h - k_0$ and $k_0 \in M^{\perp}$, we have

$$\hat{H}(a_i) = \hat{\Phi}_0(a_i) = \sum_{j=0}^{\infty} (\phi_0)_j (a_i)^j \quad i=1,2,..,n \tag{3.4}$$

As noted above, since $r_0 \in c^0$, it follows that $|(r_0)_i| < 1 \ \forall i > N$ so $(\phi_0)_i = 0 \ \forall i > N$ from the first alignment condition and we have discovered that optimal solutions $\hat{\Phi}$ are finite degree polynomials. Since $\hat{H}$ is rational, this means $\hat{K}$ is rational so the solution we have obtained is also a solution to (*OPT*).

We collect the above results into the following.

Theorem 3.1

Given $\hat{H}$ and $\hat{U}$ in *RA* and $\hat{U}$ with n distinct real zeros a_i in the unit disk with no zeros on the unit circle. There exists an optimal solution $\hat{K}_0$ to (*OPT*). Furthermore, $\hat{\Phi}_0 = \hat{H} - \hat{K}_0$ is a finite degree polynomial. $\hat{K}_0$ is constructed as follows.

i) Solve the finite LP problem (3.2) for j=0,1,2,...,N (the determination of N will be discussed below).

ii) Use the conditions (3.3) and (3.4) to determine

$$\hat{\Phi}_0 = \sum_{i=0}^{N} (\phi_0)_i \lambda^i$$

iii) The optimal $\hat{K}_0$ is

$$\hat{K}_0 = \hat{H} - \hat{\Phi}_0$$

Remarks:

1. Once we have $\hat{K}_0$, we can find $\hat{Q}$ from

$$\hat{Q} = \hat{K}_0 / \hat{U} \in RA$$

2. The easiest way to determine N is by trial and error. Begin with $N=n$ and increase it until there is no change in the maximum in (3.2).

4. l^1-Optimization – MIMO Systems – Good Rank

In order to develop the notation for the MIMO case, consider again the system shown in Figure 1. G is described by a real rational transfer function matrix as follows

$$\begin{bmatrix}\hat{z}\\ \hat{y}\end{bmatrix} = \begin{bmatrix}\hat{G}_{11} & \hat{G}_{12}\\ \hat{G}_{21} & \hat{G}_{22}\end{bmatrix}\begin{bmatrix}\hat{w}\\ \hat{u}\end{bmatrix} \tag{4.1}$$

The controller C is represented by a transfer function matrix

$$\hat{u} = \hat{C}\hat{y} \tag{4.2}$$

and the closed loop system is represented by

$$\hat{\Phi} = \hat{H} - \hat{U}\hat{Q}\hat{V} \tag{4.3}$$

where

$$\hat{H} = \hat{G}_{11} + \hat{G}_{12}\hat{M}_2\hat{Y}_1\hat{G}_{21}$$

$$\hat{U} = \hat{G}_{12}\hat{M}_2$$

$$\hat{V} = \hat{M}_1\hat{G}_{21}$$

$$\hat{Q} \in RA$$

and $\hat{G}_{22} = \hat{M}_1^{-1}\hat{N}_1 = \hat{N}_2\hat{M}_2^{-1}$

are arbitrary left and right stable coprime factorizations of $\hat{G}_{22}$ (i.e. coprime over RA). $\hat{Y}_1$ is determined from the Bezout identity.

$$\begin{bmatrix}\hat{X}_1 & -\hat{Y}_1\\ -\hat{N}_1 & \hat{M}_1\end{bmatrix}\begin{bmatrix}\hat{M}_2 & \hat{Y}_2\\ \hat{N}_2 & \hat{X}_2\end{bmatrix} = I$$

We assume that $\hat{G}$ is admissible [6], hence $\hat{H}$, $\hat{U}$, and $\hat{V} \in RA$. The parameter $\hat{Q}$ determines the controller $\hat{C}$ by means of [1]

$$\hat{C} = (\hat{Y}_2 - \hat{M}_2\hat{Q})(\hat{X}_2 - \hat{N}_2\hat{Q})^{-1} \tag{4.4}$$

$$= (\hat{X}_1 - \hat{Q}\hat{N}_1)^{-1}(\hat{Y}_1 - \hat{Q}\hat{M}_1)$$

When $\hat{Q} \in RA$ then $\hat{C} \in RA$ and this is the desired situation in the design problem. When $\hat{Q} \in A$, $\hat{C}$ is then in the fraction field of A and the problem still makes sense in that the closed loop system is stable and all stabilizing controllers are parametrized by (4.4) [7].

In the following we will assume that $\hat{U}$ and $\hat{V}$ have full normal rank, that is, full rank for almost all λ.

Consider the vector sequence

$$f = \begin{bmatrix} f_1(k)\\ \vdots \\ f_n(k)\end{bmatrix} \qquad k = 0,1,2,\ldots$$

When each of the elements is bounded, we define the l_n^∞ norm as

$$\|f\|_\infty = \max_j \sup_k |f_j(k)| \tag{4.5}$$

$$= \max_j \|f_j\|_\infty$$

Assume that $\hat{H}$ is an mxn matrix of transfer functions

$$\hat{H}(\lambda) = \sum_{i=0}^{\infty} H(i)\lambda^i$$

$\hat{H} \in A_{mn}$ will mean that $\hat{H}_{ij}(\lambda) \in A$ and so the sequence of mxn matrices

$$H = \{H(0), H(1),...\} \in l^1_{mn}$$

and $\{H_{ij}(k)\} \in l^1$

We will consider $\hat{H}$ to be a mapping

$$\hat{H} : l^\infty_n \to l^\infty_m$$

$$\hat{H}f = H*f$$

with norm

$$\|\hat{H}\|_A := \sup_{\|f\|_\infty \le 1} \|H*f\|_\infty \tag{4.6}$$

$$= \max_i \sum_{j=1}^{n} \sum_{k=0}^{\infty} |H_{ij}(k)|$$

$$= \max_i \sum_{j=1}^{n} \|H_{ij}\|_1$$

$$= \|H\|_1$$

In the MIMO case, we need the following result.

Theorem 4.1 [8]

1) Every bounded linear functional on l^1_{mn} is represented uniquely by

$$f(H) = \sum_{i=1}^{m} \sum_{j=1}^{n} \sum_{k=0}^{\infty} Y_{ij}(k)\, H_{ij}(k)$$

where $\{Y(k)\} \in l^\infty_{mn}$. Furthermore every element in l^∞_{mn} defines a member of $\left(l^1_{mn}\right)^*$ in this way and we have

$$\|f\| = \sum_{i=1}^{m} \max_j \|Y_{ij}\|_\infty =: \|Y\|_\infty$$

2) $f(H) = \langle H,Y \rangle \le \|Y\|_\infty \|H\|_1$

3) Equality in 2) holds if H is aligned with Y. Then H,Y satisfy the following conditions.

a) $H_{ij}(k) = 0$ when $|Y_{ij}(k)| < \max_j \|Y_{ij}\|_\infty$

b) $H_{ij}(k)Y_{ij}(k) \ge 0$

c) $\sum_{j=1}^{n} \|H_{ij}\|_1 = \text{constant} \ \forall\ i$

d) If $Y_{ij}(k)$ is identically zero for a fixed i, then $H_{ij}(k)$ can be any sequence such that

$$\sum_{j=1}^{n} \|H_{ij}\|_1 \le \text{constant}$$

We are now ready to consider our first MIMO problem which corresponds to the one-block problem in H_∞ theory (we will call this the "good" rank case).

Consider

$$\hat{\Phi} = \hat{H} - \hat{K}$$

where

$$\hat{H} \in RA$$

and

$$\hat{K} \in S_A$$

Therefore

$$\hat{K} = \hat{U}\hat{Q}\hat{V} \tag{4.7}$$

and the one-block problem corresponds to $\hat{U}$ having full row rank and $\hat{V}$ having full column rank. In this case, using the results given above concerning the Smith-McMillan factorization of $\hat{U}$ and $\hat{V}$, we can consider $\hat{U}$ and $\hat{V}$ to be square polynomial matrices with zeros inside the unit disc. For simplicity, we will assume that $\hat{U}$ has N_1 zeros, a_i , and $\hat{V}$ has N_2 zeros b_j that are real and distinct.

The following theorem will characterize solutions of (4.7).

Theorem 4.2 [8]

1) Given $\hat{U}$ and $\hat{V}$ as above; there exist nonzero vectors α_i, β_j such that

$$\alpha_i^T\hat{U}(a_i) = 0 \qquad i=1,2,...,N_1$$

$$\hat{V}(b_j)\beta_j = 0 \qquad j=1,2,...,N_2$$

2) Let $\hat{K} = \hat{U}\hat{Q}\hat{V}$, with $\hat{K} \in A$; then $\hat{Q} \in A$ if and only if

$$\text{i) } \alpha_i^T\hat{K}(a_i) = 0 \qquad i=1,2,...,N_1$$

$$\text{ii) } \hat{K}(b_j)\beta_j = 0 \qquad j=1,2,...,N_2$$

When i) and ii) hold, we say that $\hat{K}$ interpolates $\hat{U}$ and $\hat{V}$.

Proof:

Let $\hat{U} = \hat{S}_1\hat{T}\hat{S}_2$ and $\hat{V} = \hat{R}_1\hat{Y}\hat{R}_2$ where $\hat{S}_1,\hat{S}_2,\hat{R}_1,\hat{R}_2$ are unimodular and $\hat{T}$ and $\hat{Y}$ are the Smith forms of $\hat{U}$ and $\hat{V}$ respectively.

$$\hat{T} = \begin{bmatrix} \sigma_1 & 0 \\ 0 & \sigma_m \end{bmatrix} \qquad \hat{Y} = \begin{bmatrix} \tau_1 & 0 \\ 0 & \tau_n \end{bmatrix}$$

Since the zeros of $\hat{U}$ are distinct, $\sigma_i = 1$ for $i = 1,...,m-1$ and σ_m contains each of the zeros a_i with multiplicity one.

Define

$$\gamma_i^T(\lambda) = (0...\underset{i^{th}entry}{1}\ 0...0)\, \hat{S}_1^{-1} \qquad i=1,...,m$$

This is the ith row of $\hat{S}_1^{-1}$. At each zero of σ_m, it follows that

$$\gamma_m^T(a_i)\ \hat{U}\ (a_i) = 0$$

So define

$$\alpha_i^T := \gamma_m^T(a_i)$$

and we have satisfied the first condition of 1).

The same procedure can be used to define

$$\beta_j := \eta_n(b_j)$$

where $\eta_n(\lambda)$ is the nth column of $\hat{R}_2^{-1}$

Now for part 2). Assume $\hat{K} = \hat{U}\hat{Q}\hat{V}$ with $\hat{K} \in A$
(If) Use the Smith factorizations of part 1 to write

$$\hat{T}(\hat{S}_2\hat{Q}\hat{R}_1)\hat{Y} = \hat{S}_1^{-1}\hat{K}\hat{R}_2^{-1}$$

Let

$$\hat{Q}_1 = \hat{S}_2\hat{Q}\hat{R}_1$$

Clearly $\hat{Q} \in A$ if and only if $\hat{Q}_1 \in A$

Now

$$\hat{S}_1^{-1}\hat{K}\hat{R}_2^{-1} = \begin{bmatrix} \gamma_1^T \\ \vdots \\ \gamma_m^T \end{bmatrix} \hat{K}[\eta_1 .... \eta_n]$$

Since $\hat{Q}_1 = \hat{T}^{-1}(\hat{S}_1^{-1}\hat{K}\hat{R}_2^{-1})\hat{Y}^{-1}$ the (i,j) element can be written as

$$(\hat{Q}_1)_{ij} = \frac{\gamma_i^T\hat{K}\eta_j}{\sigma_i\tau_j}$$

and each element is in A for $i=1,..,m-1$ and $j=1,..,n-1$. When $i=m$, σ_m has N_1 zeros in the unit disk and $\tau_j \neq 0$ at each of these zeros. $\gamma_m^T\hat{K} = 0$ at each of these zeros and this is what is needed to claim $(\hat{Q}_1)_{ij} \in A$ for $i=m$ and $j=1,...,n$. For $j=n$ and $i=1,...,m$, the same argument applies, so $\hat{Q}_1 \in A$.
(Only if) when $\hat{Q} \in A$, $\hat{Q}_1 \in A$ and

$$\sigma_i\tau_j(\hat{Q}_1)_{ij} = \gamma_i^T\hat{K}\eta_j \in A \qquad \forall i,j$$

This means that

$$\alpha_i^T\hat{K}(a_i)\eta_j = 0 \qquad j=1,...,n$$

so

$$\alpha_i^T\hat{K}(a_i) = 0$$

A similar argument establishes

$$\hat{K}(b_j)\beta_j = 0$$

QED

We next show how the interpolation conditions can be used to define the space of admissible K's.

Define

$$F(l,r) = \begin{bmatrix} 0 & \alpha_{r1}\,\underline{a}_r & 0 \\ \vdots & \vdots & \vdots \\ 0 & \alpha_{rm}\,\underline{a}_r & 0 \end{bmatrix} \qquad \begin{matrix} l = 1,...,n \\ r = 1,...,N_1 \end{matrix}$$

l^{th} column

and

$$J(i,j) = \begin{bmatrix} 0 & .... & 0 \\ \vdots & & \vdots \\ \beta_{j1}\,\underline{b}_j & .... & \beta_{jn}\,\underline{b}_j \\ \vdots & & \vdots \\ 0 & .... & 0 \end{bmatrix} \qquad \begin{matrix} i = 1,...,m \\ j = 1,...,N_2 \end{matrix} \tag{4.8}$$

(i^{th} row $\rightarrow$)

where

$$\alpha_r^T = [\alpha_{r1} \cdots \alpha_{rm}] \quad \beta_j = \begin{bmatrix} \beta_{j1} \\ \vdots \\ \beta_{jn} \end{bmatrix}$$

$$\underline{a}_r = (1, a_r, a_r^2, \ldots\ldots)$$

$$\underline{b}_j = (1, b_j, b_j^2, \ldots\ldots)$$

Then by straightforward calculation it follows that

$$\alpha_r^T \hat{K}(a_r) = 0$$

is equivalent to

$$< K, F(l,r) > = 0 \qquad l=1,\ldots,n$$

and

$$\hat{K}(b_j)\beta_j = 0$$

is equivalent to

$$< K, J(i,j) > = 0 \qquad i=1,\ldots,m$$

From (4.8), notice that the matrix sequences are in c_{mn}^0, so we will formulate our problem in $l_{mn}^1 = (c_{mn}^0)^*$ to guarantee the existence of a solution.

Define

$$M = \left\{ G \in c_{mn}^0 \mid G \subset Span\{F(l,r) J(i,j)\} \right\}$$

Then

$$S_A = M^\perp = \left\{ K \in l_{mn}^1 \mid <F(l,r),K> = 0 \text{ and } <J(i,j),K> = 0 \text{ for all } i,j,l,r \right\}$$

and we are guaranteed that our problem has a solution. In fact, as in the SISO case, our problem has a solution $\hat{K} \in RA$. We establish this as follows:

Theorem 4.3 [9]

Assume $\hat{U}$ and $\hat{V}$ have no zeros on the unit circle, and $\hat{H} \in RA$. Then the problem (OPT) has a solution $\hat{K}_0$. Moreover $\hat{\Phi}_0 = \hat{H} - \hat{K}_0$ is a polynomial matrix.

Proof: Existence was established by constructing $M^\perp = S_A$. We use the alignment condition of Theorem 2.3 to show that for any minimizer $\hat{K}_0$ of $(OPTA)$, $\hat{\Phi}_0$ has at least one row which is polynomial. Next we show that given any minimizer of $(OPTA)$ for which p rows of the corresponding Φ are polynomial (where $p < m$), there exists another minimizer for which at least $p+1$ rows of the corresponding Φ are polynomial. Hence there is at least one minimizer $\hat{K}_0$ for which all rows of $\hat{\Phi}_0$ are polynomial.

First, assume $K_0 \in l_{mn}^1$ is a minimizer for $(OPTA)$ and $G_0 \in c_{mn}^0$ is a maximizer of the dual problem. Then by Theorem 2.3, Φ_0 and G_0 are aligned. If G_0 is zero, then $\mu_A = 0$ and $K_0 = H$ is a minimizer for (OPT) and Φ_0 is a polynomial. If G_0 is nonzero, there is a row, say the ith, such that $\max_j \|G_{0ij}\|_\infty > 0$. Since $G_0 \in c_{mn}^0$ there exists N such that

$$| G_{0ij}(k) | < \max_j \|G_{0ij}\|_\infty$$

$\forall j$ and $k > N$, and hence the ith row of Φ_0 is a polynomial.

Now, suppose $p < m$ and K_0 is any minimizer of $(OPTA)$ such that the first p rows of Φ are polynomial. Partition after the pth row as follows.

$$\Phi_0 = \begin{bmatrix}\Phi_{01}\\ \Phi_{02}\end{bmatrix} \quad H = \begin{bmatrix}H_1\\ H_2\end{bmatrix} \quad K_0 = \begin{bmatrix}K_{01}\\ K_{02}\end{bmatrix}$$

and consider the problem

$$\inf_{R\in S_2} \|(H_2 - K_{02}) - R\|_1$$

where

$$S_2 = \{R \in l^1_{(m-p)n} \mid [K_{01}^T \ (K_{02}+R)^T]^T \in S_A\}$$

Clearly if R_0 is any minimizer for this problem $[K_{01}^T \ (K_{02}+R_0)^T]^T$ is a minimizer for $(OPTA)$. Moreover, applying Theorem 4.2, we see that $R \in S_2$, if and only if

$$\alpha_{i2}^T R = 0 \qquad i=1,\dots,N_1$$

where α_{i2}^T denotes the last $m-p$ entries of α_i^T. Thus the same argument given above implies the existence of a minimizer R_0 such that $H_2 - \hat{K}_{02} - \hat{R}_0$ has at least one polynomial row. Hence $[K_{01}^T \ (K_{02}+R_0)^T]^T$ is a minimizer for $(OPTA)$ such that Φ has at least $p+1$ polynomial rows. So $\hat{\Phi}_0$ is a polynomial and since $\hat{H} \in RA$, $\hat{K}_0 \in RA$ solves (OPT).

QED

Now, how do we construct solutions in this case? We consider the primal problem instead of the dual problem as follows:

Recall that our problem is

$$(OPT) \qquad \min_{K\in M^\perp} \|H-K\|_1 = \min_{K\in M^\perp} \|\Phi\|_1 = \mu_0$$

where

$$\|\Phi\|_1 = \max_i \sum_{j=1}^{n} \|\Phi_{ij}\|_1$$

and

$$\|\Phi_{ij}\|_1 = \sum_{k=0}^{\infty} |\Phi_{ij}(k)|$$

Now $K \in M^\perp$ implies

$$\alpha_i^T \hat{K}(a_i) = 0 \qquad i=1,\dots,N_1$$

$$\hat{K}(b_j)\beta_j = 0 \qquad j=1,\dots,N_2$$

Since $\hat{K} = \hat{H} - \hat{\Phi}$, we can write

$$\alpha_i^T\hat{\Phi}(a_i) = \alpha_i^T \sum_{k=0}^{\infty} \Phi(k)(a_i)^k = \alpha_i^T \hat{H}(a_i)$$

and

$$\hat{\Phi}(b_j)\beta_j = \left[\sum_{k=0}^{\infty}\Phi(k)(b_j)^k\right]\beta_j = \hat{H}(b_j)\beta_j$$

Theorem 4.3 establishes the fact that the optimal $\hat{\Phi}$ is a finite degree polynomial, so we can write

$$\|\Phi_{ij}\|_1 = \sum_{k=0}^{N} |\Phi_{ij}(k)| \quad \forall i,j$$

and our problem can be stated as

$$\mu_0 = \min \gamma (= \max_i \sum_{k=0}^{N} |\Phi_{ij}(k)|)$$

such that

$$\sum_{j=1}^{n}\sum_{k=0}^{N} |\Phi_{ij}(k)| \le \gamma \qquad i=1,\ldots,m$$

and

$$\alpha_i^T \sum_{k=0}^{N} \Phi(k)(a_i)^k = \alpha_i^T \hat{H}(a_i) \qquad i=1,..,N_1$$

$$\left[\sum_{k=0}^{N} \Phi(k)(b_j)^k\right]\beta_j = \hat{H}(b_j)\beta_j \qquad j=1,\ldots,N_2$$

Using the standard methods for setting up LP problems [10], we define

$$\Phi_{ij}(k) = \Phi_{ij}^{+}(k) - \Phi_{ij}^{-}(k)$$

where

$$\Phi_{ij}^{+}(k) \ge 0, \Phi_{ij}^{-}(k) \ge 0$$

Then our problem becomes the following LP problem.

Theorem 4.4

A solution to (OPT) can be found from the solution of the LP problem

$$\mu_0 = min\ \gamma$$

such that

$$\sum_{j=1}^{n}\sum_{R=0}^{N} (\Phi_{ij}^{+}(k) + \Phi_{ij}^{-}(k)) \le \gamma \quad i=1,..,m$$

and

$$\alpha_i^T \sum_{k=0}^{N} (\Phi^{+}(k) - \Phi^{-}(k))(a_i)^k = \alpha_i^T \hat{H}(a_i)$$

$$\left[\sum_{k=0}^{N} (\Phi^{+}(k) - \Phi^{-}(k))(b_j)^k\right]\beta_j = \hat{H}(b_j)\beta_j$$

$$\Phi_{ij}^{+}(k) \ge 0,\ \Phi_{ij}^{-}(k) \ge 0,\ \gamma \ge 0$$

Remark: An alternate formulation of the LP problem in the good rank case was given by Mendlovitz [11].

5. MIMO Systems – The General Case (Bad Rank)

In this case $\hat{U}$ and $\hat{V}$ no longer have full row and full column rank respectively. We will consider the case where $\hat{U}$ has full column rank and $\hat{V}$ has full row rank. This corresponds to the 4-block problem and is the most general case. The 2-block problems occur when both $\hat{U}$ and $\hat{V}$ have full column rank or full row rank and are special cases of the 4-block problem. We refer to these cases as "bad" rank.

In this section, we will not distinguish between H and $\hat{H}$ in order to avoid becoming buried in notation. It will be clear from the context when we are discussing transfer functions $\hat{H}$ or sequences H.

Refer to (4.1) and (4.3) in order to determine the dimensions of the transfer function matrices.

Let n_z, n_y, n_w and n_u be the dimensions of the output and input vectors respectively. Then U is $n_z \times n_u$ with rank n_u, V is $n_y \times n_w$ with rank n_y and H is $n_z \times n_w$.

We make the additional assumption

Assumption 5.1. There exists n_u rows of U and n_w columns of V which are linearly independent for all λ on the unit circle.

This is slightly stronger than assuming U and V have no zeros on the unit circle. For example

$$U = \begin{bmatrix} \lambda(\lambda+1) \\ (1-\lambda)^2 \\ \lambda(1-\lambda) \end{bmatrix}$$

has no zeros on the unit circle, yet does not satisfy assumption 5.1. This assumption is imposed in order to guarantee the existence of solutions to the 4-block problem.

Now, it is possible to reorder the rows of U and the columns of V if necessary in order to insure that

$$U = \begin{bmatrix} U_1 \\ U_2 \end{bmatrix}$$

$$V = [V_1 \; V_2]$$

with U_1 $n_u \times n_u$ and invertible, V_1 $n_y \times n_y$ and invertible and neither has zeros on the unit circle.

Then we can write

$$K = UQV = \begin{bmatrix} U_1 \\ U_2 \end{bmatrix} Q[V_1 \; V_2] \tag{5.1}$$

$$= \begin{bmatrix} K_{11} & K_{12} \\ K_{21} & K_{22} \end{bmatrix}$$

Define polynomial coprime factorizations as follows.

$$\begin{aligned} U_2U_1^{-1} &= \tilde{D}_U^{-1}\tilde{N}_U \\ V_1^{-1}V_2 &= N_V D_V^{-1} \end{aligned} \tag{5.2}$$

We now characterize the feasible set S_A for our problem as follows:

Theorem 5.2 [9]

Given U and V as above and $K \in A$, there exists $Q \in A$ satisfying $K = UQV$ if and only if

i) $[-\tilde{N}_U \; \tilde{D}_U]\, K = 0$

ii) $[K_{11} \; K_{12}] \begin{bmatrix} -N_V \\ D_V \end{bmatrix} = 0$

iii) K_{11} interpolates U_1 and V_1

Proof: Assume $K = UQV$ with $K \in A$ and $Q \in A$. Then $K_{11} = U_1QV_1$ which is the good rank case, so *iii)* is established by Theorem 4.2.

From (5.1)

$$K_{21} = U_2QV_1$$

$$K_{12} = U_1QV_2$$

$$K_{22} = U_2QV_2$$

and

$$Q = U_1^{-1}K_{11}V_1^{-1}$$

Then from (5.2)

$$K_{12} = K_{11}V_1^{-1}V_2 = K_{11}N_V D_V^{-1}$$

So

$$K_{12}D_V = K_{11}N_V$$

which is *ii*).

The same manipulation will yield *i*) as well.

On the other hand, assume $K \in A$ and *i*), *ii*), *iii*) are true. From *iii*)

$$K_{11} = U_1 Q V_1$$

and $Q \in A$ from Theorem 4.2.

Reversing the calculation procedure used above and using (5.2) results in

$$K_{12} = U_1 Q V_2$$

$$K_{21} = U_2 Q V_1$$

$$K_{22} = U_2 Q V_2$$

so

$$K = UQV \text{ with } Q \in A.$$

QED

We now translate these conditions into a dual space in order to consider the existence question. Clearly $S_A = S_i \cap S_{ii} \cap S_{iii}$ where each subspace satisfies the appropriate condition in Theorem 5.2. We will characterize each of these subspaces in the following:

S_{iii} is characterized in terms of the sequences of (4.8) for the 1-block problem.

If we define

$$M_3 = \begin{bmatrix} span\{F,J\} & 0 \\ 0 & 0 \end{bmatrix} \in c^0_{n_z n_w}$$

Then

$$M_3^{\perp} = S_{iii} = \left\{ K \in l^1_{n_z n_w} \mid \langle M,K \rangle = 0 \; \forall \, M \in M_3 \right\}$$

Note that this will take care of the K_{ij} where $i = 1,\dots,n_u$ and $j = 1,\dots,n_y$

Now consider S_i which is defined by

$$[-\tilde{N}_U \, \tilde{D}_U] \, K = 0$$

Define the linear operator

$$F : l^1_{n_z n_w} \to l^1_{(n_z - n_u) n_w}$$

as

$$FK := T^*K$$

where

$$\hat{T} = [-\tilde{N}_U \tilde{D}_U]$$

Then

$$S_i = Ker\, F = \{K \in l^1_{n_z n_w} \mid T^*K = 0\}$$

Now define a bounded linear operator $G : c^0_{(n_z - n_u) n_w} \to c^0_{n_z n_w}$ as follows:

Given $\alpha \in c^0_{(n_z - n_u) n_w}$

$$(G\alpha)_{ij}(k) := \sum_{l=0}^{\infty} \sum_{m=1}^{n_z - n_u} T^T_{im}(l-k)\alpha_{mj}(l) \qquad \begin{matrix} i=1,\dots,n_z \\ j=1,\dots,n_w \\ k=0,1,2,\dots \end{matrix}$$

Then $F=G^*$ since for every $\alpha \in c^0_{(n_z-n_u)n_w}$ and $K \in l^1_{n_z n_w}$ it follows that

$$<G\alpha,K> = \sum_{i=1}^{n_z}\sum_{j=1}^{n_w}\sum_{k=0}^{\infty} K_{ij}(k)\left[\sum_{l=0}^{\infty}\sum_{m=1}^{n_z-n_u} T^T_{im}(l-k)\alpha_{mj}(l)\right]$$

$$= \sum_{m=1}^{n_z-n_u}\sum_{j=1}^{n_w}\sum_{l=0}^{\infty} \alpha_{mj}(l)\left[\sum_{i=1}^{n_z}\sum_{k=0}^{\infty} K_{ij}(k)T_{mi}(l-k)\right]$$

$$= <\alpha,FK>$$

From our previous discussion of the adjoint, recall that

$$Ker(G^*) = R(G)^{\perp} = S_i$$

Define $M_1 := R(G) \subset c^0_{n_z n_w}$
Then

$$M_1^{\perp} = S_i$$

and the same procedure can be used to define M_2 such that

$$M_2^{\perp} = S_{ii}$$

Then with $M = M_1 + M_2 + M_3$, it follows that $M^{\perp} = S_i \cap S_{ii} \cap S_{iii} = S_A$ and again we are guaranteed a solution to $(OPTA)$.

We summarize this as follows:

Theorem 5.3

Given U and V as above with assumption 5.1, the problem $(OPTA)$ has a solution K_0.

We now turn to the problem of constructing solutions.

6. Truncated Problems

In this section, we will be constructing approximate solutions in the bad rank case, so it is possible to drop our previous assumption of no zeros on the unit circle and deal with $K \in RA$. In this case our previous results that characterized admissible $K \in A$, Theorems 4.2 and 5.2, are still valid when U or V have zeros on the unit circle. We still assume, for simplicity, that the zeros are real and distinct. The general case is treated in [9].

Our problem is

$$(OPT) \qquad \mu_o = inf\left[\max_i \sum_{j=1}^{n_w}\sum_{k=0}^{\infty} |\Phi_{ij}(k)|\right]$$

Subject to

$$H - \Phi \in S_A$$

which we write as

$$(LP) \qquad \mu_0 = inf\, \gamma$$

Subject to

$$\Phi^+_{ij}(k) - \Phi^-_{ij}(k) = \Phi_{ij}(k) \tag{6.1}$$

$$\Phi^+_{ij}(k) \geq 0,\ \Phi^-_{ij}(k) \geq 0,\ \ \gamma \geq 0 \qquad \begin{matrix} i=1,\ldots,n_z \\ j=1,\ldots,n_w \\ k=0,1,\ldots \end{matrix}$$

$$\sum_{j=1}^{n_w}\sum_{k=0}^{\infty}(\Phi_{ij}^{+}(k)+\Phi_{ij}^{-}(k))\le\gamma \qquad i=1,\ldots,n_z$$

and

$$H-\Phi\in S_A$$

This is an infinite LP problem in that there are infinitely many variables and constraints and cannot be solved using standard LP techniques.

We therefore define a problem with a finite number of variables which we call a *truncated* problem as follows:

(OPT_δ) $$\mu_\delta=\min_{K\in S_\delta}\|H-K\|_A$$

where

$$S_\delta=\{K\in S: H-K \text{ is a polynomial of degree } \le\delta\}$$

Then (LP_δ) is the same as (LP) above except that $k=0,1,\ldots,\delta$ and $H-\Phi\in S_\delta$

Clearly S_δ may be empty, although it is not empty in the good rank case. In order to avoid this problem, temporarily, we assume

Assumption 6.1: There exists δ^* such that S_{δ^*} is not empty, i.e. (OPT) has a feasible point for which $H-K=\Phi$ is a polynomial of degree δ^*.

When this holds, we can construct a sequence $\{\delta_i\}$ such that with $\delta_0\ge\delta^*$, the corresponding sequence of problems $\{(OPT_{\delta i})\}$ has well defined norms $\mu_{\delta i}$ and $\mu_{\delta i}\to\mu_0$ as $i\to\infty$.

Theorem 6.2 [9]

Assume 6.1. $|\mu_\delta-\mu_0|$ can be made arbitrarily small by taking δ sufficiently large.

Proof: For $K_{\delta^*}\in S_{\delta^*}$, write

$$\|H-K\|_A=\|H-K_{\delta^*}+K_{\delta^*}-K\|_A$$

$$=\|\Phi_{\delta^*}-(K-K_{\delta^*})\|_A$$

Since $K-K_{\delta^*}\in S$ if and only if $K\in S$, we have

$$\mu_0=\inf_{K\in S}\|H-K\|_A=\inf_{K\in S}\|\Phi_{\delta^*}-K\|_A$$

Therefore, we can find $K'\in RA$, with $K'=UQ'V$ such that, for any $\varepsilon>0$

$$\|\Phi_{\delta^*}-K'\|_A=\mu_0+\varepsilon$$

Since U and V are polynomials and the polynomials are dense in RA, we can approximate K' as closely as desired by a polynomial

$$K_p=UQ_pV$$

Then

$$\Phi_p=\Phi_{\delta^*}-UQ_pV$$

is a polynomial whose norm is arbitrarily close to μ_0.

With $\delta=$ *degree* Φ_p, we have the result. QED

In the good rank case, when there are no zeros on the unit circle, we know that $\delta^*=\text{degree }\Phi_0$ where Φ_0 is an optimal solution. Even if there is a zero on the unit circle, 6.1 is still true in the good rank case. This is stated as follows.

Lemma 6.3 [9] Given U and V with good rank, there exists δ^* such that S_{δ^*} is not empty.

Proof: Consider $\Phi = H - UQV$ and use the Smith decomposition to write

$$\Phi = H - S_1 T S_2 Q R_1 Y R_2$$

and

$$S_1^{-1}\Phi R_2^{-1} = S_1^{-1} H R_2^{-1} - T S_2 Q R_1 Y$$

or

$$\Phi' = H' - U'QV'$$

Now reconstruct the interpolation conditions for the good rank case.

$$U' = TS_2 = \begin{bmatrix} 1 & 0 \\ 0 & \sigma_m \end{bmatrix} S_2$$

where

$$\sigma_m = \prod_{i=1}^{N_1} (\lambda - a_i)$$

$$V' = R_1 Y = R_1 \begin{bmatrix} 1 & 0 \\ 0 & \tau_n \end{bmatrix}$$

$$\tau_n = \prod_{j=1}^{N_2} (\lambda - b_j)$$

So

$$\alpha_i^T U'(a_i) = 0$$

and

$$V'(b_j)\beta j = 0$$

yields

$$\alpha_i^T = (0,\ldots,1)$$

$$\beta_j = \begin{bmatrix} 0 \\ \vdots \\ 1 \end{bmatrix}$$

and the interpolation conditions are

$$\alpha_i^T K'(a_i) = 0$$

$$K'(b_j)\beta_j = 0$$

which result in

$$K'_{mj}(a_i) = 0 \qquad \begin{matrix} j=1,\ldots n \\ i=1,\ldots,N_1 \end{matrix}$$

and (6.2)

$$K'_{in}(b_j) = 0 \qquad \begin{matrix} i=1,\ldots,m \\ j=1,\ldots,N_2 \end{matrix}$$

Notice that this involves only the last row and last column of K'.

Define a polynomial matrix $\Phi'(\lambda)$ as follows

$$\Phi'_{ij}(\lambda) := H'_{ij}(\lambda) - K'_{ij}(\lambda)$$

such that the conditions (6.2) are satisfied. The unspecified entries of Φ' are arbitrary.

Then K' interpolates U' and V' so there exists $Q \in RA$ such that

$$\Phi' = H' - U'QV'$$

is a polynomial, and so is

$$\Phi = S_1 \Phi' R_2.$$

Thus

$$\delta^* = degree \text{ of } \Phi$$

QED

This allows us to construct approximate solutions to good rank problems when there is a zero on the unit circle by solving a sequence of problems, each with increasing δ. Now when does 6.1 hold in the bad rank case? Refer to (5.1) and (5.2). Partition H conformally with K as

$$H = \begin{bmatrix} H_{11} & H_{12} \\ H_{21} & H_{22} \end{bmatrix}$$

Then we have

Theorem 6.4 [9]

Given U and V with bad rank. 6.1 is satisfied if and only if

$$T_{UH} := (-\tilde{N}_U \, \tilde{D}_U) \begin{bmatrix} H_{11} & H_{12} \\ H_{21} & H_{22} \end{bmatrix}$$

and

$$T_{HV} := \begin{bmatrix} H_{11} & H_{12} \\ H_{21} & H_{22} \end{bmatrix} \begin{bmatrix} -N_V \\ D_V \end{bmatrix}$$

are polynomials.

Proof: Assume 6.1 holds, and Φ is a polynomial. Since $\Phi = H - K$ and

$$[-\tilde{N}_U \, \tilde{D}_U]\, K = 0, \qquad K \begin{bmatrix} -N_V \\ D_V \end{bmatrix} = 0$$

it is clear that T_{UH} and T_{HV} are polynomials since $[-\tilde{N}_U \tilde{D}_U]$ and $[-N_V^T \, D_V^T]^T$ are polynomials.

Now assume that T_{UH} and T_{HV} are polynomials and recall that from (5.2)

$$U_2 U_1^{-1} = \tilde{D}_U^{-1} \tilde{N}_U$$

$$V_1^{-1} V_2 = N_V D_V^{-1}$$

Bring in right and left coprime polynomial factorizations

$$\tilde{D}_U^{-1} \tilde{N}_U = N_U D_U^{-1}$$

$$N_V D_V^{-1} = \tilde{D}_V^{-1} \tilde{N}_V$$

and construct the double Bezout identity in each case.

$$\begin{bmatrix} \tilde{X}_U & -\tilde{Y}_U \\ -\tilde{N}_U & \tilde{D}_U \end{bmatrix} \begin{bmatrix} D_U & Y_U \\ N_U & X_U \end{bmatrix} = I$$

$$\begin{bmatrix} \tilde{D}_V & \tilde{N}_V \\ \tilde{Y}_V & \tilde{X}_V \end{bmatrix} \begin{bmatrix} X_V & -N_V \\ -Y_V & D_V \end{bmatrix} = I$$

Each block entry is a polynomial matrix.

Define

$$B_U := \begin{bmatrix} \tilde{X}_U & -\tilde{Y}_U \\ -\tilde{N}_U & \tilde{D}_U \end{bmatrix}$$

$$B_V := \begin{bmatrix} X_V & -N_V \\ -Y_V & D_V \end{bmatrix}$$

Note that B_U^{-1} and B_V^{-1} are polynomial matrices.

Now consider (OPT') defined by

$$H' = B_U H B_V \qquad U' = B_U U \qquad V' = V B_V$$

Then

$$U' = B_U \begin{bmatrix} U_1 \\ U_2 \end{bmatrix} = \begin{bmatrix} U'_1 \\ 0 \end{bmatrix}$$

and

$$V' = [V_1 V_2] \ B_V = [V'_1 0]$$

where U'_1 and V'_1 are square and invertible. Now

$$H' = B_U H B_V = \begin{bmatrix} H'_{11} & H'_{12} \\ H'_{21} & H'_{22} \end{bmatrix}$$

where

$$[H'_{21} \ H'_{22}] = T_{UH} \, B_V$$

and

$$\begin{bmatrix} H'_{12} \\ H'_{22} \end{bmatrix} = B_U \, T_{HV}$$

and these are polynomials, by assumption.

Clearly H'_{11}, U'_1, V'_1 define a good rank subproblem which has a polynomial feasible point, i.e. $\Phi'_{11} = H'_{11} - U'_1 Q V'_1$ is polynomial. Then

$$K' = \begin{bmatrix} K'_{11} & 0 \\ 0 & 0 \end{bmatrix}$$

is a feasible point for (OPT') for which

$$\Phi' = \begin{bmatrix} \Phi'_{11} & H'_{12} \\ H'_{21} & H'_{22} \end{bmatrix}$$

is a polynomial.

Finally $K = B_U^{-1} K' B_V^{-1} = UQV$ is a feasible point of (OPT) for which

$$\Phi = H - UQV = B_U^{-1} \Phi' B_V^{-1}$$

is polynomial. QED

A remaining question is, what happens when T_{UH} and T_{HV} are not polynomials?

In this case, we can approximate H by a polynomial H_P in such a way that we can get arbitrarily close to μ_0.

Since the polynomials are dense in RA, given any $\varepsilon > 0$, we can choose H_P such that $\|H - H_P\|_A < \varepsilon$.

Now solve the problem

$$\mu_a = \inf \|H_P - K\|_A$$

Then it is easy to show that $\mu_0 < \mu_a + \varepsilon$ and $\mu_a < \varepsilon + \mu_0$ so $|\mu_0 - \mu_a| < \varepsilon$.

7. Constrained Problems

Suppose that in a particular case of (OPT) we achieved a small value of $\|z\|_\infty$, but $\|u\|_\infty$ is unacceptably large. A natural objective in this case is to reformulate (OPT) so that we want to minimize the maximum $\|z\|_\infty$ subject to the constraint that the maximum $\|u\|_\infty$ remains within a certain bound.

In H_∞ theory, this problem is quite difficult, but we will see that when the induced norm is the l^1 norm, the problem becomes quite simple and our previous method of constructing approximate solutions can be used.

Using our previous notation, we can establish the following:

Refer to Figure 1. The transfer function from w to z can be written as

$$\Phi_{zw} = H_1 - U_1 QV$$

and that from w to u as

$$\Phi_{uw} = H_2 - U_2 QV$$

where

$$H = \begin{bmatrix} H_1 \\ H_2 \end{bmatrix} \quad U = \begin{bmatrix} U_1 \\ U_2 \end{bmatrix}$$

Now suppose we want to minimize $\|\Phi_{zw}\|_A$ subject to $\|\Phi_{uw}\|_A \le 1$.

We set up the problem

(OPTC) $\quad \mu_C := \inf_{K \in S_C} \|\Phi_{zw}\|_A$

where

$$S_C = \{K \in RA : \exists\, Q \in RA \text{ satisfying } K = UQV \text{ and } \|\Phi_{uw}\|_A \le 1\}$$

Of course, this problem may not have a solution since the constraint set may be empty, i.e. there may be no controller that will satisfy $\|\Phi_{uw}\|_A \le 1$. The constraint set will be non-empty if and only if the standard problem defined by H_2, U_2, V satisfies

$$\inf \|H_2 - U_2 QV\|_A \le 1$$

When this is true, we can use the iteration procedure previously set up in order to construct arbitrarily good suboptimal solutions.

Let's set up the problem as follows:

$$(OPTC_\delta) \quad \mu_{C\delta} := \inf_{K \in S_{C\delta}} \|H_1 - K_1\|_A$$

where

$$S_{C\delta} = \{K \in S_C : H - K \text{ is a polynomial of } degree \le \delta\}$$

and

$$H - K = \begin{bmatrix} H_1 \\ H_2 \end{bmatrix} - \begin{bmatrix} K_1 \\ K_2 \end{bmatrix}$$

If (OPT) satisfies 6.1 and

$$\mu_2 := \inf \|H_2 - K_2\|_A < 1$$

then we can show that $(OPTC_\delta)$ has a polynomial feasible point for sufficiently large δ and we can construct a sequence of norms which approach μ_C. We show this as follows:

Assume (OPT) satisfies 6.1 and there exists K_2 such that $\|H_2 - K_2\|_A < 1$.

Let

$$\Phi_{\delta *} = H - K_{\delta *}$$

be the polynomial feasible point and consider (as in the proof of Theorem 6.2)

$$\|H_1 - K_1\| = \|(H_1 - K_{1\delta*}) - (K_1 - K_{1\delta*})\| = \|\Phi_{1\delta*} - \bar{K}_1\| \ .$$

Since $K_1 \in S_C$ if and only if $(K_1 - K_{1\delta*}) \in S_C$

$$\mu_C = \inf_{K \in S_C} \|H_1 - K_1\| = \inf_{K \in S_C} \|\Phi_{1\delta*} - K_1\|$$

Now, we can show that there exists

$$K' = \begin{bmatrix} K'_1 \\ K'_2 \end{bmatrix} \in S_C$$

such that

$$\|H_1 - K'_1\| \simeq \mu_C$$

and

$$\|H_2 - K'_2\| < 1$$

Therefore, we can approximate

$$\begin{bmatrix} K'_1 \\ K'_2 \end{bmatrix} \cong \begin{bmatrix} U_1 \\ U_2 \end{bmatrix} Q_P V$$

where Q_P is a polynomial.

Then

$$\Phi_{1P} = \Phi_{1\delta*} - U_1 Q_P V$$

is a polynomial whose norm can be made arbitrarily close to μ_C and

$$\Phi_{2P} = \Phi_{2\delta*} - U_2 Q_P V$$

has norm less than 1.

The LP problem to be solved here is the same as (6.1) with additional constraints as follows.

$$\sum_{j=1}^{n_w} \sum_{k=0}^{\delta} (\Phi_{ij}^{+}(k) + \Phi_{ij}^{-}(k)) \leq 1$$

where i takes on values corresponding to the constrained rows of Φ. The right hand side of the above inequality is γ for those rows corresponding to the regulated variables.

The search for δ is straightforward. Start with a value δ_0 and increase it until the LP solver finds a feasible point. If a feasible point can't be found, then the constraint is too tight and should be relaxed.

8. Robust Stability and Performance

So far, our discussion has centered on systems in which the only unknowns are the exogenous signals. This is only a portion of the design problem, and in fact a minor portion. Plant uncertainty and the ability to treat it effectively is the major problem in the design of control systems. To my knowledge, the 1963 text by Horowitz [12] was the only attempt to treat plant uncertainty until the recent text by Morari and Zafiriou [13]. All of the text books that have been used for the past 45 years have failed to treat the problem of plant uncertainty effectively and it is only recently that methods have been developed that seem to offer some hope in these difficult problems.

The problems can be stated briefly as follows: First the problem of robust stability is concerned with guaranteeing that the system will remain stable regardless of our uncertainty in the plant description.

This problem, at least, was briefly encountered in the past and the methods used to treat it were gain margin, phase margin, and M-peak. These methods were effective for single loop scalar systems and are major reasons for the popularity of Nyquist and Bode diagrams and the so-called loop shaping design methods as typified by the classical text of Bower and Schultheiss [14].

These methods left unresolved the main design problem. Specifications must be met regardless of plant uncertainty. Stability is certainly necessary but not sufficient for satisfaction of specifications. In the classical texts, specifications were usually so vaguely stated that it was never clear what the design problem was all about. Bower and Schultheiss was a notable exception, but still an adequate statement of the problem was never made and most results were ad hoc.

It is clear that in order to make some progress on such problems, we must first decide on an effective way to describe plant uncertainty. This is a very difficult practical problem and any description is subject to criticism.

The description that we adopt here is to view an uncertain plant as a nominal plant (i.e. known) plus unknown, but norm bounded perturbations. For example, we will consider

$$\Pi_a = \{P = P_0 + \Delta : \Delta \text{ causal}, \|\Delta\| \leq 1\}$$

$$\Pi_m = \{P = (I+\Delta)P_0 : \Delta \text{ causal}, \|\Delta\| \leq 1\}$$

as a nominal plant perturbed by additive and multiplicative perturbations. There are many other ways to describe classes of uncertain plants, but for simplicity we will consider only these two.

The first problem of interest is that of robust stability and we have the following results.

Assume that Δ is a linear, time-varying, strictly causal operator mapping $l^\infty \to l^\infty$, and consider the system shown in Figure 2.

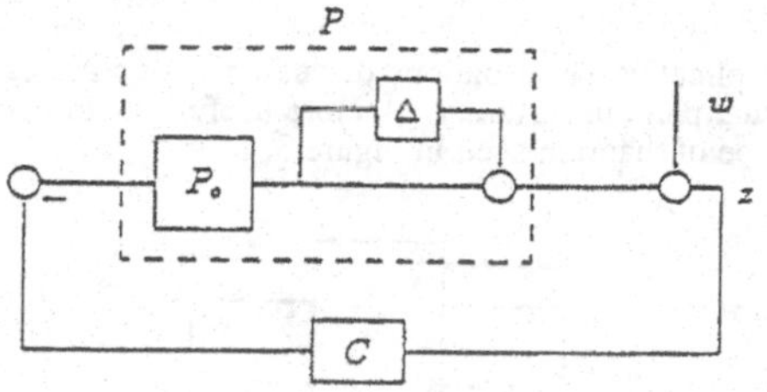

Figure 2

Theorem 6.1 [15]

Given a linear shift invariant plant P_0 and a perturbation Δ as described above with $\|\Delta\| \leq 1$. Assume C stabilizes P_0. Then C stabilizes all $P \in \Pi_m$ if and only if

$$\|P_0C(I+P_0C)^{-1}\|_A < 1 \tag{6.1}$$

The condition for additive perturbations is

$$\|C(I+P_0C)^{-1}\|_A < 1 \tag{6.2}$$

Now for the performance specification. Refer to Figure 2 and assume that the objective is to make $\|z\|$ small for all $\|w\| \leq 1$ and all $P \in \Pi_m$. More precisely, our problem is:

Find a controller C such that
1) C stabilizes all $P \in \Pi_m$.
2) $\sup_{P \in \Pi_m} \|(I+PC)^{-1}\|$ is minimized.

If we define

$$\Psi := \{C \mid C \text{ stabilizes all } P \in \Pi_m\}$$

then our problem can be stated as

$$\gamma_0 = \inf_{C \in \Psi} \sup_{P \in \Pi_m} \|(I+PC)^{-1}\| \tag{6.3}$$

Define $S_0 := (I+P_0C)^{-1}$ and $T_0 := P_0C(I+P_0C)^{-1}$. Then we have the following result.

Theorem 6.2 [16]

Let $C \in \Psi$. Then

$$\sup_{P \in \Pi_m} \|(I+PC)^{-1}\| = \frac{\|S_0\|_A}{1 - \|T_0\|_A} \tag{6.4}$$

In the scalar case we can solve (6.3) very easily as follows:

Theorem 6.3

Assume P_0 is a scalar, linear, shift-invariant plant, C stabilizes P_0 and $\gamma > 0$. Then C stabilizes all $P \in \Pi_m$ and $sup_{P \in \Pi_m} \|(1+PC)^{-1}\| < \gamma$ if and only if

$$\|S_0 \quad \gamma T_0\|_A < \gamma$$

Clearly, this involves an iterative process of choosing γ, solving a minimum norm problem, increasing or decreasing γ and resolving.

This same procedure applies in the MIMO case when the perturbation occurs at the plant input rather than the output [16].

The additive and multiplicative perturbations discussed above are called unstructured and are not very useful in modeling actual plant uncertainty. A more useful model is called structured perturbations and generally leads to the type of diagram seen in Figure 3.

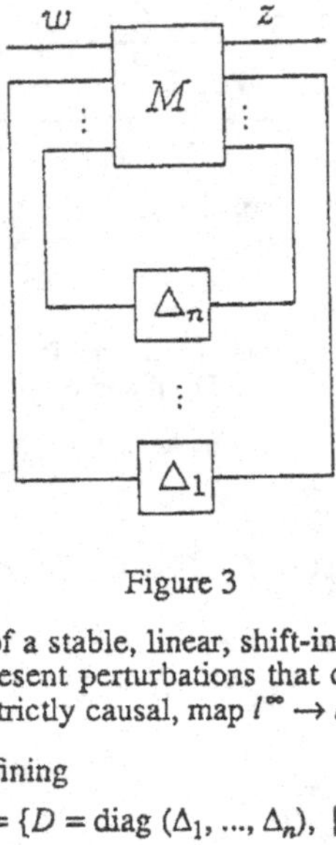

Figure 3

Here, M is the transfer function of a stable, linear, shift-invariant system which is composed of the plant and compensator and the Δ_i represent perturbations that can occur in various parts of the system. We will assume that the Δ_i are linear, strictly causal, map $l^\infty \rightarrow l^\infty$ and $\|\Delta_i\| \leq 1$.

Using these perturbations, and defining

$$D(n) := \{D = \text{diag}\,(\Delta_1, ..., \Delta_n),\ \|\Delta_i\| \leq 1\}$$

we will define the following problems.

Robust Stability Problem: Determine conditions under which the system in Figure 3 is stable for all $D \in D(n)$.

Robust Performance Problem: Determine conditions under which the system in Figure 3 is stable for all $D \in D(n)$ and $\|\Phi\| < 1$ where Φ is the system gain between w and z.

The first result here shows that the two problems are equivalent in a certain sense. Call the system of Figure 3, System II. Connect z to w through a fictitious perturbation Δ_0 and define

$$D(n+1) := \{\tilde{D} = diag(\Delta_0, \ldots, \Delta_n),\ \|\Delta_i\| \leq 1\}$$

Call this System I. Then we have

Theorem 6.4 [17]

The following are equivalent.

i) System I achieves robust stability

ii) $(I - M\tilde{D})^{-1}$ is l^∞–stable $\forall \tilde{D} \in D(n+1)$

iii) System II achieves robust performance.

Therefore we need to consider only stability robustness.

Write

$$M = \begin{bmatrix} M_{11} & \cdots & M_{1n} \\ \vdots & & \\ M_{n1} & \cdots & M_{nn} \end{bmatrix}$$

The main result here is

Theorem 6.5 [17]

$(I - MD)^{-1}$ is l^∞–stable $\forall D \in D(n)$ if and only if the system of inequalities

$$x_i \le \sum_{j=1}^{n} \|M_{ij}\|_A x_j \quad i=1,\ldots,n \tag{6.5}$$

has no solutions in $(\mathbb{R}^+)^n \setminus \{0\}$.

A more useful characterization of robust stability is the following.

Theorem 6.6

The following are equivalent

i) (6.5) has no solution in $(\mathbb{R}^+)^n \setminus \{0\}$

ii) $\rho(|M|) < 1$

iii) $\inf_R \|RMR^{-1}\|_A < 1$

where

$$|M| = \begin{bmatrix} \|M_{11}\|_A & \cdots & \|M_{1n}\|_A \\ \vdots & & \\ \|M_{n1}\|_A & \cdots & \|M_{nn}\|_A \end{bmatrix}$$

$\rho(\cdot)$ is the spectral radius

and

$$R := \{diag(r_1,\ldots,r_n),\ r_i > 0\}$$

Analysis problems are quite easy using these results. We simply compute $|M|$ for a given system and determine its spectral radius, which is its largest eigenvalue. The system is robustly stable if and only if $\rho(|M|) < 1$.

Robust synthesis problems can be approached as follows: See Figure 4

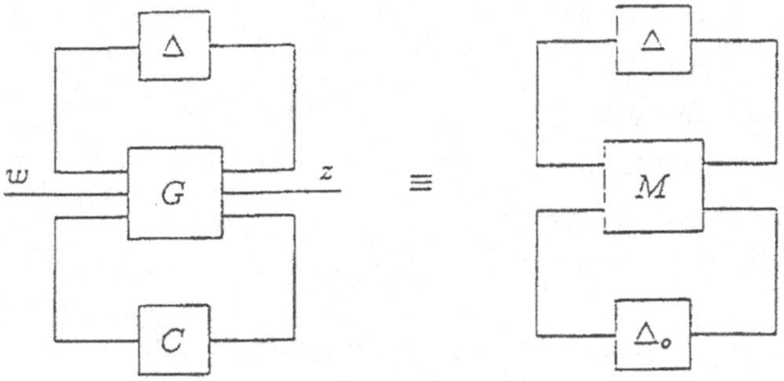

Figure 4

where we can write

$$M = H_1 - U_1 Q V_1$$

We then pose the problem

$$\inf_Q \inf_R \|RMR^{-1}\|_A = \gamma_0$$

and begin with $R = I$ say and solve $\inf_Q \|M\|_A$. Then for fixed M solve

$$\inf_R \|RMR^{-1}\|_A$$

and keep iterating. Unfortunately we have no experience yet with this procedure and don't anticipate that the problem will be convex in both variables.

This same iterative approach is used in μ-synthesis [18] and clearly, the procedure shown here is much simpler since R is a constant matrix (i.e. not a function of frequency).

9. Acknowledgement

The work discussed here was done in collaboration with Munther A. Dahleh, James S. McDonald and Mustafa Khammash and with the support of the National Science Foundation and Air Force Office of Scientific Research.

10. References

[1] Youla, D.C., Bongiorno, J.J., and Jabr, H.A., "Modern Wiener-Hopf design of optimal controllers – Part II: The multivariable case," *IEEE Trans. on Automatic Control*, Vol.AC-21, pp. 319-338, Jun. 1976.

[2] Kucera, V., *Discrete Linear Control.* John Wiley and Sons, New York, N.Y. , 1979.

[3] Luenberger, D.G., *Optimization by Vector Space Methods.* New York: John Wiley and Sons, 1969.

[4] Kailath, T., *Linear Systems.* Prentice-Hall, Inc., Englewood Cliffs, NJ, 1980.

[5] Rudin, W., *Functional Analysis.* McGraw-Hill Book Co., New York, N.Y. (Theo. 4.12), 1973.

[6] Cheng, L. and Pearson, J.B., "Frequency domain synthesis of linear multivariable regulators," *IEEE Trans. on Automatic Control*, Vol.AC-23, pp. 3-15, 1978.

[7] Vidyasagar, M., "Control Systems Synthesis: A Factorization Approach," *MIT Press*, Cambridge, MA, 1985.

[8] Dahleh, M.A. and Pearson, J.B., "l^1-optimal feedback controllers for MIMO discrete-time systems," *IEEE Trans A-C*, Vol.AC-32, no.4, pp. 314-322, Apr. 1987.

[9] McDonald, J.S. and Pearson, J.B., "l^1-optimal control of multivariable systems with output norm contraints," *Automatica* (to appear).

[10] Chvatal, V., *Linear Programming*. W.H. Freeman and Co., New York, N.Y., 1980.

[11] Mendlovitz, M., "A simple solution to the l^1 optimization problem," *Systems and Control Letters*, Vol.12, no.5, pp. 461-463, 1989.

[12] Horowitz, I.M., *Synthesis of Feedback Systems*. New York: Academic Press, 1963.

[13] Morari, M. and Zafiriou, E., *Robust Process Control*. Prentice Hall, Englewood Cliffs, N.J., 1989.

[14] Bower, J.L. and Schultheiss, P.M., *Introduction to the Design of Servomechanisms*. John Wiley & Sons, New York, N.Y, 1958.

[15] Dahleh, M.A. and Ohta, Y., "A necessary and sufficient condition for robust BIBO stability," *Systems and Control Letters*, Vol.11, pp. 271-275, 1988.

[16] Khammash, M. and Pearson, J.B., "Robust disturbance rejection in l^1-optimal control systems," *Systems and Control Letters*, Vol.14, pp. 93-101, 1990.

[17] Khammash, M. and Pearson, J.B., "Performance robustness of discrete-time systems with structured uncertainty," *IEEE Trans. A-C* (to appear).

[18] Doyle, J.C., "Structured uncertainty in control design," Proceedings of 24th CDC, Ft. Lauderdale, FL pp. 260-265, Dec. 1985.

ON THE HAMILTONIAN STRUCTURE IN THE COMPUTATION OF SINGULAR VALUES FOR A CLASS OF HANKEL OPERATORS

P. A. Fuhrmann
Department of Mathematics
Ben-Gurion University of the Negev
Beer Sheva, Israel

Abstract

The paper deals with the problem of computing the singular values of a Hankel operator with a symbol of the type m^*w with m, $w \in H_+^\infty$, m inner and w rational. In this case the problem reduces to the computation of zeroes of some functional determinant. This has been shown in Foias, Tannenbaum and Zames [1988], Lypchuk, Smith and Tannenbaum [1988], Smith [1989], Ozbay [1990]. The emphasis in this paper is to give a more illuminating and geometrical proof of the conjecture in Zhou and Khargonekar [1987]. This proof is based on the theory of polynomial models developed by the author, and it leans heavily on Smith [1989].

1 INTRODUCTION

The study of Hankel operators played always a central role in system theory, historically mainly due to its role in realization theory. This changed drastically with a shift of emphasis towards H^∞ control. With the formulation of new problems, highlighted by Zames [1981], a well as the assimilation of the Adamjan, Arov and Krein [1968a,1968b,1971,1978] results into system theory, the use of Hankel operators widened and includes now the optimal control, sensitivity minimization, robust control, balancing and model reduction. In all of these problems the computation of Hankel singular values plays a significant role.

Since our interest is in the structured result conjectured by Zhou and Khargonekar [1987] we will work solely with the Hardy spaces of the left and right half planes. The simplification, due to the greater symmetry, is significant.

The approach presented here is based on the theory of polynomial models developed by the author. The particular problem studied in this paper, i.e. that of computing the singular values of a special class of Hankel operators, is in the intersection of algebra and analysis. This is an extremely rich area and algebra provides both powerful methods and illuminating insights. The results presented here are but one instance of how polynomial algebra can be used in the study of Hankel operators. Much more comprehensive studies in the same area are Fuhrmann [1990] and Fuhrmann and Ober [1991].

The properties of Hankel operators are completely determined by its symbol. If the symbol is, as we postulate, of the form $\phi = m^*w$ with m, $w \in H_+^\infty$ with m inner and w rational then all such properties, including the computation of singular values and singular vectors, should in principle be directly computable from m and w. Now w determines a unique, up to isomorphism, minimal realization. Thus one should be able to compute all the required information from the inner function m and a minimal realization (A, B, C)

of w. That this is possible and how to do it was conjectured and proved in a special case in Zhou and Khargonekar [1987]. Proofs of the conjecture followed by Foias, Tannenbaum and Zames [1988], Lypchuk, Smith and Tannenbaum [1988], Smith [1989], and Ozbay [1990]. The method presented here is of use mostly in the case that the rational function w is of low McMillan degree.

Of course one is interested in Hankel singular values and singular vectors mostly for reasons of Norm determination and optimal approximation by finite rank operators. In this connection the work of Adamjan, Arov and Krein [1971] is pioneering and still highly relevant.

2 HANKEL OPERATORS

We will study Hankel operators defined on half plane Hardy spaces rather than on those of the unit disc as was done by Adamjan, Arov and Krein [1971]. In this we follow the choice of Glover [1984]. This choice seems to be a very convenient ones as all results on duality simplify significantly, due to the greater symmetry between the two half planes in comparison to the unit disc and its exterior.

The setting is the Hardy spaces. Thus H^2_+ is the Hilbert space of all analytic functions in the open right half plane with

$$\|f\|^2 = \sup_{x>0} \frac{1}{\pi} \int_{-\infty}^{\infty} |f(x+iy)|^2 dy.$$

The space H^2_- is similarly defined in the open left half plane. It is a theorem of Fatou that guarrantees the existence of boundary values of $H^2_\pm$-functions on the imaginary axis. Thus $H^2_\pm$ can be considered as closed subspaces of $L^2(i\mathbf{R})$, the space of Lebesgue square integrable functions on the imaginary axis. It follows from the Fourier-Plancherel and Paley-Wiener theorems that

$$L^2(i\mathbf{R}) = H^2_+ \oplus H^2_-,$$

with H^2_+ and H^2_- the Fourier-Plancherel transforms of $L^2(0,\infty)$ and $L^2(-\infty,0)$ respectively. Also H^∞_+ and H^∞_- will denote the spaces of bounded analytic functions on the open right and left half planes respectively. These spaces can be considered as subspaces of $L^\infty(i\mathbf{R})$, the space of Lebesgue measurable and essentially bounded functions on the imaginary axis. An extensive discussion of these spaces can be found in Hoffman [1962], Duren [1970] and Garnett [1981]. We will use the following conjugation $f^*(s) = f(-\bar{s})^*$.

We proceed to define Hankel operators and we do this directly in the frequency domain. Readers interested in the time domain definition and the details of the transformation into frequency domain are refered to Fuhrmann [1981], Glover [1984].

In the algebraic theory of Hankel operators the kernel and image of a Hankel operator are directly related to the coprime factorization of the symbol over the ring of polynomials. The details can be found for example in Fuhrmann [1983]. In the same way the kernel and image of a large class of Hankel operators are related to a coprime facorization over H^∞. This theme, originating in the work of Douglas, Shapiro and Shields [1971] and that of D. N. Clark, see Helton [1974], is developed extensively in Fuhrmann [1981]. Of course if the symbol of the Hankel operator is rational and in H^∞_- these two coprime factorizations are easily related.

We proceed to define Hankel operators.

Definition 2.1 *Given a function* $\phi \in L^\infty(i\mathbb{R})$ *the* **Hankel operator** $H_\phi : H^2_+ \longrightarrow H^2_-$ *is defined by*

$$H_\phi f = P_-(\phi f), \quad for\ f \in H^2_+. \tag{1}$$

The adjoint operator $(H_\phi)^* : H^2_- \longrightarrow H^2_+$ *is given by*

$$(H_\phi)^* f = P_+(\phi^* f), \quad for\ f \in H^2_-. \tag{2}$$

Here P_-, P_+ are the orthogonal projections of $L^2(i\mathbb{R})$ onto H^2_- and H^2_+ respectively.

Of course the way we defined the Hankel operator was highly arbitrary. We can define also $\hat{H}_\phi : H^2_- \longrightarrow H^2_+$ by

$$\hat{H}_\phi f = P_+ \phi f, \quad for\ f \in H^2_-. \tag{3}$$

The operator $\hat{H}_\phi$ is sometimes referred to as the involuted Hankel operator. Notice that $\hat{H}_\phi$ and H^*_ϕ are different operators. Still, in the situation we study, the involuted Hankel operator provides useful information for the study of the Hankel operator itself.

The next theorem summarizes the basic facts concerning Hankel operators. Detailed information can be found in Fuhrmann [1981] and Nikolskii [1985].

Theorem 2.1

1. *For every* $\psi \in H^\infty_+$ *the Hankel operator* H_ϕ *satisfies the functional equation*
$$P_- \psi H_\phi f = H_\phi \psi f, \quad f \in H^2_+. \tag{4}$$
2. $Ker\, H_\phi$ *is an invariant subspace, i.e. for* $f \in Ker\, H_\phi$ *and* $\psi \in H^\infty_+$ *we have* $\psi f \in Ker\, H_\phi$.
3. $Ker\, H_\phi = m H^2_+$ *with* m *inner if and only if* $\phi = m^* \psi$ *for some* $\psi \in H^\infty_+$ *and* m, ψ *have no common inner factor.*

Although we will not go into the details, it is of interest to point out the relation between certain Hankel operators and copressions of shifts.

Let m be inner. Then mH^2 is a right invariant subspace of H^2_+, i.e. it is invariant under multiplication by all H^∞_+ functions. We will put $H(m) = H^2 \ominus mH^2 = \{mH^2\}^\perp$.

Theorem 2.2 *Let* m *be inner in* H^∞_+ *and* $\psi \in H^\infty_+$. *Assume* ψ *and* m *have no common inner factor. Then,* $T_\psi : H(m) \longrightarrow H(m)$ *is defined by*

$$T_\psi = P_{H(m)} \psi | H(m).$$

1. *For every* $\psi \in H^\infty_+$ *the Hankel operator* $H_{m^*\psi}$ *satisfies*
$$T_\psi P_{H(m)} = m H_{m^* \psi}.$$

2.

$$||T_\psi|| = ||H_{m^*\psi}|| = inf\{||\psi - mq||_\infty \mid q \in H^\infty\}.$$

3. The nontrivial singular values of T_ψ and H_{m^*w} are equal.

∎

3 SINGULAR VALUES OF HANKEL OPERATORS

We consider the class of Hankel operators with symbol given by $\psi = m^*w$, where $m \in H^\infty_+$ is inner and $w \in H^\infty_+$ is rational. Since w is rational it has a unique representation of the form $w = \frac{b}{a}$, with b, a coprime and a monic, stable of degree n. Clearly the coprimeness assumption is no restriction on the problem. We do however make the following assumption

Assumption 3.1 *If ς is a pole of w then $-\varsigma$ is not a zero of w.*

Assumtion 3.1 means that $a \wedge b^* = 1$. The two coprimeness conditions are equivalent to $b \wedge aa^* = 1$.

Given a bounded operatot $T : H_1 \longrightarrow H_2$ between two Hilbert spaces, then a vector $f \in H_1$ is a singular vector associated with the singular value σ if

$$T^*Tf = \sigma^2 f$$

Given a singular vector f then for $g = \frac{1}{\sigma}Tf$ we have $T^*g = \frac{1}{\sigma}T^*Tf = \sigma f$. Hence

$$TT^*g = \sigma Tf = \sigma^2 g.$$

Thus g so defined is a singular vector of T^*. Moreover if

$$\left\{ \begin{array}{rcl} Tf & = & \sigma g \\ T^*g & = & \sigma f \end{array} \right. \tag{5}$$

then necessarily $||g|| = ||f||$, for

$$||g||^2 = (g,g) = (g, \frac{1}{\sigma}Tf) == (\frac{1}{\sigma}T^*g, f) = (f,f) = ||f||^2.$$

A pair (f,g) such that (5) holds is called a Schmidt pair of T, associated with the singular value σ. In much the same way as the singular vector determines a Schmidt pair so does a Schmidt pair deterrmine the singular vectors. Indeed if f, g satisfy (5) then

$$T^*Tf = T^*\sigma g = \sigma T^*g = \sigma^2 f$$

Thus the determination of all Schmidt pairs is equivalent to the determination of all singular vectors.

The next lemma determines the kernel and image of the involuted Hankel operator. This will be of use in the sequel.

Lemma 3.1 *Let $\phi = m^* w$ with $m, w \in H_+^\infty$, m inner and w rational. Let $w = \frac{b}{a}$ with a, b coprime. Then*

1. *$Ker\hat{H}_\phi = \frac{a}{a^*}H_-^2$.*
2. *$Im\hat{H}_\phi = X^a = \{\frac{r}{a} \mid \deg r < \deg a\}$*

Proof:

1. Let $g \in Ker\hat{H}_\phi \subset H_-^2$. Then $P_+ m^* \frac{b}{a} g = 0$, i.e. $m^* \frac{b}{a} g \in H_-^2$, or $\frac{b}{a} g = mh$, for some $h \in H_-^2$. This implies that g has zeroes coinciding with those of a. This means that $g = \frac{a}{a^*} g'$ for some $g' \in H_-^2$.
   Conversely if $g \in \frac{a}{a^*} H_-^2$ then $g = \frac{a}{a^*} g'$ and
   $$\hat{H}_\phi g = P_+ m^* \frac{b}{a}\frac{a}{a^*} g' = P_+ m^* \frac{b}{a^*} g' = 0$$
   as m^*, $\frac{b}{a^*} \in H_+^\infty$, and $g' \in H_+^2$,
2. From 1 it follows that
   $$\{Ker\hat{H}_\phi\}^\perp = H_-^2 \ominus \frac{a}{a^*} H_-^2 = X^{a^*}$$
   and hence $Im\hat{H}_\phi = X^a$.

∎

In a way this simple lemma is the key to all that follows. The spaces $Im\hat{H}_\phi$ and $\{Ker\hat{H}_\phi\}^\perp$, although derived from analytic considerations, have been identified as well studied algebraic objects. Thus the road to using algebraic tools is opened.

We proceed now to the computation of singular values of H_{m^*w}. Since in general this Hankel operator is infinite dimensional it has more than just pure point spectrum. We will denote by ρ_∞ the essential spectral radius of H_{m^*w}, but we will assume that $\rho_\infty < ||H_{m^*w}||$. Thus the norm is achieved at the largest singular value.

Lemma 3.2 (Smith) *Let $\rho > \rho_\infty$. Then ρ is a singular value of H_{m^*w} if and only if there exist polynomials p, q, of degree $\leq n-1$ and not both zero, such that*

$$f = \frac{mb^*p + \rho a q^*}{\rho^2 aa^* - bb^*} \in H_+^2 \tag{6}$$

$$g = \frac{mb^*q + \rho a p^*}{\rho^2 aa^* - bb^*} \in H_+^2 \tag{7}$$

Proof: Now ρ is a singular value of $H_{m^* w}$ if and only if there exist two H_+^2 functions f and g for which:

$$\begin{cases} H_{m^* w} f &= \rho g^* \\ H_{m^* w}^* g^* &= \rho f \end{cases} \tag{8}$$

This in turn is equivalent to the existence of two H_+^2 functions u and v for which:

$$m^* w f + u = \rho g^* \tag{9}$$

$$m w^* g^* + v^* = \rho f \tag{10}$$

Solving for f we have

$$\rho f = v^* + m w^* (\frac{m^* w f + u}{\rho})$$

and hence

$$\rho^2 f = \rho v^* + w w^* f + m w^* u$$

i.e.

$$f = \frac{m w^* u + \rho v^*}{\rho^2 - w w^*} \tag{11}$$

Similarly for g we get

$$g^* = \frac{m^* w v^* + \rho u}{\rho^2 - w w^*} \tag{12}$$

Thus ρ is a singular value if and only if there exist $u, v \in H_+^2$ for which

$$f = \frac{m w^* u + \rho v^*}{\rho^2 - w w^*} \in H_+^2 \tag{13}$$

and

$$g^* = \frac{m^* w v^* + \rho u}{\rho^2 - w w^*} \in H_-^2 \tag{14}$$

Applying the projection P_+ to equation (9) yields

$$u = -P_+ m^* w f$$

Now $w = \dfrac{b}{a}$ and so $u \in Im\hat{H}_{m^* w} = X^a$. Similarly from equation (10)

$$v^* = -P_- m w^* g^* \in Im\hat{H}_{m^* w}^* = X^{a^*}$$

Thus there exist polynomials p, q of degree $< n = \deg a$ such that $u = \dfrac{p}{a}$ and $v^* = \dfrac{q^*}{a^*}$. Substituting back into equations (13) and (14) respectively we get

$$f = \frac{m \frac{b^*}{a^*} \frac{p}{a} + \rho \frac{q^*}{a^*}}{\rho^2 - \frac{b b^*}{a a^*}} \in H_+^2 \tag{15}$$

$$g^* = \frac{m^* \frac{b}{a} \frac{q^*}{a^*} + \rho \frac{p}{a}}{\rho^2 - \frac{bb^*}{aa^*}} \in H^2_- \tag{16}$$

or equivalently

$$f = \frac{mb^* p + \rho q^* a}{\rho^2 aa^* - bb^*} \in H^2_+ \tag{17}$$

and

$$g^* = \frac{m^* bq^* + \rho pa^*}{\rho^2 aa^* - bb^*} \in H^2_- \tag{18}$$

Of course the last condition is equivalent to

$$g = \frac{mb^* q + \rho ap^*}{\rho^2 aa^* - bb^*} \in H^2_+ \tag{19}$$

We note that if there exist polynomials p and q of degree $< n$ such that (17) and (18) hold then one easily checks that f and g so defined are singular vectors corresponding to the singular value ρ. ∎

The conditions given by equations (6) and (7) can be characterized in terms of the zeroes of the polynomial Ω_ρ defined next. To clarify notation we will say, given an analytic function f and a polynomial Ω, that f is divisible by Ω, and write $f = 0 \; mod(\Omega)$, if f is analytic at all zeroes of Ω and the order of its zeroes is not less than the order of the zeroes of Ω.

Theorem 3.1 (Smith) *Let $\rho > \rho_\infty$. Let*

$$\Omega_\rho(z) = \rho^2 a(z) a^*(z) - b(z) b^*(z).$$

Assume m has no zeroes coinciding with the zeroes of Ω_ρ. Then ρ is a singular value of $H_{m^* w}$ if and only if there exist polynomials p, q, of degree $\leq n-1$ and not both zero, such that

$$mb^* p + \rho aq^* = 0 \; mod(\Omega_\rho) \tag{20}$$

and

$$mb^* q + \rho ap^* = 0 \; mod(\Omega_\rho) \tag{21}$$

If the latter condition is satisfied the corresponding singular vectors are given by (6) and (7).

Proof: Put $\Omega_\rho(z) = \rho^2 a(z) a^*(z) - b(z) b^*(z)$. Let ς be an arbitrary right half plane zero of Ω_ρ. Then if $u = \frac{p}{a}$ and $v = \frac{q}{a}$ are related by the singular value/singular vector equation (17) and (18) , then clearly $f, g^* \in H^2_+$ implies that the polynomial $mb^* p + \rho aq^*$ vanish at each zero ς of Ω_ρ in the closed right half plane at least to the order of ς as a zero of Ω_ρ.

Conversely, if $mb^*p+\rho aq^*$ vanish at each right half plane zero ς at least to the order of ς as a zero of $\Omega\rho$ then $f = \dfrac{m\frac{b^*}{a^*}\frac{p}{a}+\rho\frac{q^*}{a^*}}{\rho^2-\frac{bb^*}{aa^*}} \in H^2_+$ and hence ρ is a singular value.

However the theorem claims more. Assume there exist p, q such that (20) and (21) hold. Then f and g defined by (17) and (19) respectively are in $\{KerH_{m^*w}\}^\perp = \{mH^2_+\}^\perp = H(m)$. Now m has a meromorphic continuation into the left half plane, unless the imaginary axis is its natural boundary in which case one needs to consider pseudo-meromorphic continuations, see Douglas, Shapiro and Shields [1971] or Fuhrmann [1981]. By our assumtion m is analytic at all zeroes of Ω_ρ. Also $g^* = \dfrac{m^*bq^*+\rho pa^*}{\Omega_\rho}$ is in $\in H^2_-$. Therefore $\dfrac{bq^*+\rho ma^*p}{\Omega_\rho}$ is analytic at all zeroes of Ω_ρ. Now $\rho\frac{a}{b}$ is, by the coprimeness of b and Ω_ρ, also analytic at all left half plane zeroes of Ω_ρ. Therefore $\rho\frac{a}{b}\left\{\dfrac{bq^*+\rho ma^*p}{\Omega_\rho}\right\}$ is analytic at all left half plane zeroes of Ω_ρ. However this function can be rewritten as

$$\frac{\rho aq^*+\rho^2 m\frac{aa^*}{b}p}{\Omega_\rho} = \frac{\rho aq^*+mb^*p}{\Omega_\rho}+\frac{mp}{b}$$

The right term is clearly analytic at zeroes of Ω_ρ so we have $\dfrac{mb^*p+\rho aq^*}{\Omega_\rho}$ analytic at all left half plane zeroes of Ω_ρ.

The sufficiency of the condition is obvious, in view of Lemma 3.2. ∎

The interest of this result is in the fact that the computation of the singular values of the Hankel operator has been reduced to a problem in functional arithmetic modulo the polynomial Ω_ρ. Our aim is a direct interpretation of equations (20) and (21) in state space terms. To this end we make, in the next section, a digression into polynomial model theory.

4 POLYNOMIAL MODELS

In this section we summarize the main results concerning polynomial models. Although we will use the results solely in the scalar case, we will state them, for extra generality, in their multivariable form.

Our starting point is this basic result about free modules. A proof can be found in Van der Waerden [1949].

Theorem 4.1 *Let R be a principal ideal domain and M a free left $R-$module with n basis elements. Then every $R-$submodule N of M is free and has at most n basis elements.*

If V is a finite dimensional vector space over a field F then $V[z]$, the space of vector polynomials is a free finitely generated module over the polynomial ring $F[z]$. Throughout we will assume a basis has been chosen and thus V will be identified with F^n and similarly $V[z]$ with $F^n[z]$. Also we will identify $F^n[z]$ and $(F[z])^n$ and speak of its elements as polynomial vectors. Similarly elements of $F^{m\times n}[z]$ will be referred to as polynomial matrices. Because of the nature of the factorization results we are interested in, and for the consistency of notation, we will identify the field F with the real field $\mathbf{R}$, noting that some of the results hold in greater generality.

The next theorem is the basic representation theorem for submnodules.

Theorem 4.2 *A subset M of $\mathbf{R}^n[z]$ is a submodule of $\mathbf{R}^n[z]$ if and only if $M = D\mathbf{R}^n[z]$ for some D in $\mathbf{R}^{n\times n}[z]$.*

The following is the basic theorem that relates submodule inclusion to factorization.

Theorem 4.3 *Let $M = D\mathbf{R}^n[z]$ and $N = E\mathbf{R}^n[z]$ then $M \subset N$ if and only if $D = EG$ for some G in $\mathbf{R}^{n\times n}[z]$.*

Let π_+ and π_- denote the projections of $\mathbf{R}^m((z^{-1}))$ the space of truncated Laurent series on $\mathbf{R}^m[z]$ and $z^{-1}\mathbf{R}[[z^{-1}]]$, the space of formal power series vanishing at infinity, respectively. Since

$$\mathbf{R}^m((z^{-1})) = \mathbf{R}^m[z] \oplus z^{-1}\mathbf{R}^m[[z^{-1}]] \tag{22}$$

π_+ and π_- are complementary projections.

Given a nonsingular polynomial matrix D in $\mathbf{R}^{m\times m}[z]$ we define two projections π_D in $\mathbf{R}^m[z]$ and π^D in $z^{-1}\mathbf{R}^m[[z^{-1}]]$ by

$$\pi_D f = D\pi_- D^{-1} f \qquad for\ f \in \mathbf{R}^m[z] \tag{23}$$

$$\pi^D f = \pi_- D^{-1}\pi_+ Dh \qquad for\ h \in z^{-1}\mathbf{R}^m[[z^{-1}]] \tag{24}$$

and define two linear subspaces of $\mathbf{R}^m[z]$ and $z^{-1}\mathbf{R}^m[[z^{-1}]]$ by

$$X_D = Im\pi_D. \tag{25}$$

and

$$X^D = Im\pi^D. \tag{26}$$

An element f of $\mathbf{R}^m[z]$ belongs to X_D if and only if $\pi_+ D^{-1} f = 0$, i.e. if and only if $D^{-1}f$ is a strictly proper rational vector function.

We turn X_D into an $\mathbf{R}[z]$–module by defining

$$p \cdot f = \pi_D p f \qquad for\ p \in \mathbf{R}[z],\ f \in X_D. \tag{27}$$

Since $Ker\pi_D = D\mathbf{R}^m[z]$ the following isomorphism result holds.

Theorem 4.4 *With the previously defined module structure X_D is isomorphic to $\mathbf{R}^m[z]/D\mathbf{R}$*

Similarly, we turn X^D into an $\mathbf{R}[z]$–module by defining

$$p \cdot h = \pi_- ph \qquad for\ p \in \mathbf{R}[z],\ h \in X^D. \tag{28}$$

In X_D we will focus on a special map S_D, a generalization of the classical companion matrix, which corresponds to the action of the identity polynomial z, i.e.,

$$S_D f = \pi_D z f \qquad for \qquad f \in D.$$

Thus the module structure in X_D is identical to the module structure induced by S_D through $p \cdot f = p(S_D)f$. With this definition the study of S_D is identical to the study of the module structure of X_D. In particular the invariant subspaces of S_D are just the submodules of X_D which we proceed to investigate.

The interpretation of factorization of polynomial matrices on the level of polynomial models is described next.

Theorem 4.5 *A subset M of X_D is a submodule, or equivalently an S_D invariant subspace, if and only if $M = D_1 X_{D_2}$ for some factorization $D = D_1 D_2$ with $D_i \in \mathbf{R}^{n\times n}[z]$.*

We summarize now the connection between the geometry of invariant subspaces and the arithmetic of polynomial matrices.

Theorem 4.6 *Let M_i, $i = 1, \dots, s$ be submodules of X_D, having the representations $M_i = E_i X_{F_i}$, that correspond to the factorizations*

$$D = E_i F_i.$$

Then the following statements are true.

(i) $M_1 \subset M_2$ if and only if $E_1 = E_2 R$, i.e. if and only if E_2 is a left factor of E_1.

(ii) $\cap_{i=1}^{s} M_i$ has the representation $E_\nu X_{F_\nu}$ with E_ν the l.c.r.m. of the E_i and F_ν the g.c.r.d. of the F_i.

(iii) $M_1 + \dots + M_s$ has the representation $E_\mu X_{F_\mu}$ with E_μ the g.c.l.d. of the E_i and F_μ the l.c.l.m. of all the F_i.

Corollary 4.1 *Let $D = E_i F_i$, for $i = 1, \dots, s$. Then*

(i) We have

$$X_D = E_1 X_{F_1} + \dots + E_s X_{F_s}$$

if and only if the E_i are left coprime.

(ii) We have $\cap_{i=1}^{s} E_i X_{F_i} = 0$ if and only if the F_i are right coprime.

(iii) The decomposition

$$X_D = E_1 X_{F_1} \oplus \dots \oplus E_s X_{F_s}$$

is a direct sum if and only if $D = E_i F_i$ for all i, the E_i are left coprime and the F_i are right coprime.

The next result summarizes the relation between factorization and the spectral decomposition of linear maps.

Theorem 4.7 *Let $D(z) \in \mathbf{R}^{n\times n}[z]$ be nonsingular and let $d(z) = \det D(z)$ be its characteristic polynomial. Suppose d has a factorization $d = e_1 \cdots e_s$ with the e_i pairwise coprime. Then D admits factorizations*

$$D = D_i E_i$$

with $\det D_i = d_i$, $\det E_i = e_i$ and such that

$$X_D = D_1 X_{E_1} \oplus \cdots \oplus D_s X_{E_s} \tag{29}$$

Moreover

$$\det(S_D | D_i X_{E_i}) = e_i. \tag{30}$$

Denoting by $\tilde{T}$ the transpose of the matrix T we define, for an element $A(z) = \sum_{j=-\infty}^{\infty} A_j z^j$ of $\mathbf{R}^{m\times m}((z^{-1}))$, $\tilde{A}$ by

$$\tilde{A}(z) = \sum_{j=-\infty}^{\infty} \tilde{A}_j z^j. \tag{31}$$

In $\mathbf{R}^m((z^{-1})) \times \mathbf{R}^m((z^{-1}))$ we define a symmetric bilinear form $[f, g]$ by

$$[f, g] = \sum_{j=-\infty}^{\infty} \tilde{f}_j g_{-j-1} \tag{32}$$

where $f(z) = \sum_{j=-\infty}^{\infty} f_j z^j$ and $g(z) = \sum_{j=-\infty}^{\infty} g_j z^j$. It is clear that, as both f and g are truncated Laurent series, that the sum in (32) is well defined, containing only a finite number of nonzero terms. We denote by T^* the adjoint of a map T relative to the bilinear form of (32), i.e. T^* is the unique map that satisfies

$$[Tf, g] = [f, T^* g] \tag{33}$$

for all $f, g \in \mathbf{R}^m((z^{-1}))$. We use this global bilinear form to obtain a concrete representation of X_D^* the dual space of X_D.

Theorem 4.8 *X_D^* The dual space of X_D can be identified with $X_{\tilde{D}}$ under the pairing*

$$< f, g > = [D^{-1} f, g] \tag{34}$$

for $f \in X_D$ and $g \in X_{\tilde{D}}$. Moreover the module structures of X_D and $X_{\tilde{D}}$ are related through

$$S_D^* = S_{\tilde{D}} \tag{35}$$

The following result shows how the characterization of submodules of X_D and their relation to factorizations is reflected by duality.

Theorem 4.9 *Let $M \subset X_D$ be a submodule, represented as $M = E X_G$ for some factorization $D = EG$ into nonsingular factors. Then $M^\perp$ is a submodule of $X_{\tilde{D}}$ and is given by $M^\perp = \tilde{G} X_{\tilde{E}}$.*

The next two theorems are the central tools in polynomial model theory. They provide the characterization of all module homomorphisms and the condition for their invertibility. For the relation to the commutant lifting theorem and the spectral mapping theorem see Fuhrmann [1981] and Nikolskii [1985].

Theorem 4.10 *Let $Z : X_D \longrightarrow X_{D_1}$ be the module homomorphism defined by*

$$Zf = \pi_{D_1} N f$$

with

$$ND = D_1 M$$

holding. Then

(i) $\mathrm{Ker} Z = EX_G$, where $D = EG$ and G is a g.c.r.d. of D and N.

(ii) $\mathrm{Im} Z = E_1 X_{G_1}$, where $D_1 = E_1 G_1$ and E_1 is a g.c.l.d. of D_1 and M.

(iii) Z is surjective if and only if N and D_1 are left coprime.

(iv) Z is injective if and only if D and M are right coprime.

∎

Recall that a vector space V is **symplectic** if it is equipped with a nondegenerate, alternating bilinear form. A linear map H in a symplectic space V is **Hamiltonian** if $H^* = -H$ relative to this form. A map R in V is a **symplectic** if it is invertible and leaves the form invariant.

The canonical example of a symplectic space is $\mathbf{R}^{2n}$ with the bilinear form induced by

$$J = \begin{pmatrix} 0 & I \\ -I & 0 \end{pmatrix}. \tag{36}$$

A Hamiltonian map H in this case is given by

$$H = \begin{pmatrix} A & P \\ Q & -\tilde{A} \end{pmatrix} \tag{37}$$

with P, Q symmetric. R is symplectic if and only if $\tilde{R}JR = J$.

Contrary to inner product spaces, there are self orthogonal elements and subspaces in symplectic spaces. A subspace $\mathcal{L}$ of a symplectic space V is called **Lagrangian** if it is a maximal self orthogonal subspace.

We modify now the previous approach to duality to accomodate the study of symplectic spaces. To this end we introduce now a global alternating form on $\mathbf{R}^m((z^{-1})) \times \mathbf{R}^m((z^{-1}))$ in the following way. We define now a new bilinear form on $\mathbf{R}^m((z^{-1})) \times \mathbf{R}^m((z^{-1}))$ by

$$\{f, g\} = [\tau f, g] \quad for \quad f, g \in \mathbf{R}^m((z^{-1})) \tag{38}$$

where $\tau : \mathbf{R}^m((z^{-1})) \longrightarrow \mathbf{R}^m((z^{-1}))$ is defined by

$$(\tau f)(z) = f(-z). \tag{39}$$

Lemma 4.1 *The bilinear form defined by (38) on* $\mathbf{R}^m((z^{-1})) \times \mathbf{R}^m((z^{-1}))$ *is alternating, i.e.*

$$\{f, g\} = -\{g, f\}. \tag{40}$$

Given a rational matrix function Φ we define its **Hamiltonian or parahermitian** conjugate, Φ^* by

$$\Phi^*(z) = \Phi(-\bar{z})^*. \tag{41}$$

This implications of this symmetry were originally studied by Brockett and Rahimi [1972]. An extensive study of Hamiltonian symmetry and its associated realization theory can be found in Fuhrmann [1984].

Given a map $Z : \mathbf{R}^m((z^{-1})) \to \mathbf{R}^m((z^{-1}))$ we will denote by Z^∇ the map, assuming it exists, which satisfies

$$\{Zf, g\} = \{f, Z^\nabla g\} \tag{42}$$

for all $f, g \in \mathbf{R}^m((z^{-1}))$. Since the forms $[\ ,\ ]$, and hence also $\{\ ,\ \}$, are nondegenerate the map Z^∇, if it exists, is unique. The map Z^∇ will be called the **Hamiltonian adjoint** of Z.

Assume now that we have two nondegenerate bilinear forms on $V \times V^*$ and $V^* \times V$ which satisfy $<< x, y >> = - << y, x >>$ for all $x \in V$ and $y \in V^*$. Note that while the two forms are distinct we do not distinguish between them, as it is always clear from the context which form we use. We say x is **orthogonal to** y if $<< x, y >> = 0$. Given a subset $M \subset V$ we define $M^\perp$ as usual by

$$M^\perp = \{y \in V^* \mid << x, y >> = 0 \ \ for\ all\ x \in M\}. \tag{43}$$

Since full submodules of $\mathbf{R}^m[z]$ are of the form $D\mathbf{R}^m[z]$ with D a nonsingular polynomial matrix it is of interest to characterize its orthogonal complement. The result should be compared with Theorem 4.9.

We can use these isomorphisms to define a pairing between X_D and $X_{(D^*)}$ by

Theorem 4.11 X_D^∇ *The dual space of* X_D *can be identified with* $X_{(D^*)}$ *under the pairing*

$$<< f, g >> = \{D^{-1}f, g\} = [JD^{-1}f, g] \tag{44}$$

for all $f \in X_D$ *and* $g \in X_{(D^*)}$. *Moreover the module structures of* X_D *and* $X_{(D^*)}$ *are related through*

$$(S_D)^* = -S_{(D^*)}. \tag{45}$$

We note at this point that if D is a nonsingular Hamiltonian symmetric polynomial matrix, i.e. if $D^* = D$, then X_D with the metric of (44) is a symplectic space. Much use of this will be made later.

The pairing of elements of X_D and X_{D^*} given by (44) allows us to compute, for a subset V of X_D, the annihilator $V^\perp$, i.e., the set of all $g \in X_{D^*}$ such that $<< f, g >> = 0$ for all $f \in V$. Since $(S_D)^* = -S_{D^*}$, it follows that if $V \subset X_D$ is a submodule, then so is $V^\perp$ in X_{D^*}. However, we know that submodules are related to factorizations of D into nonsingular factors. The annihilator of V can be concretely identified.

Theorem 4.12 *Let $V \subset X_D$ be a submodule. Then $V^\perp$ is a submodule of $X_{(D^*)}$. Moreover if $V = EX_F$ where $D = EF$ is a factorization of D into nonsingular factors, then*

$$V^\perp = F^* X_{E^*}. \tag{46}$$

If we assume $\tilde{D}(-z) = D(z)$, i.e. $D^* = D$, then X_D is a symplectic space and S_D a Hamiltonian map. For details see Fuhrmann [1984]. One way to obtain Hamiltonian symmetric polynomial matrices is to study those which have a factorization of the form $D(z) = \tilde{E}(-z)E(z)$ or $D = E^*E$.

Theorem 4.13 *Let $D = E^*E$. Then E^*X_E is a Lagrangian S_D-invariant subspace of X_D.*

5 REALIZATION THEORY

As usual, given a proper rational matrix G we will say a system (A, B, C, D) is a realization of G if

$$G = D + C(zI - A)^{-1}B.$$

We will use the notation $G = [A, B, C, D]$. We will be interested in realizations associated with rational functions having the following representations

$$G = VT^{-1}U + W. \tag{47}$$

Our approach to the analysis of these systems is to associate with each representation of the form (47), a state space realization in the following way. We choose X_T as the state space and define the triple (A, B, C), with $A : X_T \longrightarrow X_T$, $B : \mathbf{R}^m \longrightarrow X_T$, and $C : X_T \longrightarrow \mathbf{R}^p$ by

$$\begin{cases} A = S_T \\ B\xi = \pi_T U\xi, \\ Cf = (VT^{-1}f)_{-1} \\ D = G(\infty). \end{cases} \tag{48}$$

We call this the associated realization to the polynomial matrix P,

$$P = \begin{pmatrix} T & U \\ -V & W \end{pmatrix}.$$

Theorem 5.1 *The system given by (48) is a realization of $G = VT^{-1}U + W$. This realization is reachable if and only if T and U are left coprime and observable if and only if T and V are right coprime.*

Theorem 5.2 *Let G be a normalized bicausal isomorphismi, and let $G = T^{-1}D$ be a left coprime factorization. Then*

1. *The polynomial models X_D and X_T contain the same elements.*
2. *Let $\mathcal{B}$ be a common basis of the two polynomial models. Let (A_T, B_T, C_T, I) and (A_D, B_D, C_D, I) be the matrix representation of the shift realizations of G and G^{-1} respectively in the given basis. Then*

$$\begin{aligned} A_D &= A_T - B_T C_T \\ B_D &= -B_T \\ C_D &= C_T \end{aligned} \tag{49}$$

∎

Let $G_1 = [A_1, B_1, C_1, D_1]$ and $G_2 = [A_2, B_2, C_2, D_2]$ be two transfer functions realized in the state spaces X_1 and X_2 respectively. If the number of inputs of the second system equals the number of outputs of the first we can feed those outputs to the second system. This gives rise to the series coupling and the corresponding transfer function is

$$G_2G_1 = \left[\begin{pmatrix} A_1 & 0 \\ B_2C_1 & A_2 \end{pmatrix}, \begin{pmatrix} B_1 \\ B_2D_1 \end{pmatrix}, \begin{pmatrix} D_2C_1 & C_2 \end{pmatrix}, D_2D_1 \right].$$

We will use also the notation $G_2G_1 = [A_2, B_2, C_2, D_2] \times [A_1, B_1, C_1, D_1]$.

Definition 5.1 *We say that (A, B, C, D) acting in $\mathbf{R}^{2n}$ is a* **Hamiltonian system** *if, with J defined by (36), we have*

$$\begin{aligned} \tilde{A}J &= -JA \\ JB &= \tilde{C} \\ D &= \tilde{D} \end{aligned} \tag{50}$$

The main result concerning Hamiltonian realizations is the following.

Theorem 5.3 *(i) A transfer function G has a Hamiltonian realization if and only if it is Hamiltonian symmetric, i.e. if $G(-\bar{z})^* = G(z)$.*

(ii) Two minimal Hamiltonian realizations of G are symplectically equivalent.

∎

For an extensive study of Hamiltonian symmetry see Fuhrmann [1984].

6 BASES AND THE BEZOUTIAN

In the previous section all development has been basis free. Once a choice of basis is made for a polynomial model then all operators, and so realizations, have associated matrix representations. Different choices of basis highlight different properties of operators and systems. We will not go into all such possibilities and will restrict ourselves to the few instances needed in the sequel. Some more complex results in this direction, and relevant to the topic of this paper can be found in Fuhrmann [1990] and Fuhrmann and Ober [1991].

X finite dimensional vector space over F and let X^* be its dual space. $\{e_1, \ldots, e_n\}$ a basis for X then $\{f_1, \ldots, f_n\}$ in X^* is the dual basis if

$$< e_i, f_j >= \delta_{ij} \qquad 1 \leq i, j \leq n.$$

We will investigate next the implications of duality in the context of polynomial models. In particular we will focus on an important pair of dual bases. Let X_q be the polynomial model associated with the polynomial $q(z) = z^n + q_{n-1}z^{n-1} + \cdots + q_0$.

We consider the following very natural pair of bases in X_q. The first one we will refer to as the **standard basis** $B_{st} = \{1, z, \ldots, z^{n-1}\}$ whereas the second, the **control basis** is given by $B_{co} = \{e_1, \ldots, e_n\}$ where

$$e_i(z) = \pi_+ z^{-i} q = q_i + q_{i+1}z + \cdots + z^{n-i}, \qquad i = 1, \ldots, n.$$

These bases are important in the study of Bezoutians, introduced next.

The Bezoutian is a well studied quadratic form associated with two polynomials. It has many applications in root location, coprimeness characterization, the Cauchy index, characterization of output feedback invariants, etc. A comprehensive exposition and further references can be found in Helmke and Fuhrmann [1989].

Given $p(z), q(z)$ of polynomials with real coefficients, $q(z)$ monic of degree n and p of degree $\leq n$, the **Bezoutian** $B(q,p)$ of q and p is defined

$$\begin{aligned} B(q,p)(z,w) &= \frac{q(z)p(w) - p(z)q(w)}{z - w} \\ &= \sum_{i=1}^{n}\sum_{j=1}^{n} b_{ij} z^{i-1} w^{j-1} \end{aligned}$$

$B = (b_{ij})$ is a symmetric matrix.

The following theorem is central for the understanding of Bezoutians inasmuch as it reduces its study to a basis free object, namely a module homomorphism of which the Bezoutian is just a matrix representation. Proofs can be found in Fuhrmann [1981] and Fuhrmann and Helmke [1989].

Theorem 6.1 *1. The standard basis and the control basis are a dual pair of bases for X_q, i.e.*

$$B_{co}^* = B_{st}$$

2. $p(S_q)$ is a self adjoint operator in the indefinite metric $<\ ,\ >$, i.e.

$$p(S_q)^* = p(S_q)$$

3. The Bezoutian is the matrix representation of the self adjoint operator $p(S_q)$ with respect to the dual pair of bases, the standard and the control, i.e.

$$B(q,p) = [p(S_q)]_{co}^{st}.$$

∎

Since $p(S_q)$ is invertible whenever p and q are coprime, it follows that, coprimeness assumed, the columns of $B(q,p)$ are linearly independent, Thus a basis of R^n. Of course this basis has a functional representation in the polynomial model X_q. This is considered next.

Lemma 6.1 *Let $p, q \in \mathbf{R}[z]$ be coprime, with $\deg p < \deg q$. Then the polynomials*

$$p(z)q_k(z) - q(z)p_k(z) = \sum_{i=1}^{n} b_{ik} z^{i-1} \tag{51}$$

constitute a basis for X_q. Here

$$p_k = \pi_+ z^{-k} p, \quad q_k = \pi_+ z^{-k} q.$$

Proof: Rewrite the definition of the Bezoutian as

$$q(z)p(w) - p(z)q(w) = \sum_{i=1}^{n}\sum_{j=1}^{n} b_{ij} z^{i-1}(z-w)w^{j-1}.$$

Applying the operation $\pi_+ w^{-k}$ to both sides and equating $w = z$ we get

$$\begin{aligned} & q(z)p_k(w) - p(z)q_k(w) \\ &= \pi_+ \sum_{i=1}^{n}\sum_{j=1}^{n} b_{ij} z^{i-1} w^{-k}(z-w)w^{j-1}|_{w=z} \\ &= \sum_{i=1}^{n} z^{i-1} \sum_{j=1}^{n} b_{ij} \pi_+ w^{-k}(z-w)w^{j-1}|_{w=z} \end{aligned} \tag{52}$$

Now

$$\pi_+ w^{-k}(z-w)w^{j-1}|_{w=z} = \begin{cases} 0 & j \neq k \\ -1 & j = k \end{cases} \tag{53}$$

and hence the result follows. ∎

Note that if we choose $p(z) = 1$ then this basis reduces to the control basis of X_q.

7 HAMILTONIAN STRUCTURE

In this section we will give our proof of the Zhou-Khargonekar result. Before proceeding to that we will present in the next theorem a summary of relevant results on various Hamiltonian realizations.

Theorem 7.1 *Let $w = a^{-1}b$ be a strictly proper, stable rational function, with a and b coprime polynomials. Thus a is stable and we assume it to be monic of degree n.*

1. *X_{aa^*} is a symplectic space, and the following are direct sum representations for it.*
$$X_{aa^*} = X_a \oplus aX_{a^*} \tag{54}$$
and
$$X_{aa^*} = a^*X_a \oplus aX_{a^*}. \tag{55}$$
Whereas the first representation is a direct sum of two subspaces only one of which is invariant, the second representation is as a direct sum of two Lagrangian invariant subspaces of S_{aa^*}.

2. *Assume the polynomial b is coprime with both a and a^*. Then the following is a direct sum representation of the polynomial model*
$$X_{aa^*} = b^*X_a \oplus aX_{a^*}.$$
Since X_{aa^*} and X_{Ω_ρ} contain the same elements, this is also a direct sum representation of X_{Ω_ρ}, i.e.
$$X_{\Omega_\rho} = b^*X_a \oplus aX_{a^*}.$$

3. *Let $\{e_1,\ldots,e_n\}$ be the control basis associated with the monic polynomial a. Then the set $\mathcal{B} = \{1, s, \ldots, s^{n-1}, ae_1^*, \ldots, ae_n^*\}$ is a symplectic basis for X_{aa^*}, i.e.*
$$[I]_{\mathcal{B}}^{\mathcal{B}^*} = \begin{pmatrix} 0 & I \\ -I & 0 \end{pmatrix} = J.$$
Let f_i be defined, as in Lemma 6.1, by $f_i(z) = b(z)a_k(z) - a(z)b_k(z)$. Then $\mathcal{B}_1 = \{b^*, \ldots, b^*z^{n-1}, af_1^*, \ldots, af_n^*\}$ is also a basis for X_{aa^*}.

4. *Let $w = a^{-1}b$. Then the shift realization of w has the following matrix representation with respect to the standard basis of X_a*
$$A = \begin{pmatrix} 0 & . & . & . & -a_0 \\ 1 & & & & . \\ & . & & & . \\ & & . & & . \\ & & & 1 & -a_{n-1} \end{pmatrix}, B = \begin{pmatrix} b_1 \\ . \\ . \\ . \\ b_n \end{pmatrix}, C = \begin{pmatrix} 0 & . & . & 0 & 1 \end{pmatrix}, \tag{56}$$

5. *$w(-z)$ has the following realization*
$$A = \begin{pmatrix} 0 & -1 & & & \\ & & . & & \\ & & & . & \\ & & & & -1 \\ a_0 & . & . & . & a_{n-1} \end{pmatrix}, B = \begin{pmatrix} 0 \\ . \\ . \\ . \\ 1 \end{pmatrix}, C = \begin{pmatrix} b_1 & . & . & . & b_n \end{pmatrix}. \tag{57}$$

6. $\gamma(z) = 1 - \frac{w(-z)w(z)}{\rho^2}$ *has the following Hamiltonian realization*

$$A = \begin{pmatrix} 0 & . & . & . & -a_0 & & & & \\ 1 & & & & . & & & & \\ & . & & & . & & & & \\ & & . & & . & & & & \\ & & & 1 & -a_{n-1} & & & & \\ & & & & & 0 & -1 & & \\ & & & & & & & . & \\ & & 0 & & & & & & . \\ & & & & & & & & -1 \\ & & & & 1 & a_0 & . & . & . \; a_{n-1} \end{pmatrix}$$

$$B = \frac{1}{\rho}\begin{pmatrix} b_1 \\ . \\ . \\ . \\ b_n \\ 0 \\ . \\ . \\ . \\ 0 \end{pmatrix}, C = \frac{1}{\rho}\begin{pmatrix} 0 & . & . & . & 0 & b_1 & . & . & . & b_n \end{pmatrix}, D = 1. \tag{58}$$

This is the matrix representation of the shift realization of γ *with respect to the basis* $\mathcal{B} = \{1, s, \dots, s^{n-1}, ae_1^*, \dots, ae_n^*\}$*. It has the same representation with respect to the basis* $\mathcal{B}_1 = \{b^*, \dots, b^* z^{n-1}, af_1^*, \dots, af_n^*\}$

7. $\gamma(z)^{-1} = \left(1 - \frac{w(-z)w(z)}{\rho^2}\right)^{-1}$ *has the following minimal realization given by*

$$\begin{pmatrix} 0 & . & . & . & -a_0 & & & & & \\ 1 & & & & . & & & & & \\ & . & & & . & & & -\frac{1}{\rho^2}(b_i b_j) & & \\ & & . & & . & & & & & \\ & & & 1 & -a_{n-1} & & & & & \\ & & & & & 0 & -1 & & & \\ & & & & & & & . & & \\ & & 0 & & & & & & . & \\ & & & & & . & & & & -1 \\ & & & & 1 & a_0 & . & . & . & a_{n-1} \end{pmatrix} \tag{59}$$

$$B = \frac{1}{\rho}\begin{pmatrix} b_1 \\ . \\ . \\ . \\ b_n \\ 0 \\ . \\ . \\ . \\ 0 \end{pmatrix}, C = \frac{1}{\rho}\begin{pmatrix} 0 & . & . & . & 0 & b_1 & . & . & . & b_n \end{pmatrix}, D = 1.$$

This is the matrix representation of the shift realization of $\gamma(z)^{-1}$ *in the polynomial model* X_{Ω_ρ} *with respect to the basis* $\{b^*, \ldots, b^* z^{n-1}, af_1^*, \ldots, af_n^*\}$.

8. *Let* (A, B, C) *be an arbitrary minimal realization of the rational function* $w = \frac{b}{a}$. *Then the transfer function* $\gamma(z) = 1 - \frac{w(-z)w(z)}{\rho^2}$ *has a minimal realization given by*

$$\left[\begin{pmatrix} A & 0 \\ \frac{1}{\rho^2}\tilde{C}C & -\tilde{A} \end{pmatrix}, \begin{pmatrix} B \\ 0 \end{pmatrix}, \begin{pmatrix} 0 & -\tilde{B} \end{pmatrix}, 1\right]. \tag{60}$$

This realization and the shift realization, in the polynomial model X_{aa^*}, *based on the representation* $\gamma(z) = 1 - \frac{1}{\rho^2}\frac{b^*b}{a^*a}$ *are symplectically equivalent.*

9. *If* π_1 *and* π_2 *are the parallel projections on the two direct summands of*

$$X_{aa^*} = X_a \oplus aX_{a^*}$$

then

$$\pi_1 S_{aa^*} \mid X_a = S_a \tag{61}$$

$$\pi_2 S_{aa^*} \mid X_a = S_{a^*} \tag{62}$$

The same holds true for the direct sum representation

$$X_{aa^*} = b^* X_a \oplus aX_{a^*}. \tag{63}$$

10. $\gamma(z)^{-1} = \left(1 - \frac{w(-z)w(z)}{\rho^2}\right)^{-1}$ *has a minimal realization given by*

$$\left[\begin{pmatrix} A & -\frac{1}{\rho}B\tilde{B} \\ \frac{1}{\rho}\tilde{C}C & -\tilde{A} \end{pmatrix}, \begin{pmatrix} \tilde{B} \\ 0 \end{pmatrix}, \begin{pmatrix} 0 & -\tilde{B} \end{pmatrix}, 1\right].$$

It also has a minimal shift realization, in the polynomial model X_{Ω_ρ}, based on the representation $\gamma(z)^{-1} = 1 + \frac{\rho^2 b^* b}{\Omega_\rho}$. *The two realizations are symplectically equivalent. In particular S_{Ω_ρ} and F_ρ defined by*

$$F_\rho = \begin{pmatrix} A & -\frac{1}{\rho}B\tilde{B} \\ \frac{1}{\rho}\tilde{C}C & -\tilde{A} \end{pmatrix}$$

are similar.

11. The polynomial Ω_ρ is, up to a constant factor, the characteritic polynomial of F_ρ.

Proof:

1. That X_{aa^*} is a symplectic space follows from Theorem 4.13. The direct sum representation (55) follows from Corollary 4.1. (54) can be checked directly.

2. We consider the module homomorphism Z in X_{aa^*} induced by b^*, i.e. $b^*(S_{aa^*})$

$$Zf = b^*(S_{aa^*})f = \pi_{aa^*}b^*f$$

This is an isomorphism, i.e. $b^*(S_{aa^*})$ is invertible if and only if $b^* \wedge aa^* = 1$, and this was assumed.

Next we study how X_a and aX_{a^*} transform under this map. Clearly $X_a \longrightarrow b^*X_a$ by $f \mapsto b^*f$, whereas, since aX_{a^*} is a submodule,

$$b^*(S_{aa^*})aX_{a^*} = ab^*(S_{a^*})X_{a^*}.$$

Also note that $a^* \wedge b^* = 1$ implies the invertibility of $b^*(S_{a^*})$. Thus the direct sum representation follows.

$$X_{aa^*} = b^*X_a \oplus aX_{a^*}$$

Consider the basis $\mathcal{B}_1$ which is the image under the map $b^*(S_{aa^*})$. As $b^*(S_{aa^*})$ commutes with S_{aa^*} it follows that S_{aa^*} has the same matrix representation as before.

3. We use the following computational rules

$$\{f^*, g^*\} = -\{f, g\}$$

$$<< f, g >> = - << g, f >>$$

and

$$<< af,g >>=<< f,a^*g >>$$

Then

$$\begin{aligned} << s^i, ae_j^* >> &= \{\frac{s^{i-1}}{aa^*}, ae_j^*\} = \{s^{i-1}, \frac{e_j^*}{a^*}\} = -\{\frac{e_j^*}{a^*}, s^{i-1}\} \\ &= \{\frac{e_j}{a}, (-s)^{i-1}\} = [\frac{e_j}{a}, s^{i-1}] =< e_j, s^{i-1} >= \delta_{ij} \end{aligned}$$

The basis $\mathcal{B}_1 = \{b^*, \ldots, b^* z^{n-1}, af_1^*, \ldots, af_n^*\}$ is the image, under the invertible map $b^*(S_{aa^*})$, of the basis $\mathcal{B}$.

4. This is a simple computation.

5. Follows from the previous realization by Hamiltonian duality. This is also the matrix representation of the shift realization of $w(-z) = b(-z)a(-z)$ in the basis $\{e_1^*, \ldots, e_n^*\}$ of X_{a^*}.

6. This can be seen in several ways. This realization is the series connection of the realizations (56) and (57), with the factor ρ taken into consideration. This is also the matrix representation of the shift realization of $\gamma(z) = 1 - \frac{b^*}{\rho}(aa^*)^{-1}\frac{b}{\rho}$ in the basis $\mathcal{B} = \{1, z, \ldots, z^{n-1}, ae_1^*, \ldots, ae_n^*\}$

7. Follows from the realization (59) and Theorem 5.2.

8. From $w(z) = C(zI - A)^{-1}B$ follows $w(-z) = \tilde{B}(zI + \tilde{A})^{-1}\tilde{C}$. The series connection of the two realizations implies the result.

9. Given two pairs of systems $[A_1, B_1, C_1, D_1]$, $[A_2, B_2, C_2, D_2]$ as well as $[A_1', B_1', C_1', D_1']$, $[A_2', B_2', C_2', D_2']$ which are isomorphic in pairs, then the series connection of the first pair is isomorphic to the series connection of the later pair. We also use the fact that the series connection of the product of two shift realizations, under coprimeness assumptions, is isomorphic to the shift realization of their product. The proof given in Fuhrmann [1976] can be adapted to cover this case.

10. Recall that, for $f \in X_a$ we have $S_a f = zf - a\xi$, where $\xi = (a^{-1}f)_{-1}$. So

$$S_{aa^*} f = zf = zf - a\xi + a\xi = S_a f + a\xi,$$

therefore

$$\pi_1 S_{aa^*} f = S_a f$$

Equation (62) follows from the invariance of aX_{a^*}.

Similarly, if $f \in X_a$ we have

$$\begin{aligned} S_{aa^*} b^* f &= \pi_{aa^*} z b^* f = z b^* f \\ &= b^*\{zf - a\xi + a\xi\} = b^* S_a f + ab^* \xi. \end{aligned}$$

11. Follows by a simple result on inverse systems.

12. Follows from the similarity of F_ρ and S_{Ω_ρ} and the fact that Ω_ρ is the characteristic polynomial of S_{Ω_ρ}.

∎

Before we state the next theorem we clarify how analytic functions act on matrices. Here we follow the Riesz-Dunford functional calculus, see Dunford and Schwartz [1957]. Thus assume we are given a matrix A and a function m analytic in a neighbourhood of the spectrum of A. Let μ be a polynomial such that $m - \mu$ has at each eigenvalue of A a zero of order not less than the order of the zero of the minimal polynomial of A at that point. Then we define $m(A) = \mu(A)$. Clearly this definition is independent of the polynomial μ.

We can pass on now to the proof of the main result.

Theorem 7.2 **(Zhou-Khargonekar)** *Let (A, B, C) be a minimal realization of $w \in H_+^\infty$. Let $m(z)$ be an inner function in the right half plane Π_+ and assume that m is analytic at all zeroes of Ω_ρ. Let F_ρ be the Hamiltonian matrix*

$$F_\rho = \begin{pmatrix} A & \frac{1}{\rho} B\tilde{B} \\ -\frac{1}{\rho}\tilde{C}C & -\tilde{A} \end{pmatrix}.$$

Then $\rho > \rho_\infty$ is a singular value of $H_{m^* w}$ if and only if

$$det[m(F_\rho)]_{11} = 0.$$

In particular

$$inf\{||w - mq||_\infty \mid q \in H^\infty\}_+ = sup\{\rho \mid det[m(F_\rho)]_{11} = 0\}$$

Proof: It suffices to show that the statement is equivalent to that of Theorem 3.1.

The statement of the theorem is independent of the particular realization of w taken. Indeed if R is a similarity in the state space then

$$\begin{pmatrix} R & 0 \\ 0 & \tilde{R}^{-1} \end{pmatrix}$$

is a symplectic map. In fact

$$\begin{pmatrix} \tilde{R} & 0 \\ 0 & R^{-1} \end{pmatrix}\begin{pmatrix} 0 & I \\ -I & 0 \end{pmatrix}\begin{pmatrix} R & 0 \\ 0 & \tilde{R}^{-1} \end{pmatrix} = \begin{pmatrix} 0 & I \\ -I & 0 \end{pmatrix}$$

and

$$\begin{pmatrix} R^{-1}AR & \frac{1}{\rho}R^{-1}B\tilde{B}\tilde{R}^{-1} \\ \frac{1}{\rho}\tilde{R}\tilde{C}CR & -\tilde{R}\tilde{A}\tilde{R}^{-1} \end{pmatrix} = \begin{pmatrix} R^{-1} & 0 \\ 0 & \tilde{R} \end{pmatrix}\begin{pmatrix} A & \frac{1}{\rho}B\tilde{B} \\ \frac{1}{\rho}\tilde{C}C & -\tilde{A} \end{pmatrix}\begin{pmatrix} R & 0 \\ 0 & \tilde{R}^{-1} \end{pmatrix}$$

As by our assumption m is analytic at the zeroes of Ω_ρ then there exists a unique polynomial μ of degree $< deg(\Omega_\rho)$ such that $m - \mu = 0 mod(\Omega_\rho)$. In particular equation (20) transforms into the polynomial equation

$$\mu^* p + \rho a q* = 0 mod(\Omega_\rho)$$

or equivalently

$$\pi_{\Omega_\rho}\mu^* p = \mu(S_{\Omega_\rho})b^* p = -\rho a q*.$$

Let π_1 be the projection on the first component relative to the direct sum $X_{\Omega_\rho} = b^* X_a \oplus aX_{a^*}$. Thus $\pi_1 p(S_{\Omega_\rho})|b^* X_a$ is a singular map. But S_{Ω_ρ} is isomorphic to F_ρ and so $\pi_1 f(S_{\Omega_\rho})|b^* X_a$ is isomorphic to $[F_\rho]_{11}$. Hence the result follows. ∎

EXAMPLE:

We conclude by computing the special case:

$$w(s) = \frac{1}{\alpha s + 1} \qquad \phi(s) = e^{-hs}$$

w is realized by $(-\frac{1}{\alpha}, \frac{1}{\sqrt{\alpha}}, \frac{1}{\sqrt{\alpha}}, 0)$ and hence

$$F_\rho = \begin{pmatrix} -\frac{1}{\alpha} & \frac{1}{\alpha\rho} \\ -\frac{1}{\alpha\rho} & \frac{1}{\alpha} \end{pmatrix}.$$

So

$$e^{-hF_\rho} = e^{-h\begin{pmatrix} -\frac{1}{\alpha} & \frac{1}{\alpha\rho} \\ -\frac{1}{\alpha\rho} & \frac{1}{\alpha} \end{pmatrix}} = e^{-\frac{h}{\alpha}\begin{pmatrix} -1 & \frac{1}{\rho} \\ -\frac{1}{\rho} & 1 \end{pmatrix}}.$$

The eigenvalues of F_ρ are

$$\lambda = \frac{\pm i\sqrt{1-\rho^2}}{\rho}$$

The corresponding eigenvectors are

$$\begin{pmatrix} 1 \\ \rho \pm i\sqrt{1-\rho^2} \end{pmatrix}.$$

Since

$$\begin{pmatrix} 1 & 1 \\ \rho + i\sqrt{1-\rho^2} & \rho - i\sqrt{1-\rho^2} \end{pmatrix}^{-1} =$$

$$\frac{i}{2\sqrt{1-\rho^2}}\begin{pmatrix} \rho - i\sqrt{1-\rho^2} & -1 \\ -(\rho + i\sqrt{1-\rho^2}) & 1 \end{pmatrix}$$

we have

$$\begin{pmatrix} -1 & \frac{1}{\rho} \\ -\frac{1}{\rho} & 1 \end{pmatrix} = \frac{i}{2\sqrt{1-\rho^2}}\begin{pmatrix} 1 & 1 \\ \rho + i\sqrt{1-\rho^2} & \rho - i\sqrt{1-\rho^2} \end{pmatrix}$$

$$\times \begin{pmatrix} \frac{i\sqrt{1-\rho^2}}{\rho} & 0 \\ 0 & \frac{-i\sqrt{1-\rho^2}}{\rho} \end{pmatrix}\begin{pmatrix} \rho - i\sqrt{1-\rho^2} & -1 \\ -(\rho + i\sqrt{1-\rho^2}) & 1 \end{pmatrix}$$

This allows us to compute e^{-hF_ρ}. In particular

$$[e^{-hF_\rho}]_{11} =$$
$$\frac{i}{2\sqrt{1-\rho^2}}\left[(\rho - i\sqrt{1-\rho^2})e^{-\frac{ih\sqrt{1-\rho^2}}{\alpha\rho}} - (\rho + i\sqrt{1-\rho^2})e^{\frac{ih\sqrt{1-\rho^2}}{\alpha\rho}}\right]$$

Thus

$$det[e^{-hF_\rho}]_{11} = 0 \Longleftrightarrow$$

$$e^{\dfrac{2ih\sqrt{1-\rho^2}}{\alpha\rho}} = \frac{\rho - i\sqrt{1-\rho^2}}{\rho + i\sqrt{1-\rho^2}} = (\rho - i\sqrt{1-\rho^2})^2$$

Or

$$e^{\dfrac{ih\sqrt{1-\rho^2}}{\alpha\rho}} = \pm(\rho - i\sqrt{1-\rho^2}). \tag{64}$$

Putting $y = \dfrac{h\sqrt{1-\rho^2}}{\alpha\rho}$, then equation (64) reduces to

$$\cos y = \rho, \quad \sin y = -\sqrt{1-\rho^2}$$

and hence to

$$\tan y = -\frac{-\sqrt{1-\rho^2}}{\rho}.$$

Finally we get

$$\tan y + \frac{\alpha y}{h} = 0.$$

So if y is the smallest zero of this equation, namely the zero that satisfies $\pi/2 < y < \pi$, then

$$\rho = \frac{1}{\sqrt{1+\frac{\alpha^2 y^2}{h^2}}}.$$

References

[1968a] V. M. Adamjan, D. Z. Arov and M. G. Krein, "Infinite Hankel matrices and generalized problems of Caratheodory-Fejer and F. Riesz", *Funct. Anal. Appl.* 2, 1-18.

[1968b] V. M. Adamjan, D. Z. Arov and M. G. Krein, "Infinite Hankel matrices and generalized problems of Caratheodory-Fejer and I. Schur", *Funct. Anal. Appl.* 2, 269-281.

[1971] V. M. Adamjan, D. Z. Arov and M. G. Krein, "Analytic properties of Schmidt pairs for a Hankel operator and the generalized Schur-Takagi problem", *Math. USSR Sbornik* 15 (1971), 31-73.

[1978] V. M. Adamjan, D. Z. Arov and M. G. Krein, "Infinite Hankel block matrices and related extension problems", Amer. Math. Soc. Transl., series 2, Vol. 111, 133-156.

[1971] R. G. Douglas, H. S. Shapiro and A. L. Shields, "Cyclic vectors and invariant subspaces for the backward shift", *Ann. Inst. Fourier, Grenoble*, 20, 37-76.

[1957] N. Dunford and J. T. Schwartz, *Linear Operators*, vol. 1, Interscience, New York.

[1987] H. Bercovici, C. Foias & A. Tannenbaum, "On skew Toeplitz operators, I", *Operator Theory: Advances and Applications*, 29, 21-43.

[1987] C. Foias & A. Tannenbaum, "On the Nehari problem for a certain class of L^∞-functions appearing in control, II", *Journal of Functional Analysis*, 74, 146-159.

[1986] C. Foias, A. Tannenbaum & G. Zames, "Some explicit formulae for the singular values of certain Hankel operators with factorizable symbol", *SIAM J. Mathematical Analysis*, 19, 1081-1091

[1986] C. Foias, A. Tannenbaum & G. Zames, *IEEE Trans. Aut. Contr.* 31, 763-767.

[1976] P. A. Fuhrmann, "On series and parallel coupling of a class of discrete time infinite dimensional linear systems", *SIAM J. Contr. Optim.* 14, 339-358.

[1981] P. A. Fuhrmann, *Linear Systems and Operators in Hilbert Space*, McGraw-Hill, New York.

[1981] P. A. Fuhrmann, "Polynomial models and algebraic stability criteria", *Proceedings of Joint Workshop on Synthesis of Linear and Nonlinear Systems*, Bielefeld June 1981, 78-90.

[1983] P. A. Fuhrmann, "On symmetric rational transfer functions", *Linear Algebra and Appl.*, 50, 167-250.

[1984] P. A. Fuhrmann, "On Hamiltonian transfer functions", *Lin. Alg. Appl.*, 84, 1-93.

[1990] P. A. Fuhrmann, "A polynomial approach to Hankel norm and balanced approximations", to appear, *Lin. Alg. Appl.*, 101 pp.

[1991] P. A. Fuhrmann and R. Ober, "A functional approach to Riccati balancing", to appear.

[1984] K. Glover, "All optimal Hankel-norm approximations and their L^∞-error bounds", *Int. J. Contr.* 39, 1115-1193.

[1989] U. Helmke and P. A. Fuhrmann, "Bezoutians", *Lin. Alg. Appl.*, vols. 122-124, 1039-1097.

[1988] T. A. Lypchuk, M. C. Smith &, A. Tannenbaum, "Weighted sensitivity minimization: general plants in H^∞ and rational weights,", *Lin. Alg. Appl.* vol. 109, 71-90.

[1985] N. K. Nikolskii, *Treatise on the Shift Operator*, Springer Verlag, Berlin.

[1990] H. Ozbay "A simpler formula for the singular values of a certain Hankel operator", *Sys. and Contr. Lett.*.

[1989] M. C. Smith, "Singular values and vectors of a class of Hankel operators", *Sys. and Contr. Lett.* 12, 301-308.

[1949] B. L. Van der Waerden, *Modern Algebra*, F. Ungar, New York.

[1981] G. Zames, *IEEE Trans. Aut. Contr.* 16, 301,320.

[1987] K. Zhou and P. P. Khargonekar, "On the weighted sensitivity minimization problem for delay systems", *Sys. and Contr. Lett.* 8, 307-312.

NEHARI INTERPOLATION PROBLEM FOR RATIONAL MATRIX FUNCTIONS: THE GENERIC CASE

Joseph A. Ball*
Department of Mathematics
Virginia Tech
Blacksburg, Virginia 24061

Israel Gohberg
School of Mathematical Sciences
Tel-Aviv University
Tel-Aviv, Ramat-Aviv, Israel

Leiba Rodman**
Department of Mathematics
College of William and Mary
Williamsburg, Virginia 23187-8795

Abstract

We study here a matrix Nehari problem, regarded as an interpolation problem, for the class of rational matrix functions. The interpolation data are assumed to be in a simple, but generic form. This allows us to present a self-contained exposition of the matrix Nehari problem, including criteria for solvability and description of all solutions. The proofs are based on the state-space method.

Contents

---

* Partially supported by the NSF Grant DMS-8701615-02.
** Partially supported by the NSF Grant DMS-9000839.

1. INTRODUCTION

The usual formulation of the matrix Nehari problem is as a distance problem. Let Π^- be the open left half plane $\{z \in \mathbb{C} : \mathrm{Re} z < 0\}$ with the boundary $\partial\Pi^- = \{z \in \mathbb{C} : \mathrm{Re} z = 0\} \cup \{\infty\}$. We are given an $M \times N$ matrix function $K(z)$ uniformly bounded on Π^- and wish to compute the L_∞-distance (i.e. the distance with respect to the sup norm on $\partial\Pi^-$) of $K(z)$ to the H_∞-space of all $M \times N$ matrix valued functions on $\partial\Pi^-$ having uniformly bounded analytic continuation to Π^-, and to find the best matrix H_∞ approximants. We shall assume that the original function $K(z)$ is rational; then without loss of generality we may assume that the analytic approximants are rational as well. The distance in question then is

$$\inf\{\|K - R\|_\infty : R \in \mathcal{R}_{M\times N}\{\Pi^- \cup \partial\Pi^-)\}$$

where we denote by $\mathcal{R}_{M\times N}(\sigma)$ the set of $M \times N$ rational matrix functions with no poles in σ, and $\|\cdot\|_\infty$ is defined by

$$\|F\|_\infty = \sup\{\|F(z)\| : z \in \partial\Pi^-\}.$$

The Nehari distance problem can be thought of as an interpolation problem in the following way. Let $\{z_1, \dots, z_k\}$ be the poles of $K(z)$ in Π^- and let

$$K(z) = (z - z_i)^{-k_i} K^{(i)}_{-k_i} + (z - z_i)^{-k_i+1} K^{(i)}_{-k_i+1} + \cdots + (z - z_i)^{-1} K^{(i)}_{-1} + [\text{analytic at } z_i]$$

be the Laurent expansion for $K(z)$ at $z_i (i = 1, \dots, k)$, where $K^{(i)}_{-1} \neq 0$. Then it is easily seen that a rational matrix function F has the form $F = K + R$ with R a rational matrix function analytic on $\overline{\Pi^-} = \Pi^- \cup \partial\Pi^-$ if and only if

$$F \text{ analytic on } \overline{\Pi^-}\setminus\{z_1, \dots, z_k\}$$

and

$$F(z) = (z - z_i)^{-k_i} K^{(i)}_{-k_i} + \cdots + (z - z_i)^{-1} K^{(i)}_{-1} + [\text{analytic at } z_i]$$

Thus, the unknown function $F(z)$ has prescribed poles in $\overline{\Pi^-}$ with prescribed singular parts of the local Laurent series. One controls the norm of the interpolant on the boundary while maintaining the local interpolation conditions conditions on the interior.

In this paper we consider only the generic case when the poles of $K(z)$ in $\overline{\Pi^-}$ are simple. Namely, the local Laurent series of $K(z)$ in a neighborhood of every pole $z_0 \in \Pi^-$ has the form

$$K(z) = (z - z_0)^{-1} Q_0 + [\text{analytic at } z_0],$$

where Q_0, the residue of $K(z)$ at z_0, is a rank one matrix. We assume also that $K(z)$ has no poles on the boundary $\partial\Pi^-$. The suitably formulated Nehari problem is: Describe all $R \in \mathcal{R}_{M\times N}(\Pi^- \cup \partial\Pi^-)$ such that

$$\|K + R\|_\infty \leq 1,$$

in particular, give necessary and sufficient conditions for existence of such R. We state the main result of this paper.

THEOREM 1.1. *Let $z_1, \dots, z_n$ be the poles of $K(z)$ in Π^-, all simple poles, and let Q_i be the residue of $K(z)$ at z_i written as $Q_i = \gamma_i w_i$, where γ_i is a nonzero $M \times 1$ column and w_i is a nonzero $1 \times N$ row. Let*

$$P = \left[-\frac{w_i w_j^*}{z_i + \bar{z}_j} \right]_{i,j=1}^{n}$$

$$Q = \left[-\frac{\gamma_i^* \gamma_j}{\bar{z}_i + z_j} \right]_{i,j=1}^{n}$$

Assume that the matrix $I - PQ$ is nonsingular. Then there is $R \in \mathcal{R}_{M \times N}(\Pi^- \cup \partial\Pi^-)$ such that

$$\|K + R\|_\infty \leq 1 \tag{1.1}$$

if and only if the biggest eigenvalue $\lambda_1(PQ)$ of PQ is less than 1 (it turns out that both P and Q are positive definite matrices and therefore all eigenvalues of PQ are positive). If $\lambda_1(PQ) < 1$, then all rational matrix functions F of the form $F = K + R$, $R \in \mathcal{R}_{M \times N}(\Pi^- \cup \partial\Pi^-)$ with $\|F\|_\infty \leq 1$ are given by the formula

$$F = (\Theta_{11}G + \Theta_{12})(\Theta_{21}G + \Theta_{22})^{-1} \tag{1.2}$$

where $G \in \mathcal{R}_{M \times N}(\Pi^- \cup \partial\Pi^-)$, $\|G\|_\infty \leq 1$, and

$$\Theta(z) = \begin{bmatrix} \Theta_{11}(z) & \Theta_{12}(z) \\ \Theta_{21}(z) & \Theta_{22}(z) \end{bmatrix}$$

is a fixed rational matrix function given by

$$\Theta(z) = I + \begin{bmatrix} C & 0 \\ 0 & B^* \end{bmatrix} \begin{bmatrix} (zI - A)^{-1} & 0 \\ 0 & \left(zI - A^*\right)^{-1} \end{bmatrix} \Lambda^{-1} \cdot \begin{bmatrix} -C^* & 0 \\ 0 & B \end{bmatrix},$$

where

$$\Lambda = \begin{bmatrix} Q & I \\ I & P \end{bmatrix}, \; A = \begin{bmatrix} z_1 & & & \\ & z_2 & & 0 \\ & & \ddots & \\ 0 & & & z_n \end{bmatrix}; \; B = \begin{bmatrix} w_1 \\ w_2 \\ \vdots \\ w_n \end{bmatrix}; \; C = [\gamma_1 \gamma_2 \cdots \gamma_n].$$

We exclude in Theorem 1.1 the degenerate case in which $I - PQ$ is singular. In the degenerate case one still has that the existence of $R \in \mathcal{R}_{M \times N}(\Pi^- \cup \partial\Pi^-)$ with $\|K + R\|_\infty \leq 1$ is equivalent to $\lambda_1(PQ) \leq 1$ (we will not prove this statement). However, there is no linear fractional formula as in (1.2) in the degenerate case.

For the distance formulation of the Nehari problem, the following corollary is obtained immediately from Theorem 1.1 upon replacing K by a suitable multiple αK, α a positive constant.

COROLLARY 1.2. *Given* K(z), P *and* Q *as in Theorem 1.1,*

$$\inf\{\|K - R\|_\infty : R \in \mathcal{R}_{M\times N}(\Pi^- \cup \partial\Pi^-)\}$$

is equal to $\sqrt{\lambda_1(PQ)}$.

A general method for solving various interpolation problems (including the Nehari problem) was developed in [BGR1]. The method is based on reduction to so-called homogeneous interpolation problems with metric restrictions, and on the solution of the latter problem being in explicit state space form. In this paper we consider the Nehari interpolation problem only in simple, but still generic, cases. This restriction on generality allows us to present the method in a self-contained transparent way and avoid some of the more sophisticated formalism needed for the general case and for other interpolation problems.

The Nehari problem and its numerous generalizations has a long and distinguished history starting with the work of I. Schur in the early part of this century (see [G]). Many methods were developed to study this problem for the matrix case, for example, the method of Adamyan-Arov-Krein [AAL], the commutant lifting method [FF], and others. The first solution of the Nehari problem in the state space form for rational matrix functions was given by Glover [Gl]; it was motivated by control applications.

In this paper we focus mainly on the left half-plane case. An analogous Nehari problem for the unit disk can be solved using essentially the same approach.

A more general version of the Nehari problem - Nehari-Takagi problem - also can be solved for rational matrix functions using the approach presented in this paper. In the Nehari-Takagi problem one seeks to minimize $\|K + R\|_\infty$, where K is a given $M \times N$ rational matrix function, and R is any $M \times N$ rational matrix function with at most k poles in Π^- counted with multiplicities (the integer $k \geq 0$ is fixed in advance). We refer the reader to Chapter 20 in [BGR] for a complete solution of the Nehari-Takagi problem for rational matrix functions in the general case using the state-space methods.

The paper is organized as follows. In Section 2 we use heuristic arguments (as well as prove rigorously) to reduce the Nehari interpolation problem to certain homogeneous interpolation problem for rational matrix functions. The solution of the latter problem is given (in a broader context than it is actually needed for the proof of Theorem 1.1) in Section 4, based on the solution of a general interpolation problem (without norm constraints) given in Section 3. Connections with Hankel operators and their norms are developed in Section 5. In the sixth section we prove Theorem 1.1. Finally, in the last section we state (without full proof) the corresponding result for the unit disk.

Throughout the paper the following notation will be used. $\mathbb{C}^{M\times N}$ stands for the set of all $M \times N$ matrices (all matrices are assumed to have complex entries). For $A \in \mathbb{C}^{M\times N}$,

$$\operatorname{Im} A = \{Ax : x \in \mathbb{C}^{N\times 1}\}$$

$$\operatorname{Ker} A = \{x : x \in \mathbb{C}^{N\times 1} \text{ and } Ax = 0\}$$

$$\operatorname{Im}_\ell A = \{yA : y \in \mathbb{C}^{1\times M}\}$$

$$\operatorname{Ker}_\ell A = \{y : y \in \mathbb{C}^{1\times M} \text{ and } yA = 0\}$$

Block matrices

$$\begin{bmatrix} A_1 & 0 & \cdots & 0 \\ 0 & A_2 & \cdots & 0 \\ \vdots & \vdots & \ddots & \vdots \\ 0 & 0 & \cdots & A_k \end{bmatrix} \text{ and } \begin{bmatrix} Y_1 \\ Y_2 \\ \vdots \\ Y_k \end{bmatrix}$$

will be denoted $\operatorname{diag}\left[A_i\right]_{i=1}^k$ and $\operatorname{col}\left[Y_i\right]_{i=1}^k$, respectively. For hermitian matrices A and B, we use $A \le B$ to designate that $B - A$ is positive semidefinite. I_M is the $M \times M$ identity matrix. Linear span of vectors $x_1, \dots, x_p$ is denoted span $\{x_1, \dots, x_p\}$. The residue of a scalar or matrix function f(z) corresponding to $z_0 \in \mathbb{C}$ is denoted

$$\operatorname{Res}_{z=z_0} f(z) = \frac{1}{2\pi i} \int_{|z-z_0|=\varepsilon} f(z)dz$$

($\varepsilon > 0$ sufficiently small). We denote by wno(f) the winding number of a nonzero continuous scalar function f(z) defined on the imaginary axis:

$$\operatorname{wno}(f) = \frac{1}{2\pi}\left[\lim_{x\to\infty} \arg f(ix) - \lim_{x\to-\infty} \arg f(ix)\right],$$

where arg z is the argument of z chosen to depend continuously on z. The well-known logarithmic property of the winding number will be often used:

$$\operatorname{wno}(fg) = \operatorname{wno}(f) + \operatorname{wno}(g).$$

2. REDUCTION TO A HOMOGENEOUS INTERPOLATION PROBLEM.

Let Δ be either the open unit disk $\mathcal{D}$ or the open left half-plane Π^-, and let $\partial\Delta$ be the boundary of Δ (including infinity if $\Delta = \Pi^-$). Suppose that we are given an $M \times N$ rational matrix function K(z) without poles on $\partial\Delta$ and having only simple poles in Δ. We wish to describe all $R \in \mathcal{R}_{M\times N}(\Delta \cup \partial\Delta)$ such that

$$\|K + R\|_\infty \le 1, \tag{2.1}$$

where

$$\|F\|_\infty = \sup\{\|F(z)\| : z \in \partial\Delta\}.$$

Using the decomposition

$$K(z) = K_1(z) + K_2(z),$$

where $K_1(z)$ has all its poles in Δ and $K_1(\infty) = 0$, while $K_2(z)$ has all its poles outside $\overline{\Delta}$, and using the obvious fact that $R \in \mathcal{R}_{M\times N}(\Delta \cup \partial\Delta)$ if and only if $K_2 + R \in \mathcal{R}_{M\times N}(\Delta \cup \partial\Delta)$, we can assume without loss of generality that all poles of K(z) are in Δ and $K(\infty) = 0$. As these poles are assumed simple, the function K(z) has the form

$$K(z) = \sum_{i=1}^{n} \left(z - z_i\right)^{-1} \gamma_i w_i,$$

where $z_1, \dots, z_n$ are distinct points in Δ, γ_i are nonzero $M \times 1$ columns, and w_i are nonzero $1 \times N$ rows.

We now proceed with a heuristic line of argument to show how one can reduce the Nehari problem to a simpler interpolation problem for rational matrix functions.

From experience with other interpolation problems (see [BGR1]) we expect that the set of rational functions of the form $F = K + R$, where $\|F\|_\infty \le 1$, is given by a linear fractional parametrization

$$F = (\Theta_{11}G + \Theta_{12})(\Theta_{21}G + \Theta_{22})^{-1}, \tag{2.2}$$

where

$$\Theta(z) = \begin{bmatrix} \Theta_{11}(z) & \Theta_{12}(z) \\ \Theta_{21}(z) & \Theta_{22}(z) \end{bmatrix}$$

is a fixed $(M + N) \times (M + N)$ rational matrix function, and $G(z)$ is any $M \times N$ rational matrix function without poles in $\Delta \cup \partial\Delta$ and having $\|G\|_\infty \le 1$. Let us look for properties of Θ which will guarantee that (2.2) does what we want. Rewrite (2.2) as

$$\begin{bmatrix} F \\ I \end{bmatrix} = \Theta \begin{bmatrix} G \\ I \end{bmatrix} \left(\Theta_{21}G + \Theta_{22}\right)^{-1}. \tag{2.3}$$

Let us assume for the moment that it is known that $\left(\Theta_{21}G + \Theta_{22}\right)^{-1}$ is analytic in $\Delta \cup \partial\Delta$ so that the only poles in $\Delta \cup \partial\Delta$ of the expression of the right hand side of (2.3) arise from Θ. From $F = K + R$ with all poles of R outside $\Delta \cup \partial\Delta$, we see that

$$\operatorname{Im} \operatorname{Res}_{z=z_i} \begin{bmatrix} F \\ I \end{bmatrix} = \operatorname{span} \begin{bmatrix} \gamma_i \\ 0 \end{bmatrix}.$$

Thus we must have:

(I) *Θ has a simple pole at $z = z_i$ with* $\operatorname{Im} \operatorname{Res}_{z=z_i} \Theta(z) = \operatorname{span} \begin{bmatrix} \gamma_i \\ 0 \end{bmatrix}$

and $\Theta(z)$ is analytic at all other points in $\Delta \cup \partial\Delta$.

Our next goal is to understand the poles of Θ^{-1}. From (2.3) we have

$$\Theta^{-1} \begin{bmatrix} F \\ I \end{bmatrix} = \begin{bmatrix} G \\ I \end{bmatrix} \left(\Theta_{21}G + \Theta_{22}\right)^{-1} \tag{2.4}$$

which is analytic on $\Delta \cup \partial\Delta$ (again, we assume the analyticity of $(\Theta_{21}G + \Theta_{22})^{-1}$). We seek the weakest possible conditions on the poles of Θ^{-1} to insure the analyticity of (2.4) on $\Delta \cup \partial\Delta$ for all F of the form $F = K + R$, where $R(z) \in \mathcal{R}_{M\times N}(\Delta \cup \partial\Delta)$. Note that an $N \times 1$ column function $\varphi(z)$ has the property that

$\begin{bmatrix} F \\ I \end{bmatrix} \varphi$ is analytic on $\Delta \cup \partial\Delta$ if and only if:

(i) φ is analytic on $\Delta \cup \partial\Delta$
(ii) $w_i \varphi(z_i) = 0$ for $i = 1, \ldots, n$.

Consider a condition slightly weaker than the analyticity of $\Theta^{-1}\begin{bmatrix} F \\ I \end{bmatrix}$ for all F of the form $F = K + R$, where $R(z) \in \mathcal{R}_{M\times N}(\Delta \cup \partial\Delta)$, namely:

$\Theta^{-1}\begin{bmatrix} F \\ I \end{bmatrix}\varphi$ *is analytic on* $\Delta \cup \partial\Delta$ *for every* $N\times 1$ *vector function* φ *satisfying* (i) *and* (ii), *and every* F *of the form* $F = K + R$, *where* $R(z) \in \mathcal{R}_{M\times N}(\Delta \cup \partial\Delta)$. (2.5)

It follows that (2.5) holds if and only if

(II) Θ^{-1} *has a simple pole at* $z = z_i$ with $\mathrm{Im}_\ell\, \mathrm{Res}_{z=z_i}\, \Theta^{-1} = \mathrm{span}\,\{[0, w_i]\}$ *and* Θ^{-1} *is analytic at all other points in* $\Delta \cup \partial\Delta$.

Note that (II) guarantees that $\Theta^{-1}\begin{bmatrix} F \\ I \end{bmatrix}$ has at worst a simple pole for all $F = K + R$, $R(z) \in \mathcal{R}_{M\times N}(\Delta \cup \partial\Delta)$.

To guarantee analyticity of $\Theta^{-1}\begin{bmatrix} F \\ I \end{bmatrix}$ at z_i we need in addition that the residue is zero at z_i; thus, letting

$$F(z) = \sum_{i=1}^{n} \gamma_i w_i (z - z_i)^{-1} + R(z),\ R(z) \in \mathcal{R}_{M\times N}(\Delta \cup \partial\Delta),$$

we have

$$\mathrm{Res}_{z=z_i}\, \Theta_{-1}\begin{bmatrix} F \\ I \end{bmatrix} = \tilde{\Theta}_{-1}^{(i)}\begin{bmatrix} R(z_i) \\ I \end{bmatrix} + \tilde{\Theta}_{0}^{(i)}\begin{bmatrix} \gamma_i w_i \\ 0 \end{bmatrix} = 0.$$

Here $\tilde{\Theta}_{-1}^{(i)}$ and $\tilde{\Theta}_{0}^{(i)}$ are the coefficients in the Laurent series of $\Theta(z)^{-1}$ at z_i:

$$\Theta(z)^{-1} = (z - z_i)^{-1}\tilde{\Theta}_{-1}^{(i)} + \tilde{\Theta}_{0}^{(i)} + (z - z_i)\,\{\text{analytic at } z_i\}.$$

In other words,

$$u\left[\tilde{\Theta}_{-1}^{(i)}\tilde{\Theta}_{0}^{(i)}\right]\begin{bmatrix} R(z_i) \\ I \\ \gamma_i w_i \\ 0 \end{bmatrix} = 0$$

for all row vectors u. Choose u_0 so that

$$u_0\, \tilde{\Theta}_{-1}^{(i)} = [0, w_i] \tag{2.6}$$

and set

$$\left[\tilde{w}_{i1}\ \tilde{w}_{i2}\right] = u_0\, \tilde{\Theta}_{0}^{(i)}. \tag{2.7}$$

Then, since $\mathrm{Im}_\ell \tilde{\Theta}_{-1}^{(i)} = \mathrm{span}\,\{[0, w_i]\}$, we see that

$$\mathrm{Im}_\ell\left[\tilde{\Theta}_{-1}^{(i)}\tilde{\Theta}_0^{(i)}\right] = \mathrm{span}\left\{\left[0, w_i, \tilde{w}_{i1}, \tilde{w}_{i2}\right]\right\} + \left\{\left[0, 0, u\tilde{\Theta}_0^{(i)}\right] : u \in \mathrm{Ker}_\ell \tilde{\Theta}_{-1}^{(i)}\right\}$$

Thus analyticity of $\Theta^{-1}\begin{bmatrix} F \\ I \end{bmatrix}$ requires in particular

$$\begin{bmatrix} 0 & w_i & \tilde{w}_{i1} & \tilde{w}_{i2} \end{bmatrix}\begin{bmatrix} R(z_i) \\ I \\ \gamma_i w_i \\ 0 \end{bmatrix} = 0$$

i.e.

(III) $$\tilde{w}_{i1}\,\gamma_i = -1, \textit{ where } \tilde{w}_{i1} \textit{ comes from (2.6), (2.7)}$$

as well as

(III') $$u\tilde{\Theta}_0^{(i)}\begin{bmatrix} \gamma_i \\ 0 \end{bmatrix} = 0 \textit{ whenever } u\tilde{\Theta}_{-1}^{(i)} = 0$$

However, from

$$u\Theta^{-1}(z)\,\Theta(z) = u$$

together with

$$\mathrm{Im}_\ell\, \mathrm{Res}_{z=z_i}\Theta(z) = \mathrm{span}\begin{bmatrix} \gamma_i \\ 0 \end{bmatrix}$$

we see that (III') is automatic. Conversely, reversing the steps shows that if (III) is satisfied (along with (I) and (II)), then $\Theta^{-1}\begin{bmatrix} F \\ I \end{bmatrix}$ is analytic at z_i. (V)

The conditions (I) – (III) are the interpolation conditions at the points $z_1, \dots, z_n$ that the function $\Theta(z)$ will have to satisfy in order that the function F given by the equality (2.2) has the property that $F = K + R$ for some $R \in \mathscr{R}_{M\times N}(\Delta \cup \partial\Delta)$.

To take care of the second property of F, viz. $\|F\|_\infty \le 1$, we require in addition that

(IV) $$(\Theta(z))^*\begin{bmatrix} I & 0 \\ 0 & -I \end{bmatrix}\Theta(z) = \begin{bmatrix} I & 0 \\ 0 & -I \end{bmatrix}, z \in \partial\Delta.$$

It is convenient to introduce the $(M+N)\times(M+N)$ matrix $J=\begin{bmatrix} I_M & 0 \\ 0 & -I_N \end{bmatrix}$. An $(M+N)\times(M+N)$ matrix X is called *J-unitary* if $X^*JX - J$. Thus, the equality (IV) simply means that $\Theta(z)$ is J-unitary on $\partial\Delta$. By (I), $\Theta(z)$ has no poles on $\partial\Delta$; the equality (IV) then implies that $\Theta(z)^{-1}$ has no poles on $\partial\Delta$. In other words, $\Theta(z)$ has no poles and no zeros on $\partial\Delta$. Next, assume

$$\begin{bmatrix} F \\ I \end{bmatrix} = \Theta \begin{bmatrix} G \\ I \end{bmatrix}$$

where $G \in \mathcal{R}_{M\times N}(\Delta \cup \partial\Delta)$ and $\|F\|_\infty \le 1$. Then

$$0 \ge F(z)^* F(z) - I =$$

$$\begin{bmatrix} G(z)^* I \end{bmatrix} \Theta(z)^* J \Theta(z) \begin{bmatrix} G(z) \\ I \end{bmatrix} = G(z)^* G(z) - I \tag{2.8}$$

for $z \in \partial\Delta$. Therefore, $\|G\|_\infty \le 1$ as well. This shows that condition (IV) is indeed applicable for our purposes. Finally, we require also

(V) $\qquad \det\Theta_{22}(z)$ *has no zeros in* $\Delta \cup \partial\Delta$.

The existence of $\Theta(z)$ with all properties (I) – (V) is not proved yet. But if such $\Theta(z)$ exists, then we have a complete solution of the Nehari problem for rational matrix functions:

THEOREM 2.1. *Let there be given a function*

$$K(z) = \sum_{i=1}^{n} (z - z_i)^{-1} \gamma_i w_i,$$

where $z_1, \ldots, z_n$ *are distinct points in* Δ, γ_i *are nonzero* $M \times 1$ *columns and* w_i *are nonzero* $1 \times N$ *rows.*

Suppose there exists an $(M+N)\times(M+N)$ *rational matrix function* $\Theta(z) = \begin{bmatrix} \Theta_{11}(z) & \Theta_{12}(z) \\ \Theta_{21}(z) & \Theta_{22}(z) \end{bmatrix}$ *satisfying* (I) – (V). *Then all rational matrix functions* F(z) *of the form*

$$F(z) = K(z) + R(z),\ R \in \mathcal{R}_{M\times N}(\Delta \cup \partial\Delta)$$

with $\|F\|_\infty \le 1$ *are given by*

$$F = (\Theta_{11}G + \Theta_{12})(\Theta_{21}G + \Theta_{22})^{-1} \tag{2.9}$$

with $G \in \mathcal{R}_{M\times N}(\Delta \cup \partial\Delta)$, $\|G\|_\infty \le 1$.

Proof. Let $G \in \mathcal{R}_{M\times N}(\Delta \cup \partial\Delta)$, $\|G\|_\infty \le 1$. Condition (I) implies that Θ_{21} and Θ_{22} are analytic in $\Delta \cup \partial\Delta$, and therefore so is $\Theta_{21}G + \Theta_{22}$. Write

$$\Theta_{21}G + \Theta_{22} = \Theta_{22}(\Theta_{22}^{-1}\Theta_{21}G + I) \tag{2.10}$$

Since Θ is J-unitary on $\partial\Delta$ and has no poles on $\partial\Delta$, we conclude (Proposition A.2 in the appendix) that $\|\Theta_{22}^{-1}\Theta_{21}\| < 1$ on $\partial\Delta$. Hence $\|\Theta_{22}^{-1}\Theta_{21}G\| < 1$ on $\partial\Delta$, and since $\Theta_{22}^{-1}\Theta_{21}G$ is also analytic in Δ, it follows that

$$\text{wno}\det\left(\Theta_{22}^{-1}\Theta_{21}G + I\right) = 0. \tag{2.11}$$

(To verify (2.11) use the standard homotopy argument: let $E(t) = t\Theta_{22}(z)^{-1}\Theta_{21}(z)G(z) + I$ and observe that wno det $E(0) = 0$, that det $E(t) \neq 0$ for $z \in \partial\Delta$ and that the winding number of a function depends continuously on that function.) Now use (2.10), the assumption (V) and the logarithmic property of the winding number to conclude that

$$\text{wno}\det\left(\Theta_{22}^{-1}\Theta_{21}G + I\right) = 0,$$

and so $\Theta_{21}G + \Theta_{22}$ has no poles in $\Delta \cup \partial\Delta$. (Engineers will recognize this argument as a form of the Small Gain Theorem.) Now write (2.9) in the form

$$\begin{bmatrix} F \\ I \end{bmatrix} = \Theta\begin{bmatrix} G \\ I \end{bmatrix}\left(\Theta_{21}G + \Theta_{22}\right)^{-1}. \tag{2.12}$$

From the properties of Θ, $\begin{bmatrix} G \\ I \end{bmatrix}$, $\Theta_{21}G + \Theta_{22}$, we know that F is analytic on $\Delta\backslash\{z_1, \dots, z_n\}$ and has at most a simple pole at each point z_i.

We next claim that F has a pole at z_i. Suppose not. From (2.12) we know that

$$\Theta^{-1}\begin{bmatrix} F \\ I \end{bmatrix} \text{ is analytic at } z_i. \tag{2.13}$$

If F is analytic at z_i then

$$\text{Res}_{z=z_i}\, \Theta^{-1}\begin{bmatrix} F \\ I \end{bmatrix} = \tilde{\Theta}_{-1}^{(i)}\begin{bmatrix} F(z_i) \\ I \end{bmatrix} = 0,$$

where $\tilde{\Theta}_{-1}^{(i)} = \text{Res}_{z=z_i}\, \Theta(z)^{-1}$. But this is impossible, because we know that

$$\text{Im}_{\ell}\,\tilde{\Theta}_{-1}^{(i)} = \text{span}\{[0 \quad w_i]\},$$

and therefore

$$[0 \quad w_i]\begin{bmatrix} F(z_i) \\ I \end{bmatrix} = 0, \text{ i.e. } w_i = 0,$$

a contradiction. Thus necessarily F has a pole at $z_1, \dots, z_n$. From (2.12)

$$\text{Im Res}_{z=z_i}\begin{bmatrix} F(z) \\ I \end{bmatrix} \subset \text{Im Res}_{z=z_i}\Theta(z) = \text{Im}\begin{bmatrix} \gamma_i \\ 0 \end{bmatrix}.$$

Thus

$$\text{Im Res}_{z=z_i}F(z) = \text{span}\,\gamma_i,$$

and

$$F(z) = \sum_{i=1}^{n} \gamma_i \hat{w}_i \left(z - z_i\right)^{-1} + R(z) \tag{2.14}$$

with $\hat{w}_i$ a nonzero row and $R \in \mathcal{R}_{M\times N}(\Delta \cup \partial\Delta)$. Next we claim $\hat{w}_i = w_i$. To see this analyze (2.13), this time with F in the form (2.14). Since $\mathrm{Im}_\ell \tilde{\Theta}_{-1}^{(i)} = \mathrm{span}\{[0 \quad w_i]\}$ it is clear that $\Theta^{-1}\begin{bmatrix} F \\ I \end{bmatrix}$ has at worst a simple pole at z_i. Therefore we compute the residue

$$0 = \mathrm{Res}_{z=z_i}\ \Theta^{-1}\begin{bmatrix} F \\ I \end{bmatrix} = \tilde{\Theta}_{-1}^{(i)}\begin{bmatrix} F(z_i) \\ I \end{bmatrix} + \tilde{\Theta}_0^{(i)}\begin{bmatrix} \gamma_i \hat{w}_i \\ 0 \end{bmatrix} = \begin{bmatrix} \tilde{\Theta}_{-1}^{(i)} & \tilde{\Theta}_0^{(i)} \end{bmatrix}\begin{bmatrix} F(z_i) \\ I \\ \gamma_i \hat{w}_i \\ 0 \end{bmatrix} \tag{2.15}$$

Letting $\begin{bmatrix} \tilde{w}_{i1} & \tilde{w}_{i2} \end{bmatrix} = u\tilde{\Theta}_0^{(i)}$, where u is a row vector such that $\begin{bmatrix} 0 & w_i \end{bmatrix} = u\tilde{\Theta}_{-1}^{(i)}$, we obtain

$$\begin{bmatrix} 0 & w_i & \tilde{w}_{i1} & \tilde{w}_{i2} \end{bmatrix}\begin{bmatrix} F_0^{(i)} \\ I \\ \gamma_i \hat{w}_i \\ 0 \end{bmatrix} = 0$$

i.e.

$$w_i = -\tilde{w}_{i1} \gamma_i \hat{w}_i$$

But by condition (III) on Θ we have $\tilde{w}_{i1}\gamma_i = -1$ and hence $w_i = \hat{w}_i$ as required.

Finally, a computation analogous to (2.8) shows that $\|F\|_\infty \le 1$, because $\|G\|_\infty \le 1$.

Conversely let F(z) be a rational matrix function of the form $F(z) = K(z) + R(z)$, $R \in \mathcal{R}_{M\times N}(\Delta \cup \partial\Delta)$, with $\|F\| \le 1$.

Define $\begin{bmatrix} G_1 \\ G_2 \end{bmatrix}$ by

$$\begin{bmatrix} F \\ I \end{bmatrix} = \Theta\begin{bmatrix} G_1 \\ G_2 \end{bmatrix}. \tag{2.16}$$

Thus,

$$I = \Theta_{21}G_1 + \Theta_{22}G_2 = \Theta_{22}\left(\Theta_{22}^{-1}\Theta_{21}G_1G_2^{-1} + I\right)G_2 \tag{2.17}$$

From Θ being J-unitary on $\partial\Delta$ and from $\|F\|_\infty < 1$, get

$$\|G_2(z)\| \ge \|G_1(z)\|, \quad z \in \partial\Delta. \tag{2.18}$$

Indeed,

$$0 \geq F(z)^*F(z) - I = [G_1(z)^* G_2(z)^*]\Theta(z)^* J\Theta(z) \begin{bmatrix} G_1(z) \\ G_2(z) \end{bmatrix} =$$

$$G_1(z)^* G_1(z) - G_2(z)^* G_2(z) \tag{2.19}$$

for $z \in \partial\Delta$, and (2.18) follows. If for some $z \in \partial\Delta$, $G_2(z)$ had degenerate rank, then necessarily by (2.19) $\begin{bmatrix} G_1(z) \\ G_2(z) \end{bmatrix}$ would also have degenerate rank. But the left side of (2.16) is maximal rank at all points of $\partial\Delta$ and Θ is a regular on $\partial\Delta$, so $\begin{bmatrix} G_1 \\ G_2 \end{bmatrix}$ necessarily is maximal rank on $\partial\Delta$. Thus $G_2(z)$ is invertible on $\partial\Delta$. Because of (2.18), we have $\|G_1(z)G_2(z)^{-1}\| \leq 1$, $z \in \partial\Delta$, and because of Proposition A.2 in the appendix we have $\|\Theta_{22}(z)^{-1}\Theta_{22}(z)\| < 1$, $z \in \partial\Delta$. Now the standard homotopy argument (cf. the proof of (2.11)) gives

$$\text{wno det}\left(\Theta_{22}^{-1}\Theta_{21}G_1G_2^{-1} + I\right) = 0$$

By assumption (V)

$$\text{wno det}\ \Theta_{22} = 0.$$

Then these together with (2.17) imply

$$\text{wno det}\ G_2 = 0. \tag{2.20}$$

We must show that $F(z) = \sum_{i=1}^{n} \gamma_i w_i (z - z_i)^{-1} + R(z)$, $R \in \mathcal{R}_{M\times N}(\Delta \cup \partial\Delta)$, forces G_1, G_2 analytic on $\Delta \cup \partial\Delta$, i.e. $\Theta^{-1}\begin{bmatrix} F \\ I \end{bmatrix}$ analytic. This follows using the local Laurent series for Θ^{-1} and $\begin{bmatrix} F \\ I \end{bmatrix}$ and using equalities

$$\begin{bmatrix} 0 & w_i \end{bmatrix} \begin{bmatrix} \gamma_i w_i \\ 0 \end{bmatrix} = 0$$

$$\begin{bmatrix} 0 & w_i & \tilde{w}_{i1} & \tilde{w}_{i2} \end{bmatrix} \begin{bmatrix} R(w_i) \\ I \\ \gamma_i w_i \\ 0 \end{bmatrix} = 0$$

(cf. (2.15)). Now (2.20) implies that G_2^{-1} is analytic in $\Delta \cup \partial\Delta$ as well, and hence $G = G_1G_2^{-1}$ is analytic on $\Delta \cup \partial\Delta$. Now (IV) implies $\|G\|_\infty \leq 1$. □

3. THE HOMOGENEOUS INTERPOLATION PROBLEM.

This section is of preliminary character, and its results will be used in the proof of Theorem 2.1. Here we consider a basic problem of interpolation of prescribed null and pole structure for a rational matrix function. No norm inequalities will be imposed here and the interpolation points need not be in the unit disc or the right half-plane. The exposition in the section is based largely in [BGR2], [BGR3]. Again, only a simple, albeit generic, situation is studied here. For the general affine interpolation problem for rational matrix function we refer the readers to [BGR1].

We are given distinct points $z_1, z_2, \dots, z_n \in \mathbb{C}$ and nonzero $1 \times p$ row vectors $\varphi_1, \varphi_2, \dots, \varphi_n$, as well as additional distinct points $w_1, w_2, \dots, w_n \in \mathbb{C}$ and nonzero $p \times 1$ column vectors $\psi_1, \psi_2, \dots, \psi_n$. To start with, we assume that the set of points $\{z_1, \dots, z_n\}$ is disjoint from the set $\{w_1, \dots, w_n\}$. The problem is to construct a rational $p \times p$ matrix function $W(z)$ with value I at infinity such that

For $i = 1, \dots, n$, $W(z)$ *is analytic at* z_i, $\det W(z)$ *has a simple zero at* z_i *and* $\varphi_i W(z_i) = 0$ (3.1i)

For $j = 1, \dots, n$, $W(z)^{-1}$ *is analytic at* w_j, $\det W(z)^{-1}$ *has a simple zero at* w_j, *and* $W(w_j)^{-1}\psi_j = 0$. (3.1ii)

The poles of $W(z)$ *occur only at* $w_1, w_2, \dots, w_n$. (3.1iii)

The poles of $W(z)^{-1}$ *occur only at* $z_1, \dots, z_n$. (3.1iv)

It is easily seen that the interpolation conditions (2.1i) – (2.1ii) can be rewritten in the form

For $i = 1, \dots, n$, $W(z)^{-1}$ *has a simple pole at* z_i *such that* $\mathrm{Im}_\ell \mathrm{Res}_{z=z_i} W(z)^{-1} = span\ \{\varphi_i\}$. (3.2i)

For $j = 1, \dots, n$, $W(z)$ *has a simple pole at* w_j *such that* $\mathrm{Im}\mathrm{Res}_{z=w_j} W(z) = span\{\psi_j\}$. (3.2ii)

In the more delicate case when the sets $\{z_1, \dots, z_n\}$ and $\{w_1, \dots, w_n\}$ overlap, the conditions (3.1i) and (3.1ii) do not make sense (because one cannot evaluate $W(z)$ and $W(z)^{-1}$ at z_i if z_i is simultaneously a zero and a pole of $W(z)$), but the conditions (3.2i) and (3.2ii) still make sense. So we replace the interpolation conditions (3.1i), (3.1ii) by (3.2i), (3.2ii) and drop the requirement that the sets $\{z_1, \dots, z_n\}$ and $\{w_1, \dots, w_n\}$ must be disjoint. In addition, in the case $z_i = w_j = \xi_{ij}$ it turns out to be natural to impose a third condition, namely,

$$\varphi_{i1}\,\psi_j = \rho_{ij} \tag{3.3}$$

where ρ_{ij} is a given number, and φ_{i1} is a row vector such that the row vector function

$$[\varphi_i + \varphi_{i1}(z - \xi_{ij})]\ W(z)$$

is analytic at ξ_{ij} with value 0 at ξ_{ij}. In what sense the condition (3.3) is natural will become apparent later on. From the fact that $W(z)^{-1}W(z) = I$ we see that a necessary condition for the problem (3.2i), (3.2ii), (3.3) to be solvable is that $\varphi_i\psi_j = 0$ whenever $z_i = w_j$. The number ρ_{ij} is called the coupling number associated with φ_i and ψ_j (note that by definition ρ_{ij} depends on φ_{i1} as well).

The solution to the interpolation problem (3.1iii), (3.1iv), (3.2i), (3.2ii), (3.3) is given by the following theorem.

THEOREM 3.1. *Let*

$$C_\pi = [\psi_1, \dots, \psi_n],\ A_\pi = \mathrm{diag}\big[w_i\big]_{i=1}^{n}$$

$$A_\zeta = \mathrm{diag}\left[z_i\right]_{i=1}^{n}, \; B_\zeta = \mathrm{col}\left[\varphi_i\right]_{i=1}^{n}$$

and define the numbers

$$s_{ij} = \varphi_i\psi_j(w_j - z_i)^{-1} \text{ if } w_j \neq z_i$$
$$s_{ij} = -\rho_{ij} \text{ if } w_j = z_i.$$

Then there exists a $p \times p$ *rational function* $W(z)$ *with* $W(\infty) = I$ *and which satisfies the conditions* (3.1iii), (3.1iv), (3.2i), (3,2ii) *and* (3.3) *if and only if* $\varphi_i\psi_j = 0$ *whenever* $z_i = w_j$ *and the matrix*

$$S = [s_{ij}]_{1 \le i,j \le n}$$

is invertible. In this case the solution $W(z)$ *is unique and is given by*

$$W(z) = I + C_\pi(zI - A_\pi)^{-1}S^{-1}B_\zeta, \tag{3.4}$$

with the inverse $W(z)^{-1}$ *given by*

$$W(z)^{-1} = I - C_\pi S^{-1}(zI - A_\zeta)^{-1}B_\zeta. \tag{3.5}$$

The matrix S constructed in Theorem 3.1 is called the *coupling matrix* associated with the interpolation data $\{\{w_j\}_{j=1}^{n}; \{z_i\}_{i=1}^{n}; \{\varphi_i\}_{i=1}^{n}; \{\psi_j\}_{j=1}^{n}; \rho_{ij}$ for any pair (i,j) with $w_j = z_i\}$.

It will be convenient to have the following lemma first before proving Theorem 3.1. We recall that a realization

$$V(z) = I + C(zI - A)^{-1}B$$

of a $p \times p$ rational matrix function $V(z)$, which is analytic and takes value I at infinity, is called *minimal* if the size of the matrix A is minimal among all realizations of $V(z)$. Equivalently the minimality of this realization means that for m large enough the matrix

$$[B, AB, \dots, A^{m-1}B]$$

has linearly independent rows, and the matrix $\mathrm{col}\left[CA^i\right]_{i=0}^{m-1}$ has linearly independent columns.

LEMMA 3.2. *Let* $W(z)$ *be a* $p \times p$ *rational matrix function with* $W(\infty) = I$. (i) *Assume that* $w_1, \dots, w_n \in \mathbb{C}$ *are the only distinct poles of* $W(z)$ *and that all poles of* $W(z)$ *are simple. Let*

$$C_\pi = [\psi_1 \cdots \psi_n]; \; A_\pi = \mathrm{diag}\left[w_i\right]_{i=1}^{n},$$

where ψ_j is a $p \times 1$ *column vector that spans the one-dimensional subspace* $\mathrm{Im}\,\mathrm{Res}_{z=w_j} W(z)$.

Then $W(z)$ *has the form*

$$W(z) = I + C_\pi(zI - A_\pi)^{-1}\tilde{B} \tag{3.6}$$

for some $n \times p$ *matrix* $\tilde{B}$ *with nonzero rows, and the realization* (3.6) *is minimal.* (ii) *Assume that* $z_1, \dots, z_n \in \mathbb{C}$ *are the only distinct poles of* $W(z)^{-1}$ *and that all poles of* $W(z)$ *are simple. Let*

$$A_\zeta = \mathrm{diag}\left[z_i\right]_{i=1}^{n}, \; B = \mathrm{col}\left[\varphi_i\right]_{i=1}^{n},$$

where φ_i *is an* $1 \times p$ *row vector that spans the one-dimensional subspace* $\mathrm{Im}_\ell\,\mathrm{Res}_{z=z_i} W(z)^{-1}$. *Then* $W(z)^{-1}$ *has the form*

$$W(z)^{-1} = I - \tilde{C}(zI - A_\zeta)^{-1}B_\zeta \tag{3.7}$$

for some $p \times n$ *matrix* $\tilde{C}$ *with nonzero columns, and the realization* (3.7) *is minimal.*

The sign $-$ before $\tilde{C}$ in (3.7) appears as a matter of convenience in anticipation of subsequent formulas.

Proof. We prove that part (i) only; the proof of (ii) is analogous. Let τ_j be an $1 \times p$ row such that

$$\mathrm{Res}_{z=w_j} W(z) = \psi_j \tau_j,$$

and let $\tilde{B} = \mathrm{col}\left[\tau_i\right]_{i=1}^{n}$. It is easy to see that the difference

$$W(z) - (I + C_\pi(zI - A_\pi)^{-1}\tilde{B}) \tag{3.8}$$

is analytic at each point $w_1, \dots, w_n$, and therefore is analytic everywhere in the complex plane, as well as at infinity. By Liouville's theorem, the matrix function (3.8) is actually a constant, and evaluation at infinity gives the desired formula (3.6). As for the minimality of (3.6), it follows from the simple observation that in any realization of $W(z)$:

$$W(z) = I + C(zI - A)^{-1}B$$

the numbers $w_1, \dots, w_n$ must be eigenvalues of A, and since these numbers are distinct, the size of A must be at least n. □

Now let $W(z)$ be a rational matrix function satisfying the hypotheses of Lemma 3.2(i) and (ii). So $W(z)$ is given by (3.6) while its inverse is given by (3.7), and these two formulas must be related to each other directly. Such a relation (which, by the way, allows us to eliminate the matrices $\tilde{B}$ and $\tilde{C}$) is described as follows.

LEMMA 3.3. *Let*

$$W(z) = I + C_\pi(zI - A_\pi)^{-1}\tilde{B} \tag{3.9}$$

and

$$W(z)^{-1} = I - \tilde{C}(zI - A_\zeta)^{-1}B_\zeta \tag{3.10}$$

be minimal realizations. Then there exists a unique invertible matrix S *such that*

$$\tilde{B} = S^{-1}B_\zeta,\ \tilde{C} = C_\pi S^{-1}$$

and

$$SA_\pi - A_\zeta S = B_\zeta C_\pi \tag{3.11}$$

Proof. We use the following basic fact on minimal realizations: if (3.9) is a realization of $W(z)$, then

$$W(z)^{-1} = I - C_\pi(zI - (A_\pi - \tilde{B}C_\pi))^{-1}\tilde{B} \tag{3.12}$$

is a realization of $W(z)^{-1}$, and if (3.9) is minimal, then so is (3.12). Thus, under the hypotheses of Lemma 3.3, the realization (3.12) is minimal. Now, by the state space isomorphism theorem, there exists a unique invertible matrix S such that

$$C_\pi = \tilde{C}S,\ A_\pi - \tilde{B}C_\pi = S^{-1}A_\zeta S,\ \tilde{B} = S^{-1}B_\zeta.$$

The second equality here can be rewritten in the form (3.11), and we are done. □

Proof of Theorem 3.1. Suppose first that a solution W(z) with $W(\infty) = I$ of the interpolation conditions (3.1iii), (3.1iv), (3.2i), (3.2ii) and (3.3) exist. Ignoring condition (3.3) for the moment, we see from Lemma 3.3 that W(z) and $W(z)^{-1}$ necessarily have minimal realizations

$$W(z) = I + C_\pi(zI - A_\pi)^{-1}S^{-1}B_\zeta$$
$$W(z)^{-1} = I - C_\pi S^{-1}(zI - A_\zeta)^{-1}B_\zeta$$

where

$$C_\pi = [\psi_1 \cdots \psi_n],\ A_\pi = \mathrm{diag}\left[w_i\right]_{i=1}^n,\ A_\zeta = \mathrm{diag}\left[z_i\right]_{i=1}^n,\ B_\zeta = \mathrm{col}\left[\varphi_i\right]_{i=1}^n$$

and S is an invertible solution of

$$SA_\pi - A_\zeta S = B_\zeta C_\pi. \tag{3.13}$$

If we let s_{ij} denote the $(i,j)^{\text{th}}$ entry of S, we see that (3.13) is equivalent to the system of scalar equations

$$s_{ij}w_j - z_i s_{ij} = \varphi_i\psi_j,\ 1 \le i,j \le n.$$

If $w_j \neq z_i$, we can solve to get

$$s_{ij} = (w_j - z_i)^{-1}\varphi_i\psi_j\ (w_j \neq z_i).$$

For i, j such that $z_i = w_j$, we see that necessarily $\varphi_i\psi_j = 0$ for any solution to exist; (we also arrived at the necessity of this condition for W(z) earlier by a more direct argument). We now analyze the impact of the interpolation condition (3.3). For each pair of indices i and j for which $z_i = w_j$, we denote this common value by ξ_{ij}. In general define row vectors $\gamma_1, \dots, \gamma_n$ by

$$\mathrm{col}\left[\gamma_i\right]_{i=1}^n = S^{-1}\mathrm{col}\left[\varphi_i\right]_{i=1}^n \tag{3.14}$$

so W(z) has the partial fraction representation

$$W(z) = I + \sum_{\ell=1}^{n}(z - w_\ell)^{-1}\psi_\ell\gamma_\ell.$$

Then W(z) has a Laurent expansion at the point $z_i = \xi_{ij} = w_j$ of the form

$$W(z) = (z - \xi_{ij})^{-1}\psi_i\gamma_j + \left[I + \sum_{\ell,\ell\neq j}(\xi_{ij} - w_\ell)^{-1}\psi_\ell\gamma_\ell\right] + (z - \xi_{ij})\,.\,[\text{analytic at } \xi_{ij}].$$

Then the condition on φ_{i1} in (3.3) means

$$\varphi_i\left[I + \sum_{\ell,\ell\neq j}(\xi_{ij} - w_\ell)^{-1}\psi_\ell\gamma_\ell\right] + \varphi_{i1}\psi_j\gamma_j = 0. \tag{3.15}$$

The interpolation condition $\varphi_{i1}\psi_j = \rho_{ij}$ (3.3) then gives

$$\varphi_i = -\rho_{ij}\gamma_j + \sum_{\ell,\ell\neq j}(w_\ell - \xi_{ij})^{-1}\varphi_i\psi_\ell\gamma_\ell \tag{3.16}$$

If in general we define a matrix $\tilde{S} = \left[\tilde{s}_{ij}\right]_{1\le i,j\le n}$ by

$$\left[\tilde{s}_{ij}\right] = \begin{cases} (w_j - z_i)^{-1}\varphi_i\psi_j & \text{if } w_j = z_i \\ -\rho_{ij} & \text{if } z_i = \xi_{ij} = w_j \end{cases}$$

and if we recall (3.14) then (3.16) can be expressed as

$$\varphi_i = \left\{ i^{\text{th}} \text{ row of } \tilde{S} \cdot S^{-1}\text{col}\left[\varphi_i\right]_{i=1}^{n} \right\} \tag{3.17}$$

whenever i is such that $z_i = w_j$ for some j. If $z_i \neq w_j$ for any j (3.17) is trivial. Thus necessarily

$$(\tilde{S} - S)S^{-1}B_{\zeta} = 0$$

or

$$\tilde{S}\ \text{col}\left[\gamma_i\right]_{i=1}^{n} = S\ \text{col}\left[\gamma_i\right]_{i=1}^{n}$$

Since any given row $\tilde{S}$ and S differ in at most one entry and moreover each of the row vectors $\gamma_1, \ldots, \gamma_n$ are nonzero, this finally forces $\tilde{S} = S$. Hence in particular $\tilde{S}$ is invertible and the necessity direction in Theorem 3.1 follows. This also establishes uniqueness; if a solution exists, necessarily it is given by the formula in the statement of the theorem.

Conversely, suppose that $\varphi_i\psi_j = 0$ whenever $z_i = w_j$ and S defined in Theorem 3.1 is invertible. Define W(z) by (3.4). By a direct check, S satisfies the Sylvester equation (3.11); so $W(z)^{-1}$ is given by (3.5). Moreover, since the column vectors $\psi_1, \ldots, \psi_n$ and row vectors $\varphi_1, \ldots, \varphi_n$ are all nonzero, these realizations can be seen to be minimal. From all these facts, the interpolation conditions (3.1iii), (3.1iv), (3.2i) and (3.2ii) all follows. Finally to prove (3.3) begin with

$$\varphi_i = \{i^{\text{th}} \text{ row of } S \cdot S^{-1}\ \text{col}\left[\varphi_i\right]_{i=1}^{n}\} = \{i^{\text{th}} \text{ row of } S\ \text{col}\left[\gamma_i\right]_{i=1}^{n}\}$$

Now comparison of (3.15) and (3.16) proves (3.3). □

4. HOMOGENEOUS INTERPOLATION PROBLEM WITH CONSTRAINTS.

In this section we will focus on the existence and formulas for the rational matrix function Θ(z) that was used in Section 2. We consider the conditions (I) – (IV), leaving out for the time being the condition (V). Also, we consider the left half-plane case in this section.

Recall the conditions:

(I) *Θ has a simple pole at* $z = z_i$ *with* $\text{Im Res}_{z=z_i} \Theta(z) = \text{span}\begin{bmatrix}\gamma_i \\ 0\end{bmatrix}$ *and Θ(z) is analytic at all other points in* $\Pi^- \cup \partial\Pi^-$.

(II) Θ^{-1} *has a simple pole at* $z = z_i$ *with* $\text{Im}_\ell\ \text{Res}_{z=z_i} \Theta^{-1}(z) = \text{span}\{[0\ \ w_i]\}$ *and* Θ^{-1} *is analytic at all other points in* $\Pi^- \cup \partial\Pi^-$.

(III) *If* $\Theta(z)^{-1} = (z - z_i)^{-1}\ \tilde{\Theta}_{-1}^{(i)} + \tilde{\Theta}_0^{(i)} + \ldots$ *and* $\begin{bmatrix}0 & w_i\end{bmatrix} = u\ \tilde{\Theta}_{-1}^{(i)}, \begin{bmatrix}\tilde{w}_{i1} & \tilde{w}_{i2}\end{bmatrix} = u\ \tilde{\Theta}_0^{(i)}$ *for some* $1 \times (M+N)$ *row vector* u, *then* $\tilde{w}_{i1}\gamma_i = -1$.

(IV) $(\Theta(z))^* \begin{bmatrix} I_M & 0 \\ 0 & -I_N \end{bmatrix} \Theta(z) = \begin{bmatrix} I_M & 0 \\ 0 & -I_N \end{bmatrix}$, $z \in \partial\Pi^-$.

We continue to use the matrices A, B, C, P and Q introduced in Sections 1 and 2:

$$A = \operatorname{diag}\left[z_i\right]_{i=1}^{n},\ B = \operatorname{col}\left[w_i\right]_{i=1}^{n},\ C = [\gamma_1\ \gamma_2\ \cdots\ \gamma_n].$$

$$P = \left[-\frac{w_i w_j^*}{z_i+\bar{z}_j}\right]_{i,j=1}^{n}\qquad,\qquad Q = \left[-\frac{\gamma_j^*\gamma_i}{\bar{z}_i+z_j}\right]_{i,j=1}^{n}$$

THEOREM 4.1. *There exists a rational matrix function $\Theta(z)$ satisfying (I) – (IV) if and only if the matrix $\Lambda = \begin{bmatrix} Q & I \\ I & P \end{bmatrix}$ is invertible. In this case such $\Theta(z)$ is unique up to multiplication on the right by a constant J-unitary matrix, and the unique such $\Theta(z)$ with $\Theta(\infty) = I$ is given by the formula*

$$\Theta(z) = I + \begin{bmatrix} C & 0 \\ 0 & B^* \end{bmatrix}\begin{bmatrix} (zI-A)^{-1} & 0 \\ 0 & (zI+A^*)^{-1} \end{bmatrix}\Lambda^{-1}\begin{bmatrix} -C^* & 0 \\ 0 & B \end{bmatrix}. \tag{4.1}$$

The proof of Theorem 4.1 will be reduced to the homogeneous interpolation problem studied in the previous section and to the usage of Theorem 3.1.

Proof of Theorem 4.1. Assume first that Λ is invertible. We will verify that the rational matrix function given by (4.1) indeed satisfies the conditions (I) – (IV).

The equalities

$$AT + PA^* = -BB^*,\ A^*Q + QA = -C^*C$$

are straightforward using the definitions of A, B, C, P and Q. Consequently,

$$\Lambda\begin{bmatrix} A & 0 \\ 0 & -A^* \end{bmatrix} - \begin{bmatrix} -A^* & 0 \\ 0 & A \end{bmatrix}\Lambda = \begin{bmatrix} -C^* & 0 \\ 0 & B \end{bmatrix}\begin{bmatrix} C & 0 \\ 0 & B^* \end{bmatrix}. \tag{4.2}$$

On the other hand, formula (4.1) implies

$$\Theta(z)^{-1} = I - \begin{bmatrix} C & 0 \\ 0 & B^* \end{bmatrix}\left\{zI - \left\{\begin{bmatrix} A & 0 \\ 0 & -A^* \end{bmatrix} - \Lambda^{-1}\begin{bmatrix} -C^* & 0 \\ 0 & B \end{bmatrix}\begin{bmatrix} C & 0 \\ 0 & B^* \end{bmatrix}\right\}\right\}^{-1}\Theta(z)^{-1}$$

$$\times\,\Lambda^{-1}\begin{bmatrix} -C^* & 0 \\ 0 & B \end{bmatrix}.$$

In view of (4.2) this formula can be rewritten as

$$\Theta(z)^{-1} = I - \begin{bmatrix} C & 0 \\ 0 & B^* \end{bmatrix}\left\{zI - \Lambda^{-1}\begin{bmatrix} -A^* & 0 \\ 0 & A \end{bmatrix}\Lambda\right\}^{-1}\Lambda^{-1}\begin{bmatrix} -C^* & 0 \\ 0 & B \end{bmatrix} =$$

$$I - \begin{bmatrix} C & 0 \\ 0 & B^* \end{bmatrix}\Lambda^{-1}\begin{bmatrix} (zI+A^*)^{-1} & 0 \\ 0 & (zI-A)^{-1} \end{bmatrix}^{-1}\begin{bmatrix} -C^* & 0 \\ 0 & B \end{bmatrix}. \tag{4.3}$$

We are now ready to prove that $\Theta(z)$ satisfies the property (IV). It will be convenient to rewrite this property in the form

$$\begin{bmatrix} I & 0 \\ 0 & -I \end{bmatrix} \Theta(z) = \left(\Theta(-\bar{z})\right)^{-1*} \begin{bmatrix} I & 0 \\ 0 & -I \end{bmatrix}, \tag{4.4}$$

where the equality holds for every $z \in \mathbb{C}$ which is not a pole of $\Theta(z)$. By using the formulas (4.1) and (4.3) the verification of (4.4) amounts to the equality

$$\begin{bmatrix} I & 0 \\ 0 & -I \end{bmatrix} \left\{ I + \begin{bmatrix} C & 0 \\ 0 & B^* \end{bmatrix} \begin{bmatrix} (zI - A)^{-1} & 0 \\ 0 & (zI + A^*)^{-1} \end{bmatrix} \Lambda^{-1} \begin{bmatrix} -C^* & 0 \\ 0 & B \end{bmatrix} \right\} =$$

$$\left\{ I - \begin{bmatrix} -C & 0 \\ 0 & B^* \end{bmatrix} \begin{bmatrix} (-zI + A)^{-1} & 0 \\ 0 & (-zI - A^*)^{-1} \end{bmatrix} \Lambda^{-1*} \begin{bmatrix} C^* & 0 \\ 0 & B \end{bmatrix} \right\} \times \begin{bmatrix} I & 0 \\ 0 & -I \end{bmatrix}.$$

This equality is easily seen, in particular because $\Lambda = \Lambda^*$.

It will be convenient to rewrite (III) in the form as it was used in Section 3:

(III') *If* $\begin{bmatrix} \tilde{w}_{i1} & \tilde{w}_{i2} \end{bmatrix}$ *is such that the function* $\left\{ \begin{bmatrix} 0 & w_i \end{bmatrix} + \begin{bmatrix} \tilde{w}_{i1} & \tilde{w}_{i2} \end{bmatrix} (z - z_i) \right\} \Theta(z)$ *is analytic at* z_i *with value* 0 *at* z_i, *then*

$$\tilde{w}_{i1} \gamma_i = -1$$

One checks easily that (III) is equivalent to (III').

Since $\Theta(z)$ satisfies (IV), by using the formula (4.4), one easily verifies that $\Theta(z)$ is analytic and invertible in the closed right half-plane $\{z \in \mathbb{C} : \mathrm{Re} z \geq 0\} \cup \{\infty\}$ except for the points $-\bar{z}_1, \ldots, -\bar{z}_n$. At these points both $\Theta(z)$ and $\Theta(z)^{-1}$ have a simple pole, and

$$\mathrm{Im}\, \mathrm{Res}_{z=-\bar{z}_i}\, \Theta(z) = \mathrm{span}\left\{ \begin{bmatrix} 0 \\ w_i^* \end{bmatrix} \right\}. \tag{4.5}$$

$$\mathrm{Im}_\ell\, \mathrm{Res}_{z=-\bar{z}_i}\, \Theta(z)\text{-}1 = \mathrm{span}\left\{ \begin{bmatrix} -\gamma_i^* & 0 \end{bmatrix} \right\}. \tag{4.6}$$

Also, if $\tilde{\gamma}_{i1}$ and $\tilde{\gamma}_{i2}$ are such that the function

$$\left\{ \begin{bmatrix} -\gamma_i^* & 0 \end{bmatrix} + \begin{bmatrix} \tilde{\gamma}_{i1} & \tilde{\gamma}_{i2} \end{bmatrix} (z + \bar{z}_i) \right\} \Theta(z)$$

is analytic at $-\bar{z}_i$ with value 0 at $-\bar{z}_i$, then

$$\tilde{\gamma}_{i2} w_i^* = -1. \tag{4.7}$$

Consider the matrix S given by the following formula:

$$S = \begin{bmatrix} S_1 & S_2 \\ S_3 & S_4 \end{bmatrix}, \tag{4.8}$$

where

$$S_1 = \left[s_{ij}^{(1)}\right]_{i,j=1}^{n}, \quad s_{ij}^{(1)} = \begin{cases} 1 & \text{if } \quad i=j \\ \begin{bmatrix} 0 & w_i \end{bmatrix}\begin{bmatrix} \gamma_j \\ 0 \end{bmatrix}\left(z_j - z_i\right)^{-1} = 0 & \text{if } i \neq j \end{cases}$$

$$S_2 = \left[s_{ij}^{(2)}\right]_{i,j=1}^{n}, \quad s_{ij}^{(2)} = \begin{bmatrix} 0 & w_i \end{bmatrix}\begin{bmatrix} 0 \\ w_j^* \end{bmatrix}\left(-\bar{z}_j - z_i\right)^{-1}$$

$$S_3 = \left[s_{ij}^{(3)}\right]_{i,j=1}^{n}, \quad s_{ij}^{(3)} = \begin{bmatrix} -\gamma_i^* & 0 \end{bmatrix}\begin{bmatrix} \gamma_j \\ 0 \end{bmatrix}\left(z_j + \bar{z}_i\right)^{-1}$$

$$S_4 = \left[s_{ij}^{(4)}\right]_{i,j=1}^{n}, \quad s_{ij}^{(4)} = \begin{cases} 1 & \text{if } \quad i=j \\ \begin{bmatrix} -\gamma_i^* & 0 \end{bmatrix}\begin{bmatrix} 0 \\ w_j^* \end{bmatrix}\left(-\bar{z}_j + \bar{z}_i\right)^{-1} = 0 & \text{if } i \neq j \end{cases}.$$

We see that

$$S = \begin{bmatrix} I & P \\ Q & I \end{bmatrix} = \begin{bmatrix} 0 & I \\ I & 0 \end{bmatrix}\Lambda. \tag{4.9}$$

It's now clear that Θ given by (4.1) is constructed exactly as is W in Theorem 3.1 (see (3.4)) if the data in Theorem 3.1 is specialized to have the form:

2n in place of n

$$\begin{bmatrix} \gamma_1 \\ 0 \end{bmatrix}, \cdots \begin{bmatrix} \gamma_n \\ 0 \end{bmatrix}, \begin{bmatrix} 0 \\ -w_1^* \end{bmatrix}, \cdots, \begin{bmatrix} 0 \\ -w_n^* \end{bmatrix} \text{ in place of } \psi_1, \dots, \psi_n$$

$$z_1, \cdots, z_n, -\bar{z}_1, \cdots, -\bar{z}_n \text{ in place of } w_1, \cdots, w_n, w_{n+1}, \cdots, w_{2n}$$

$$-\bar{z}_1, \cdots, -\bar{z}_n, z_1, \cdots, z_n \text{ in place of } z_1, \cdots, z_n, z_{n+1}, \cdots, z_{2n}$$

$$\begin{bmatrix} -\gamma_1^* & 0 \end{bmatrix}, \dots, \begin{bmatrix} -\gamma_n^* & 0 \end{bmatrix}, \begin{bmatrix} 0 & w_1 \end{bmatrix}, \dots, \begin{bmatrix} 0 & w_n \end{bmatrix} \text{in place of } \varphi_1, \dots, \varphi_n, \varphi_{n+1}, \dots, \varphi_{2n}$$

and

$$\rho_{1,n+1} = \rho_{2,n+2} = \rho_{n,2n} = \rho_{n+1,1} = \cdots = \rho_{2n,n} = -1.$$

Hence by Theorem 3.1, $\Theta(z)$ given by (4.1) satisfies (I), (II) and (III').

Conversely, assume there exists a rational matrix function $\Theta(z)$ satisfying the properties (I), (II), (III') and (IV). Rewriting $\Theta(z)$ in the form (4.4), we see that $\Theta(z)$ satisfies also (4.5), (4.6) and (4.7) and is analytic and invertible in $\mathbb{C} \cup \{\infty\}$, except for points $z_1, \dots, z_n, -\bar{z}_1, \dots, -\bar{z}_n$. By Theorem 3.1, the matrix S defined by (4.8) is invertible. In view of (4.9) Λ is invertible as well. □

5. CONNECTION WITH THE HANKEL OPERATOR.

Another ingredient that we need for the solution of the Nehari problem is the connection with the Hankel operator. For the expert reader, this material is well known in much greater generality; see e.g. [Gl, BGR1].

As before, let Π^- be the open left half-plane, and let

$$K(z) = \sum_{i=1}^{n} (z - z_i)^{-1}\gamma_i w_i, \tag{5.1}$$

where $z_1, \dots, z_n \in \Pi^-$; γ_i are nonzero $M \times 1$ columns, and w_i are nonzero $1 \times N$ rows. In this section we express the norm of the Hankel operator associated with K(z) in therms of solution of certain Lyapunov equations associated with z_i, γ_i, w_i. Throughout this section, we let

$$A = \mathrm{diag}\left[z_i\right]_{i=1}^{n} \quad ; \quad B = \mathrm{col}\left[w_i\right]_{i=1}^{n} \quad ; \quad C = \left[\gamma_1, \dots, \gamma_n\right],$$

and we write K(z) in the form

$$K(z) = C(zI - A)^{-1}B.$$

The Hankel operator $\mathcal{H}_K$ associated with K(z) will be defined as an integral operator from $L_N^2(0,\infty)$ into $L_M^2(0,\infty)$ ($\mathbb{C}^N$ and $\mathbb{C}^M$-valued L^2 space over $(0,\infty)$, respectively, with respect to the Lebesque measure). Observe that K(z) has an inverse Laplace transform $\breve{K}(t)$ the matrix entries of which are absolutely integrable over $(0,\infty)$:

$$K(z) = \int_0^{\infty} e^{-tz}\, \breve{K}(t)dt, \ \mathrm{Re}\, z \geq 0.$$

Actually, $\breve{K}(t)$ is exponentially decaying, i.e.

$$\|\breve{K}(t)\| \leq \alpha e^{-\beta t}, \ t \geq 0, \tag{5.2}$$

where $\alpha > 0$ and $\beta > 0$ are constants independent of t (β can be taken any positive number such that all z_i have real parts smaller than $-\beta$). We define $\mathcal{H}_K : L_N^2(0,\infty) \to L_M^2(0,\infty)$ as the integral operator

$$(\mathcal{H}_K f)(t) = \int_0^{\infty} \breve{K}(s + t)f(s)ds \quad , \quad f \in L_N^2(0, \infty). \tag{5.3}$$

Because of (5.2) the linear operator $\mathcal{H}_K$ is bounded, and we will see that $\mathcal{H}_K$ is actually a finite rank operator. Let $\|\mathcal{H}_K\|$ denote the induced operator norm of $\mathcal{H}_K$ as an operator from $L_N^2(0, \infty)$ to $L_M^2(0, \infty)$:

$$\|\mathcal{H}_K\| = \sup\{\|\mathcal{H}_K u\|_2 : \in L_N^2(0, \infty), \|u\|_2 \leq 1\},$$

where $\|\cdot\|_2$ stands for the norm in the Hilbert spaces $L_N^2(0, \infty)$ and $L_M^2(0, \infty)$. Introduce the $n \times n$ hermitian matrices

$$P = \left[-\frac{w_i w_j^*}{z_i + \bar{z}_j} \right]_{i,j=1}^{n}$$

and

$$Q = \left[-\frac{\gamma_i^* \gamma_j}{\bar{z}_i + z_j} \right]_{i,j=1}^{n} .$$

The matrices P and Q are called *controllability and observability gramians*, respectively, associated with the points $z_1, \dots, z_n$, the columns $\gamma_1, \dots, \gamma_n$ and the rows $w_1, \dots, w_n$. We collect the properties of P and Q that will be used later in the following proposition.

PROPOSITION 5.1 (i).*The controllability and observability gramians satisfy the Lyapunov equations*

$$AP + PA^* = -BB^* \tag{5.4}$$

and

$$A^*Q + QA = -C^*C \tag{5.5}$$

(ii) *The formulas*

$$P = \int_0^\infty e^{sA} BB^* e^{sA^*} ds \tag{5.6}$$

$$Q = \int_0^\infty e^{sA^*} CC^* e^{sA} ds \tag{5.7}$$

hold.

(iii) *Both* P *and* Q *are positive definite.*

Proof. The equations (5.4) and (5.5) can be verified directly. Since A (and therefore also A^*) has all eigenvalues in Π^-, the solutions to (5.4) and (5.5) are unique. By direct substitution it is verified that the right hand sides of (5.6) and (5.7) satisfy the equations (5.4) and (5.5), respectively. By the uniqueness of solutions we must have equalities (5.6) and (5.7). These equalities also show that P and Q are positive semidefinite. To verify (iii) it is sufficient to show that KerQ = 0 and KerP = 0. We verify only that KerQ = 0; the proof of KerP = 0 is analogous. Let Qx = 0. The equation (5.5) implies $-x^*C^*Cx = 0$, and therefore Cx = 0. Now

$$QAx = -C^*Cx - A^*Qx = 0,$$

and therefore KerQ is A-invariant subspace such that Cx = 0 for all $x \in$ KerQ.

But the form of A (in particular, the property that $z_1, \dots, z_n$ are distinct) and of C (in particular, the property $\gamma_i \neq 0$) imply that the only A-invariant subspace on which C is zero is the zero subspace. So KerQ = {0}. □

The formulas (5.6) and (5.7) suggest that the following two operators will be useful: the controllability operator $\Xi : L_N^2(0, \infty) \to \mathbb{C}^n$ given by

$$\Xi f = \int_0^\infty e^{sA} Bf(s)ds, \qquad f \in L_N^2(0,\infty) \tag{5.8}$$

and the observability operator $\Omega : \mathbb{C}^n \to L_M^2(0,\infty)$ given by

$$(\Omega x)(t) = Ce^{tA}x,\ 0 \le t < \infty. \tag{5.9}$$

Note that these operators are well-defined because all the eigenvalues of A are in Π^-. The operator Ξ is surjective. Indeed, for a fixed $x \in \mathbb{C}^N$ and fixed $\gamma > 0$ let $c_\gamma \in L_N^2(0,\infty)$ be defined by $c_\gamma(t) = x$ if $0 \le t \le \gamma$, $c_\gamma(t) = 0$ if $t > \gamma$. One verifies that

$$\Xi c_\gamma = (e_{\gamma A} - I)A^{-1}Bx = \gamma \sum_{m=0}^{\infty} \frac{1}{(m+1)} \gamma^m A^m Bx \in \text{Range}(\Xi).$$

Consequently, all coefficients of this power series in γ are in the range of Ξ; in other words, $A^m Bx \in \text{Range}(\Xi)$ for $m = 0,1,\ldots$ and every $x \in \mathbb{C}^N$. As $z_i \neq z_j$ for $i \neq j$ and the rows of B are nonzero, we obtain $\text{Range}(\Xi) = \mathbb{C}^n$ as claimed. Also, Ω is injective since the columns of C are nonzero (and $z_i \neq z_j$ for $i \neq j$). We indicate the equality $Q = \Omega^* \Omega$, which can be verified by writing out

$$<\Omega x, \Omega x> = < Qx, x>$$

(here $<\cdot,\cdot>$ is the standard scalar product in $\mathbb{C}^n$) and using (5.7). Analogously, $P = \Xi\, \Xi^*$.

We now express the singular values of the Hankel operator in terms of the controllability and observability gramians.

THEOREM 5.2. *Let* K(z) *be a given rational matrix function in the form (5.1), where* $z_1, \ldots, z_n \in \Pi^-$; γ_i *are nonzero* $M \times 1$ *columns, and* w_i *are nonzero* $1 \times N$ *rows. Further, let* $\mathcal{H}_K$ *be the associated Hankel operator (given by (5.3)), and let* P *and* Q *be the associated controllability and observability gramians, respectively. Then:*

(i) $\mathcal{H}_K$ *has finite rank equal to* n

(ii) *the equalities*

$$s_k(\mathcal{H}_K) = \sqrt{\lambda_k(PQ)},\ k = 1, \ldots, n$$

hold, where $s_k(\mathcal{H}_K)$ *are the nonzero singular values of* $\mathcal{H}_K$ *arranged in the nonincreasing order, and* $\lambda_k(PQ)$ *are the eigenvalues of* PQ *(also arranged in the nonincreasing order).*

In particular,

$$\|\mathcal{H}_K\|^2 = \lambda_{max}(PQ),$$

the largest eigenvalue of PQ.

Observe that since both P and Q are positive definite, the eigenvalues of PQ are positive.

Proof. We check first that the factorization

$$\mathcal{H}_K = \Omega\Xi \tag{5.9}$$

holds. To see this, note that $K(z) = C(zI - A)^{-1}B$ is the Laplace transform of $\breve{K}(t) = Ce^{tA}B$. Thus

$$[\mathcal{H}_K(f)](t) = \int_0^\infty Ce^{(t+s)A}Bf(s)ds = Ce^{tA}\int_0^\infty e^{sA}Bf(s)ds = \Omega\Xi f$$

from the definitions (5.8) and (5.9).

The factorization (5.10) implies, in particular, that $\mathcal{H}_K$ is a finite rank operator. Moreover, since Ω is surjective and Ξ is injective, the rank of $\mathcal{H}_K$ is equal to the size of A. This proves (i).

For the proof of part (ii) write (for $k = 1, \ldots, n$):

$$(s_k(\mathcal{H}_K))^2 = (s_k(\Omega\Xi))^2 = \lambda_k(\Xi^*\Omega^*\Omega\Xi).$$

Use now the general result that for linear bounded operators $S : X \to Y$, $T : Y \to X$ acting between Banach spaces X and Y, the nonzero spectrum of ST coincides with the nonzero spectrum of TS (to verify this fact observe the equality

$$\begin{bmatrix} I & \lambda S \\ 0 & I \end{bmatrix}^{-1}\begin{bmatrix} I-\lambda ST & 0 \\ -T & I \end{bmatrix}\begin{bmatrix} I & \lambda S \\ 0 & I \end{bmatrix} = \begin{bmatrix} I & 0 \\ -T & I-\lambda TS \end{bmatrix}$$

where $\lambda \in \mathbf{C}$). Hence

$$(s_k(\mathcal{H}_K))^2 = \lambda_k(\Xi\Xi^*\Omega^*\Omega) = \lambda_k(PQ).$$ □

We conclude this section with a norm inequality for the Hankel operator. First, we extend the definition of the Hankel operator to the class of rational $N \times M$ matrix functions F(z) without poles on the imaginary line (including infinity). Using the partial fractions, we can write

$$F(z) = F_1(z) + F_2(z),$$

where F_1 has all poles in the open left half-plane and $F_1(\infty) = 0$, while F_2 is analytic in the closed left hand-plane an at infinity. Now the Hankel operator $\mathcal{H}_F$ is defined by the formula (5.3) applied to F_1:

$$(\mathcal{H}_F f)(t) = \int_0^\infty \check{F}_1(s+t)f(s)ds, \quad f \in L^2_N(0,\infty).$$

THEOREM 5.3. *For any* $M \times N$ *rational matrix function* F(z) *without poles on the imaginary line (including infinity) the inequality*

$$\|\mathcal{H}_F\| \le \|F\|_\infty := \sup\{\|F(z)\| : z \in \partial\Pi^-\} \tag{5.11}$$

holds, where $\partial\Pi^- = i\,\mathbf{R} \cup \{\infty\}$ *is the boundary of* Π^-.

Proof. We use the Laplace transforms. Let $H^2_M(\Pi^+)$ be the Hardy space of M-dimensional vector valued square summable functions on the imaginary line that are non-tangential limits (almost everywhere in the Lebesque measure) of analytic functions in the open right half-plane Π^+. The Hardy space $H^2_N(\Pi^-)$ is defined analogously, with respect to the open left half-plane Π^-. Introduce the Laplace transforms

$$\mathcal{F}_+ : L^2_M(0,\infty) \to H^2_M(\Pi^+)$$

$$\mathcal{F}_- : L^2_N(0,\infty) \to H^2_N(\Pi^-)$$

given by

$$(\mathcal{F}_+h)(z) = \int_0^\infty e^{-tz}h(t)dt, \qquad h \in L^2_M(0,\infty)$$

$$(\mathcal{F}_-h)(z) = \int_0^\infty e^{tz}h(t)dt, \qquad h \in L^2_N(0,\infty).$$

It is well-known that $\mathcal{F}_+$ and $\mathcal{F}_-$ are isometric automorphisms. One verifies that

$$\mathcal{F}_+\mathcal{H}_F\mathcal{F}_-^* = P_{H^2_M(\Pi^+)}\mathcal{M}_F|H^2_N(\Pi^-),$$

where $P_{H^2_M(\Pi^+)}$ is the orthogonal projection (in the Hilbert space $L^2_M(i\mathbb{R})$) onto $H^2_M(\Pi^+)$ and $\mathcal{M}_F$ is the multiplication operator induced by F:

$$(\mathcal{M}_F f)(z) = F(z)f(z),\ f \in L^2_N\ (i\mathbb{R}).$$

Now for every $h(z) \in H^2_N(\Pi^-)$ we have

$$\|\mathcal{H}_F\mathcal{F}_-^*h(z)\|^2 = \|\mathcal{F}_+\mathcal{H}_F\mathcal{F}_-^*h(z)\|^2 =$$

$$\|P_{H^2_M(\Pi^+)}\mathcal{M}_F h(z)^2\| \le \|\mathcal{M}_F h(z)^2\| = \int_{-\infty}^{\infty}\|F(iw)h(iw)\|^2dw \le$$

$$\max_{-\infty<w<\infty}\|F(iw)\|^2 \cdot \int_{-\infty}^{\infty}\|h(iw)\|^2dw = \|F\|^2_\infty\|h\|^2 = \|F\|^2_\infty\|\mathcal{F}_-^*h(z)\|^2,$$

where in the last equality we used the fact that $\mathcal{F}_-^*$ is an isometric isomorphism. The inequality (5.11) is now evident. □

6. PROOF OF THE MAIN THEOREM.

We give here the proof of Theorem 1.1. A preliminary lemma is needed. Let P and Q be defined as in Theorem 1.1.

LEMMA 6.1. *Assume that* 1 *is not an eigenvalue of* PQ, *and set* $Z = (I - PQ)^{-1}$. *Then the number of eigenvalues of* PQ *bigger than 1 (counted with multiplicities) equals the number of negative eigenvalues of* QZ. *In particular,* $\lambda_{max}(PQ) < 1$ *if and only if* QZ *is positive definite.*

Proof. By Proposition 5.1 both P and Q are positive definite matrices. So QZ is invertible, and the equality $QZ = Z^*Q$ implies that QZ is Hermitian. So the second statement in the lemma is indeed a particular case of the first statement.

Next note that

$$QZ = Q(I-PQ)^{-1} = Q^{1/2}(Q^{-1/2} - PQ^{1/2})^{-1} = Q^{1/2}(I - Q^{1/2}PQ^{1/2})^{-1}Q^{1/2}.$$

Thus the number of negative eigenvalues of QZ is the same as the number of negative eigenvalues of $I - Q^{1/2}PQ^{1/2}$, that is, the number of eigenvalues of $Q^{1/2}PQ^{1/2}$ which are larger than 1. By the general fact that XY and YX have the same nonzero eigenvalues, we see that this is the same as the number of eigenvalues of PQ larger than 1. □

Proof of Theorem 1.1. Assume there exists $R \in \mathcal{R}_{M\times N}(\Pi^- \cup \partial\Pi^-)$ such that (1.1) holds. Let $F = K + R$. Use the Hankel operators $\mathcal{H}_K$ and $\mathcal{H}_F$ introduced in Section 5. We have $\mathcal{H}_K = \mathcal{H}_F$ and

$$\|\mathcal{H}_K\| = \|\mathcal{H}_F\| \le \|F\|_\infty \le 1,$$

where Theorem 5.3 was used. By Theorem 5.2 $\lambda_1(PQ) \le 1$, and, because of our assumption that $I - PQ$ is nonsingular, actually $\lambda_1(PQ) < 1$.

Conversely, assume $\lambda_1(PQ) < 1$. By Lemma 6.1 QZ is positive definite, where $Z = (I - PQ)^{-1}$. At this point we use Theorem 4.1, which is applicable because

$$\Lambda = \begin{bmatrix} Q & I \\ I & P \end{bmatrix} = \begin{bmatrix} I & 0 \\ P & I \end{bmatrix}\begin{bmatrix} 0 & I \\ I - PQ & 0 \end{bmatrix}\begin{bmatrix} I & 0 \\ Q & I \end{bmatrix}$$

is invertible. Let $\Theta(z)$ be given by formula (4.1). By Theorem 4.1 we know that $\Theta(z)$ satisfies the properties (I) – (IV). In view of Theorem 2.1 we only have to prove that $\Theta(z)$ also satisfies (V), i.e.

$$\det\Theta_{22}(z) \textit{ has no zeros in } \Pi^- \cup \partial\Pi^-. \tag{6.1}$$

In this proof the condition $QZ > 0$ will be used.

Using formula (4.1), a straightforward calculation shows that

$$\frac{J - \Theta(z)\, J(\Theta(w))^*}{-z - \overline{w}} = -\begin{bmatrix} C & 0 \\ 0 & B^* \end{bmatrix}\begin{bmatrix} (zI - A)^{-1} & 0 \\ 0 & (zI + A^*)^{-1} \end{bmatrix}\Lambda^{-1} \times$$

$$\begin{bmatrix} \left(\overline{w}I - A^*\right)^{-1} & 0 \\ 0 & (\overline{w}I + A)^{-1} \end{bmatrix}\begin{bmatrix} C^* & 0 \\ 0 & B \end{bmatrix},$$

where, as usual, $J = \begin{bmatrix} I_M & 0 \\ 0 & -I_N \end{bmatrix}$. As $\Lambda^{-1} = \begin{bmatrix} -ZP & Z \\ Z^* & -QZ \end{bmatrix}$, it follows that

$$K(z,w) := \begin{bmatrix} 0 & I_N \end{bmatrix}\frac{J - \Theta(z)\, J(\Theta(w))^*}{-z - \overline{w}}\begin{bmatrix} 0 \\ I_N \end{bmatrix} =$$

$$\frac{1}{-z - \overline{w}} B^*(zI - A^*)^{-1}QZ(\overline{w}I + A)^{-1}B.$$

In view of the condition $QZ > 0$ we obtain

$$K(z, z) \ge 0 \tag{6.2}$$

for $z \in \Pi^-$. On the other hand, using the partition

$$\Theta(z) = \begin{bmatrix} \Theta_{11}(z) & \Theta_{12}(z) \\ \Theta_{21}(z) & \Theta_{22}(z) \end{bmatrix}$$

we have

$$K(z, w) = \frac{1}{-z - \overline{w}}\Big[-I + \Theta_{22}(z)\big(\Theta_{22}(w)\big)^* - \Theta_{21}(z)\big(\Theta_{21}(w)\big)^*\Big] =$$

$$\frac{1}{-z - \overline{w}}\Theta_{22}(z)\Big[I - \Theta_{22}(z)^{-1}\ \big(\Theta_{22}(w)\big)^{*-1} -$$

$$\Theta_{22}(z)^{-1}\Theta_{21}(z)\ \big(\Theta_{21}(w)\big)^*\big(\Theta_{22}(w)\big)^{*-1}\Big]\Theta_{22}(w)^*.$$

Comparison with (6.2) yields (for $z \in \Pi^-$)

$$I - \Theta_{22}(z)^{-1}\big(\Theta_{22}(z)\big)^{*-1} - \Theta_{22}(z)^{-1}\Theta_{21}(z)\big(\Theta_{21}(z)\big)^*\big(\Theta_{22}(z)\big)^{*-1} \geq 0.$$

In other words, for $z \in \Pi^-$ the matrix valued function $[\Theta_{22}(z)^{-1}\Theta_{21}(z), \Theta_{22}(z)^{-1}]$ is a contraction, and therefore $\Theta_{22}(z)^{-1}$ is a contraction as well. Thus, $\Theta_{22}(z)$ cannot have poles in Π^-, and the property (6.1) follows. □

7. THE DISK CASE.

In this section we review the Nehari interpolation problem for the case of the unit disk. The basic ideas and methods are the same as in the proof of Theorem 1.1, and therefore many details will be omitted here.

Let $\mathcal{D}$ be the open unit disk $\{z \in \mathbb{C} : (z) \leq 1\}$ with the boundary $\partial\mathcal{D} = \{z \in \mathbb{C} : |z| = 1\}$. Consider an $M \times N$ rational matrix function $K(z)$ which has only simple poles in $\mathcal{D}$ and has no poles on $\partial\mathcal{D}$. The Nehari interpolation problem (in the unit disk) is: Describe all $R \in \mathcal{R}_{M\times N}$ $(\mathcal{D} \cup \partial\mathcal{D})$ such that

$$\|K + R\|_\infty \leq 1,$$

where now

$$\|F\|_\infty = \sup\{\|F(z)\| : z \in \partial\mathcal{D}\}$$

for a rational matrix function $F(z)$ with no poles on $\partial\mathcal{D}$. The result analogous to Theorem 1.1 is stated as follows:

THEOREM 7.1. *Let $z_1, \dots, z_n$ be the poles of $K(z)$ in $\mathcal{D}$, and let $Q_i = \mathrm{Res}_{z=z_i} K(z)$ written as $Q_i = \gamma_i w_i$, where γ_i is a nonzero $M \times 1$ column and w_i is a nonzero $1 \times N$ row. Let*

$$P = \left[\frac{w_i w_j^*}{1 - z_i \overline{z}_j}\right]_{i,j=1}^n \quad , \quad Q = \left[\frac{\gamma_i^* \gamma_j}{1 - \overline{z}_i z_j}\right]_{i,j=1}^n \tag{7.1}$$

be an $n \times n$ matrices (both P and Q turn out to be positive definite). Then there exists $R \in \mathcal{R}_{M\times N}(\mathcal{D} \cup \partial\mathcal{D})$ such that

$$\|K + R\|_\infty \leq 1 \tag{7.2}$$

if and only if $\lambda_{max}(PQ)$, the maximal eigenvalue of PQ, does not exceed 1. If $\lambda_{max}(PQ) < 1$, then the rational matrix functions $F(z) = K(z) + R(z)$ with the property (7.1) are characterized as matrix functions of the form

$$F(z) = (\Theta_{11}(z)G(z) + \Theta_{12}(z))(\Theta_{21}(z)G(z) + \Theta_{22}(z))^{-1},$$

where G(z) *is an arbitrary rational* $M \times N$ *matrix function without poles in* $\mathcal{D} \cup \partial\mathcal{D}$ *and such that* $\|G\|_\infty \le 1$, *while*

$$\Theta(z) = \begin{bmatrix} \Theta_{11}(z) & \Theta_{12}(z) \\ \Theta_{21}(z) & \Theta_{22}(z) \end{bmatrix}$$

is a fixed $(M+N) \times (M+N)$ *rational matrix function given by the formula*

$$\Theta(z) = I + (z - z_0)\begin{bmatrix} C & 0 \\ 0 & B^* \end{bmatrix}\begin{bmatrix} (zI - A)^{-1} & 0 \\ 0 & \left(I - zA^*\right)^{-1} \end{bmatrix} \times$$

$$\begin{bmatrix} -Q & I \\ I & -P \end{bmatrix}^{-1}\begin{bmatrix} \left(I - z_0A^*\right)^{-1}C^* & 0 \\ 0 & \left(A - z_0I\right)^{-1}B \end{bmatrix}.$$

Here z_0 *is an arbitrary point chosen on the unit circle, and*

$$A = \operatorname{diag}\left[z_i\right]_{i=1}^n, \quad C = \left[\gamma_1\gamma_2\dots\gamma_n\right], \; B = \operatorname{col}\left[w_i\right]_{i=1}^n .$$

Observe that because of the equality

$$\begin{bmatrix} -Q & I \\ I & -P \end{bmatrix} = \begin{bmatrix} I & 0 \\ -P & I \end{bmatrix}\begin{bmatrix} 0 & I \\ I - PQ & 0 \end{bmatrix}\begin{bmatrix} I & 0 \\ -Q & I \end{bmatrix}$$

the matrix $\begin{bmatrix} -Q & I \\ I & -P \end{bmatrix}$ is invertible under the hypothesis $\lambda_{max}(PQ) < 1$.

The proof of Theorem 7.1 is considerably simplified if we assume that $z_i \neq 0 (i = 1, \dots, n)$. In this case $\Theta(z)$ has no poles or zeros at infinity, and (up to multiplication on the right by a constant J-unitary matrix, where $J = \begin{bmatrix} I_M & 0 \\ 0 & -I_N \end{bmatrix}$) the following formula can be given:

$$\Theta(z) = D + \begin{bmatrix} C & 0 \\ 0 & B^* \end{bmatrix}\begin{bmatrix} (zI - A)^{-1} & 0 \\ 0 & \left(I - zA^*\right)^{-1} \end{bmatrix}\begin{bmatrix} -Q & I \\ I & -P \end{bmatrix}^{-1} \cdot \begin{bmatrix} A^{*-1}C^* & 0 \\ 0 & B^* \end{bmatrix} D,$$

where

$$D = I - \begin{bmatrix} C & 0 \\ 0 & -B^*A^{*-1} \end{bmatrix}\begin{bmatrix} -Q & I \\ I & -P \end{bmatrix}^{-1}\begin{bmatrix} (I - \alpha A)^{-1} & 0 \\ 0 & (\alpha I - A)^{-1} \end{bmatrix}\begin{bmatrix} -C^* & 0 \\ 0 & B \end{bmatrix},$$

and where $\alpha \in \partial\mathcal{D}$ is a chosen point.

We indicate the main steps is the proof of Theorem 7.1. The counterpart of Theorem 4.1 (the proof of which uses idea analogous to those used in the proof of Theorem 4.1) looks as follows.

THEOREM 7.2 *There exists a rational matrix function $\Theta(z)$ satisfying (I) – (IV) of Section 2 (with Π^- replaced by $\mathcal{D}$) if and only if the matrix $\Lambda = \begin{bmatrix} -Q & I \\ I & -P \end{bmatrix}$ is invertible. In this case such $\Theta(z)$ is unique up to multiplication on the left by a constant J-unitary matrix. If $z_0 \in \partial\mathcal{D}$, then the unique rational matrix function $\Theta(z)$ with the properties (I) -(IV) and satisfying $\Theta(z_0) = I$ is given by the formula (assuming Λ is invertible)*

$$\Theta(z) = I + (z - z_0)\begin{bmatrix} C & 0 \\ 0 & B^* \end{bmatrix}\begin{bmatrix} (zI - A)^{-1} & 0 \\ 0 & (I - zA^*)^{-1} \end{bmatrix} \cdot$$

$$\Lambda^{-1}\begin{bmatrix} (I - z_0 A^*)^{-1} C^* & 0 \\ 0 & (A - z_0 I)^{-1} B \end{bmatrix}.$$

The Hankel operator has a different form for the unit disk case. Let K(z) be as before, i.e. rational $M \times N$ matrix function with only simple poles in $\mathcal{D}$ and no poles on the boundary $\partial\mathcal{D}$. For the Nehari problem we can assume without loss of generality that K(z) has no poles outside the unit disk and takes value 0 at infinity. Then

$$K(z) = C(zI - A)^{-1}B,$$

where A, B and C are as in the statement of Theorem 7.1. We can also write

$$K(z) = \sum_{j=1}^{\infty} K_j z^{-j}$$

(the Laurent series of K(z) in a neighborhood of infinity). The *Hankel matrix* induced by K(z) is defined to be

$$\mathcal{H}_K = [K_{i+j-1}]_{1 \le i,j < \infty} \cdot$$

We let $\|\mathcal{H}_K\|$ denote the induced operator norm of $\mathcal{H}_K$ as an operator mapping the space ℓ_N^2 of square summable $\mathbb{C}^N$-valued sequences into the space ℓ_M^2 of square summable $\mathbb{C}^M$-valued sequences. As is well known, the same results as in Section 6 hold true for the Hankel operator introduced here, where P and Q are the matrices introduced in Theorem 7.1. We point out here that these results (for the generic case) also can be proved by the same elementary arguments as those used in Section 6.

APPENDIX. J-UNITARY MATRICES

For the reader's convenience, we present here some well-known facts on J-unitary matrices that are used in the main text.

Let $J = \begin{bmatrix} I_M & 0 \\ 0 & -I_N \end{bmatrix}$ be an $(M + N) \times (M + N)$ matrix. An $(M + N) \times (M + N)$ matrix θ is called *J-unitary* if $\theta^* J \theta = J$. Partition θ consistently with the partition of J:

$$\theta = \begin{bmatrix} \theta_{11} & \theta_{12} \\ \theta_{21} & \theta_{22} \end{bmatrix}.$$

The J-unitary property of θ implies the relations

$$\theta_{11}\theta_{11}^* - \theta_{12}\theta_{12}^* = I_M \tag{A.1}$$

$$\theta_{11}\theta_{21}^* - \theta_{12}\theta_{22}^* = 0 \tag{A.2}$$

$$\theta_{21}\theta_{21}^* - \theta_{22}\theta_{22}^* = -I_N. \tag{A.3}$$

In particular, from (A.3) $\theta_{22}\theta_{22}^* = I + \theta_{21}\theta_{21}^*$, so θ_{22} is invertible.

Now let x_1, x_2, y_1, y_2 be variables ($x_2, y_1 \in \mathbb{C}^M$, $y_2, x_1 \in \mathbb{C}^N$) which are related according to the equality

$$\begin{bmatrix} y_1 \\ x_1 \end{bmatrix} = \begin{bmatrix} \theta_{11} & \theta_{12} \\ \theta_{21} & \theta_{22} \end{bmatrix} \begin{bmatrix} x_2 \\ y_2 \end{bmatrix}. \tag{A.4}$$

Define the matrix $U = \begin{bmatrix} U_{11} & U_{12} \\ U_{21} & U_{22} \end{bmatrix}$ by

$$\begin{bmatrix} y_1 \\ y_2 \end{bmatrix} = \begin{bmatrix} U_{11} & U_{12} \\ U_{21} & U_{22} \end{bmatrix} \begin{bmatrix} x_1 \\ x_2 \end{bmatrix} \tag{A.5}$$

Since θ_{22} is invertible, U is well-defined and is given by

$$\begin{bmatrix} U_{11} & U_{12} \\ U_{21} & U_{22} \end{bmatrix} = \begin{bmatrix} \theta_{12}\theta_{22}^{-1} & \theta_{11} - \theta_{12}\theta_{22}^{-1}\theta_{21} \\ \theta_{22}^{-1} & -\theta_{22}^{-1}\theta_{21} \end{bmatrix}. \tag{A.6}$$

When U and θ are related in this way we say that U is the Redheffer transform of θ and write $U = R[\theta]$. Since θ is J-unitary, we deduce from (A.4) that

$$\|y_1\|^2 - \|x_1\|^2 = \|x_2\|^2 - \|y_2\|^2.$$

Rewrite this equation as

$$\|y_1\|^2 + \|y_2\|^2 = \|x_1\|^2 + \|x_2\|^2.$$

As x_1 and x_2 are arbitrary, we deduce from (A.5) that U is unitary: $U^*U = I$. The converse holds by reversing the argument. We thus have established the following.

LEMMA A.1. *If* $U = R[\theta]$ *is the Redheffer transform of* θ, *then* θ *is J-unitary if and only if* U *is unitary.*

We next consider the case where the matrices θ and U are related by $U = R[\theta]$ and θ is J-*contractive*, i.e., $\theta^*J\theta \leq J$. In terms of the vectors x_2, y_2, y_1, x_1, by (A.4) this means

$$\|y_1\|^2 - \|y_2\|^2 \leq \|x_2\|^2 - \|y_2\|^2. \tag{A.7}$$

This can be rewritten as

$$\|y_1\|^2 + \|y_2\|^2 \leq \|x_1\|^2 + \|x_2\|^2. \tag{A.8}$$

As x_1 and x_2 are arbitrary, then by (A.5) we conclude that U is a contraction in the usual sense: $U^*U \leq I$. We have established the first part of the following.

PROPOSITION A.2 (a) *If* $U = R[\theta]$ *is the Redheffer transform of* θ, *then* θ *is J-contractive if and only if* U *is contractive.*

(b) *If* $\Theta = \begin{bmatrix} \theta_{11} & \theta_{12} \\ \theta_{21} & \theta_{22} \end{bmatrix}$ *is J-contractive, then*

(i) $\Theta^* = \begin{bmatrix} \theta_{11}^* & \theta_{21}^* \\ \theta_{12}^* & \theta_{22}^* \end{bmatrix}$ *is J-contractive, and*

(ii) θ_{22} *is invertible and* $\|\theta_{22}^{-1}\theta_{21}\| < 1$.

Proof. Suppose that $\Theta = \begin{bmatrix} \theta_{11} & \theta_{12} \\ \theta_{21} & \theta_{22} \end{bmatrix}$ is a J-contraction. Then in particular

$$\theta_{12}^*\theta_{12} - \theta_{22}^*\theta_{22} = [0, I]\begin{bmatrix} \theta_{11}^* & \theta_{21}^* \\ \theta_{12}^* & \theta_{22}^* \end{bmatrix}\begin{bmatrix} I & 0 \\ 0 & -I \end{bmatrix}\begin{bmatrix} \theta_{11} & \theta_{12} \\ \theta_{21} & \theta_{22} \end{bmatrix}\begin{bmatrix} 0 \\ I \end{bmatrix} \le -I.$$

Thus

$$\theta_{12}^*\theta_{12} + I \le \theta_{22}^*\theta_{22}$$

and since θ_{22} is square, necessarily θ_{22} is invertible. This means that the system of equations (A.4) as in the J-unitary case can be rearranged in the form (A.5) with $U = \begin{bmatrix} U_{11} & U_{12} \\ U_{21} & U_{22} \end{bmatrix}$ given by (A.6). By the equivalence of (A.7) and (A.8), we see that Θ is J-contractive if and only if U is contractive, and statement (a) follows. But a matrix U is contractive (in the usual sense) if and only if its adjoint U^* is contractive. Hence the J-contractiveness of $\begin{bmatrix} \theta_{11} & \theta_{12} \\ \theta_{21} & \theta_{22} \end{bmatrix}$ leads to the contractiveness of

$$\begin{bmatrix} \theta_{12}\theta_{22}^{-1} & \theta_{11} - \theta_{12}\theta_{22}^{-1}\theta_{21} \\ \theta_{22}^{-1} & -\theta_{22}^{-1}\theta_{21} \end{bmatrix}^* = \begin{bmatrix} \theta_{22}^{*-1}\theta_{12}^* & \theta_{22}^{*-1} \\ \theta_{11}^* - \theta_{21}^*\theta_{22}^{*-1}\theta_{12}^* & -\theta_{21}^*\theta_{22}^{*-1} \end{bmatrix}.$$

Then also the matrix

$$\begin{bmatrix} 0 & -I \\ I & 0 \end{bmatrix}\begin{bmatrix} \theta_{22}^{*-1}\theta_{12}^* & \theta_{22}^{*-1} \\ \theta_{11}^* - \theta_{21}^*\theta_{22}^{*-1}\theta_{12}^* & -\theta_{21}^*\theta_{22}^{*-1} \end{bmatrix}\begin{bmatrix} 0 & I \\ -I & 0 \end{bmatrix} =$$

$$\begin{bmatrix} \theta_{21}^*\theta_{22}^{*-1} & \theta_{11}^* - \theta_{21}^*\theta_{22}^{*-1}\theta_{12}^* \\ \theta_{22}^{*-1} & -\theta_{22}^{*-1}\theta_{12}^* \end{bmatrix} = R[\Theta^*]$$

is contractive. Thus by part (a) we conclude that Θ^* is J-contractive, and part (i) of (b) follows. Finally, from the J-contractiveness of $\begin{bmatrix} \theta_{11}^* & \theta_{21}^* \\ \theta_{12}^* & \theta_{22}^* \end{bmatrix}$ we get

$$\theta_{21}\theta_{21}^* - \theta_{22}\theta_{22}^* = [0\ I]\begin{bmatrix} \theta_{11} & \theta_{12} \\ \theta_{21} & \theta_{22} \end{bmatrix}\begin{bmatrix} I & 0 \\ 0 & -I \end{bmatrix}\begin{bmatrix} \theta_{11}^* & \theta_{21}^* \\ \theta_{12}^* & \theta_{22}^* \end{bmatrix}\begin{bmatrix} 0 \\ I \end{bmatrix} \le -I.$$

Therefore

$$\theta_{22}^{-1}\theta_{21}\theta_{21}^{*}\theta_{22}^{*-1} \leq I - \theta_{22}^{-1}\theta_{22}^{*-1}$$

so $\|\theta_{21}^{*}\theta_{22}^{*-1}x\| < \|x\|$ for all $x \neq 0$. Thus $\|\theta_{22}^{-1}\theta_{21}\| < 1$ as asserted. □

REFERENCES

[AAK] V. M. Adamjan, D. Z. Arov and M. G. Krein, Infinite block Hankel matrices and their connection with the interpolation problem, *Amer. Math. Soc. Transl.*(2) 111(1978), 133-156; Russian original, 1971.

[BGR1] J. A. Ball, I. Gohberg and L. Rodman, Interpolation of Rational Matrix Functions, OT45, Birkhäuser, 1990.

[BGR2] J. A. Ball, I. Gohberg and L. Rodman, Tangential interpolation problems for rational matrix functions, *Proc. Symposia in Applied Math* 40(1990), 59-86.

[BGR3] J. A. Ball, I. Gohberg and L. Rodman, Boundary Nevanlinna-Pick interpolation for rational matrix functions, preprint.

[FF] C. Foias and A. E. Frazho, The Communtant Lifting Approach to Interpolation Problems, OT44, Birkhäuser, 1990.

[G] I. Gohberg (editor), I. Schur Methods in Operator Theory and Signal Processing, OT18, Birkhäuser, 1986.

[Gl] K. Glover, All optimal Hankel-norm approximations of linear multivariable systems and their L_∞ error bounds, *International J. Control* 39(1984), 1115-1193.

Time variant extension problems of Nehari type

and the band method*

I. Gohberg[1], M.A. Kaashoek[2], H.J. Woerdeman[3]

The solutions of the Nehari problem and its four block generalization are derived for the time variant case. The maximum entropy principle is also established in this setting. The results are derived by applying the band method to contractive extension problems.

1. Introduction

An important set of mathematical techniques used in robust control, in particular, in H-infinity theory, is based on different interpolation and extension results, like those of Nehari, its four block generalization, and Nevanlinna-Pick. The aim of this paper is to develop the analogues of the Nehari theorem and the four block result for the time variant case, and to deduce the corresponding maximum entropy principle in this setting. The results are obtained by applying the band method, which provides a general scheme to deal with positive and contractive extension problems (see [8,9]).

In Section 2 we state the main results for the time variant Nehari problem and in Section 3 for the time variant four block problem. The elements of the band method which one needs for the contractive extension problems are reviewed in Section 4. The proofs of the results in Sections 2 and 3 are sketched in the last section.

The present paper concerns the discrete time case ; the continuous time versions will be dealt with in a separate publication.

2. The time variant Nehari problem

---

* This paper covers partly the seminar given by the second author for the Second 1990 C.I.M.E. Session at Como, Italy.

1 School of Mathematics, The Raymond and Beverly Sackler Faculty of Exact Sciences, Tel-Aviv University, Tel-Aviv, Israel.

2 Faculteit Wiskunde en Informatica, Vrije Universiteit, Amsterdam, The Netherlands.

3 Department of Mathematics, College of William and Mary, Williamsburg, VA 23185, U.S.A.

In this section we deal with the following time variant version of the Nehari problem:

$$\text{(2.1)} \qquad \left\| \begin{pmatrix} \ddots & & & & \\ \ddots & A_{-1,-1} & & ? & \\ \ddots & A_{0,-1} & A_{00} & ? & \\ \ddots & A_{1,-1} & A_{10} & A_{11} & \\ \ddots & \ddots & \ddots & \ddots & \ddots \end{pmatrix} \right\| < 1.$$

The given entries $A_{ij} : \mathcal{H} \to \mathcal{H}$, $-\infty < j \leq i < \infty$, are (bounded linear) operators acting on the Hilbert space $\mathcal{H}$, and the problem is to complete the operator matrix in (2.1) in such a way that the required norm condition is fulfilled.

To state the problem in a precise way, we need some notation. By $\mathcal{W}$ we denote the linear space of all double infinite operator matrices $V = [V_{jk}]_{j,k=-\infty}^{\infty}$ such that

$$\text{(2.2)} \qquad \sum_{\nu=-\infty}^{\infty} \sum_{j-i=\nu} \sup \|V_{ij}\| < \infty.$$

The entries V_{ij} of V are assumed to be operators acting on the Hilbert space $\mathcal{H}$, and $\|V_{ij}\|$ stands for the operator norm of V_{ij}. Each $V \in \mathcal{W}$ defines in a natural way (via the usual rules of matrix multiplication) a bounded linear operator (also denoted by V) on the Hilbert space

$$\widetilde{\ell_2}(\mathcal{H}) := \{ \begin{pmatrix} \vdots \\ x_{-2} \\ x_{-1} \\ x_0 \\ x_1 \\ x_2 \\ \vdots \end{pmatrix} \mid x_j \in \mathcal{H}, \sum_{j=-\infty}^{\infty} \|x_j\|^2 < \infty \}.$$

The problem (2.1) may now be stated more precisely as follows. Given operators $A_{ij} : \mathcal{H} \to \mathcal{H}$, $-\infty < j \leq i < \infty$, determine all $V \in \mathcal{W}$ such that

(i) $\|V\| < 1$,

(ii) $V_{ij} = A_{ij}, \quad j \leq i$.

Here $\|V\|$ denotes the norm of V as an operator on $\widetilde{\ell_2}(\mathcal{H})$. A semi-infinite operator matrix $V \in \mathcal{W}$ satisfying (i) and (ii) will be called a *strictly contractive extension* of the lower triangular part $\{A_{ij} \mid j \leq i\}$. In order that such an extension V exists it is necessary that the operators

$$\text{(2.3)} \qquad \Lambda_k = (\, A_{ij}\,)_{i=k, j=-\infty}^{\infty \quad k} : \ell_2^{\uparrow}(\mathcal{H}) \to \ell_2^{\downarrow}(\mathcal{H}) \;, k \in \mathbf{Z},$$

are uniformly bounded in the operator norm by a constant strictly less than one, that is,

$$\text{(2.4)} \qquad \sup_{k \in \mathbf{Z}} \|\Lambda_k\| < 1.$$

Here $\ell_2^{\uparrow}(\mathcal{H})$ and $\ell_2^{\downarrow}(\mathcal{H})$ are the Hilbert spaces

$$\ell_2^{\uparrow} := \{\begin{pmatrix} \vdots \\ x_2 \\ x_1 \\ x_0 \end{pmatrix} \mid x_j \in \mathcal{H}, \sum_{j=0}^{\infty} \|x_j\|^2 < \infty\}, \ \ell_2^{\downarrow}(\mathcal{H}) := \{\begin{pmatrix} x_0 \\ x_1 \\ x_2 \\ \vdots \end{pmatrix} \mid x_j \in \mathcal{H}, \sum_{j=0}^{\infty} \|x_j\|^2 < \infty\},$$

Since $V \in \mathcal{W}$, we must also have

$$\sum_{v=-\infty}^{0} \sup_{j-i=v} \|A_{ij}\| < \infty. \tag{2.5}$$

It turns out that the conditions (2.4) and (2.5) are not only necessary but also sufficient for the existence of a strictly contractive extension of the lower triangular part $\{A_{ij} \mid j \leq i\}$. In fact, if (2.4) and (2.5) hold, then one can construct explicitly all desired extensions. The next two theorems give the details of the construction.

THEOREM 2.1. *For $-\infty < j \leq i < \infty$ let A_{ij} be operators acting on the Hilbert space $\mathcal{H}$, and let Λ_k, $k \in \mathbf{Z}$, be defined by (2.3). Suppose that (2.4) and (2.5) hold. Put*

$$\begin{pmatrix} \hat{\alpha}_{ii} \\ \hat{\alpha}_{i+1,i} \\ \vdots \end{pmatrix} = (I - \Lambda_i \Lambda_i^*)^{-1} \begin{pmatrix} I \\ 0 \\ \vdots \end{pmatrix}, \quad i \in \mathbf{Z}, \tag{2.6}$$

$$\begin{pmatrix} \hat{\beta}_{ii} \\ \hat{\beta}_{i+1,i} \\ \vdots \end{pmatrix} = \Lambda_i (I - \Lambda_i^* \Lambda_i)^{-1} \begin{pmatrix} \vdots \\ 0 \\ I \end{pmatrix}, \quad i \in \mathbf{Z}, \tag{2.7}$$

$$\begin{pmatrix} \vdots \\ \hat{\gamma}_{i-1,i} \\ \hat{\gamma}_{ii} \end{pmatrix} = \Lambda_i^* (I - \Lambda_i \Lambda_i^*)^{-1} \begin{pmatrix} I \\ 0 \\ \vdots \end{pmatrix}, \quad i \in \mathbf{Z}, \tag{2.8}$$

$$\begin{pmatrix} \vdots \\ \hat{\delta}_{i-1,i} \\ \hat{\delta}_{ii} \end{pmatrix} = (I - \Lambda_i^* \Lambda_i)^{-1} \begin{pmatrix} \vdots \\ 0 \\ I \end{pmatrix}, \quad i \in \mathbf{Z}, \tag{2.9}$$

and let

$$\alpha := (\alpha_{ij})_{i,j=-\infty}^{\infty}, \ \alpha_{ij} = \begin{cases} \hat{\alpha}_{ij} \hat{\alpha}_{jj}^{-1/2}, & i \geq j; \\ 0, & i < j; \end{cases} \tag{2.10}$$

$$(2.11)\qquad \beta := (\beta_{ij})_{i,j=-\infty}^{\infty},\ \beta_{ij} = \begin{cases} \hat{\beta}_{ij}\hat{\delta}_{jj}^{-1/2}, & i \geq j; \\ 0, & i < j; \end{cases}$$

$$(2.12)\qquad \gamma := (\gamma_{ij})_{i,j=-\infty}^{\infty},\ \gamma_{ij} = \begin{cases} \hat{\gamma}_{ij}\hat{\alpha}_{jj}^{-1/2}, & i \leq j; \\ 0, & i > j; \end{cases}$$

$$(2.13)\qquad \delta := (\delta_{ij})_{i,j=-\infty}^{\infty},\ \delta_{ij} = \begin{cases} \hat{\delta}_{ij}\hat{\delta}_{jj}^{-1/2}, & i \leq j; \\ 0, & i > j; \end{cases}$$

Then the semi-infinite matrix G defined by

$$(2.14)\qquad G := \beta\delta^{-1} = \alpha^{*-1}\gamma^{*}$$

is the unique strictly contractive extension of the given lower triangular part $\{A_{ij} \mid i \geq j \geq 0\}$ with the additional property that $(G(I-G^*G)^{-1})_{ij} = 0$ for $j > i$.

The extension G constructed in the above thorem will be called the *triangular extension* of $\{A_{ij} \mid j \leq i\}$. Given this triangular extension, more precisely, given the factorizations in (2.14), one can describe explicitly the set of all strictly contractive extensions of $\{A_{ij} \mid j \leq i\}$. The result is the following.

THEOREM 2.2. *For $-\infty < j \leq i < \infty$ let A_{ij} be operators acting on the Hilbert space $\mathcal{H}$, and let Λ_k, $k \in \mathbb{Z}$, be defined by (2.3). Suppose that (2.5) holds. Then the lower triangular part $\{A_{ij} \mid j \leq i\}$ has a strictly contractive extension if and only if (2.4) holds. Suppose that (2.4) holds, and let α, β, γ and δ be defined by (2.6)-(2.13). Then each strictly contractive extension F of the given lower triangular part is of the form*

$$(2.15)\qquad F = (\alpha E + \beta)(\gamma E + \delta)^{-1},$$

where $E = (E_{ij})_{i,j=-\infty}^{\infty}$ is an element of $\mathcal{W}$ with $\|E\| < 1$ and $E_{ij} = 0$, $j \leq i$. Furthermore, (2.15) gives a one-one correspondence between all such E and all strictly contractive extensions F. Alternatively, each strictly contractive extension F of the given lower triangular part is of the form

$$(2.16)\qquad F = (\alpha^* + K\beta^*)^{-1}(\gamma^* + K\delta^*)\,,$$

where $K = (K_{ij})_{i,j=-\infty}^{\infty}$ is an element of $\mathcal{W}$ with $\|K\| < 1$ and $K_{ij} = 0$, $j \leq i$. Also, (2.16) parametrizes all strictly contractive extensions F of the given lower triangular part.

The triangular extension may also be identified by a maximum entropy principle. To introduce this principle in the present context we need the notion of (left) multiplicative diagonals. Let $V \in \mathcal{W}$, and suppose that V is positive definite as an operator on $\widetilde{\ell}_2(\mathcal{H})$. Then (see [10]) V admits a factorization

$$V = V_+ D V_+^*,$$

where $V_+ \in \mathcal{W}$, V_+ is invertible in $\mathcal{W}$, V_+ and V_+^{-1} are uppertriangular (that is, $(V_+)_{ij} = (V_+^{-1})_{ij} = 0$ for $j < i$) and D is a diagonal (i.e., $D_{ij} = 0$ for $j \neq i$). The operator matrix $D = \operatorname{diag}\,(D_{ii})_{i=0}^{\infty}$ is called the *(left) multiplicative diagonal* of V; its entries may be described explicitly (see [10]).

THEOREM 2.3. *For $-\infty < j \leq i < \infty$ let A_{ij} be operators acting on the Hilbert space $\mathcal{H}$, and let Λ_k, $k \in \mathbb{Z}$, be defined by (2.3). Suppose that (2.4) and (2.5) hold. Put*

$$M := \operatorname{diag}([(I - \Lambda_i \Lambda_i^*)^{-1}]_{00})_{i=-\infty}^{\infty}.$$

Then for the (left) multiplicative diagonal $D(\Psi)$ of $I - \Psi\Psi^*$, where Ψ is a strictly contractive extension of the given lower triangular part, the following inequality holds

$$D(\Psi) \leq M^{-1}. \tag{2.17}$$

Moreover, in (2.17) equality holds if and only if Ψ is the unique triangular extension of the given lower triangular part.

Here $A \leq B$ means that $B - A$ is a positive operator on $\widetilde{\ell}_2(\mathcal{H})$.

For the time invariant version of Theorems 2.1 - 2.3 the reader is referred to the papers [1] and [4].

3. The time variant four block problem

The four block problem has a relative short history. It has been introduced by J.C. Doyle [2,3] in connection with H_∞ control. For the time invariant case the conditions for the existence of a solution were obtained by A. Feintuch and B.A. Francis in [7]. The description of all solutions for the suboptimal case, together with an analysis of the maximum entropy solution, is due to H. Dym and I. Gohberg [5]. The present paper extends the Dym-Gohberg results to the time variant case. For a class of perturbations of Toeplitz matrices a related distance formula was derived in [6].

The extension problem solved in this section is a strictly contractive extension problem in the algebra $\mathcal{W}^{2\times 2}$, consisting of 2×2 block matrices with elements in $\mathcal{W}$. We view an element in $\mathcal{W}^{2\times 2}$ as an operator on $\widetilde{\ell}_2(\mathcal{H}) \oplus \widetilde{\ell}_2(\mathcal{H})$. Whenever we write the norm of an element in $\mathcal{W}^{2\times 2}$ we mean its norm as an operator. The problem is the following. Given $A = (A_{ij})_{i,j=-\infty}^{\infty}$, $B_- = (B_{ij}^-)_{i,j=-\infty}^{\infty}$, $C = (C_{ij})_{i,j=-\infty}^{\infty}$ and $D = (D_{ij})_{i,j=-\infty}^{\infty}$ in $\mathcal{W}$, with $B_{ij}^- = 0, j > i$, find $B_+ = (B_{ij}^+)_{i,j=-\infty}^{\infty} \in \mathcal{W}$ with $B_{ij}^+ = 0$ for $j \leq i$, such that

$$\left\| \begin{pmatrix} A & B_- + B_+ \\ C & D \end{pmatrix} \right\| < 1.$$

We shall call such $\begin{pmatrix} A & B_- + B_+ \\ C & D \end{pmatrix}$ a *strictly contractive extension* of $\begin{pmatrix} A & B_- \\ C & D \end{pmatrix}$. A necessary condition for the existence of a strictly contractive extension is that any part of the operator $\begin{pmatrix} A & B_- + B_+ \\ C & D \end{pmatrix}$ that is left unchanged when changing B_+, should have norm less than a fixed number smaller than one. Let us write down these restrictions explicitly. Introduce the operators K_i, $i = -\infty, \dots, \infty$, as follows:

$$K_{-\infty} := \begin{pmatrix} A \\ C \end{pmatrix} : \widetilde{\ell}_2(\mathcal{H}) \to \widetilde{\ell}_2(\mathcal{H}) \oplus \widetilde{\ell}_2(\mathcal{H}), \quad K_\infty = \begin{pmatrix} C & D \end{pmatrix} : \widetilde{\ell}_2(\mathcal{H}) \oplus \widetilde{\ell}_2(\mathcal{H}) \to \widetilde{\ell}_2(\mathcal{H}),$$

and

$$K_i = \begin{pmatrix} (A_{jk})_{j=i,\ k=-\infty}^{\infty\quad \infty} & (B_{jk}^-)_{j=i,\ k=-\infty}^{\infty\quad i} \\ C & (D_{jk})_{j=-\infty,\ k=-\infty}^{\infty\quad i} \end{pmatrix} : \widetilde{\ell}_2(\mathcal{H}) \oplus \ell_2^{\uparrow}(\mathcal{H}) \to \ell_2^{\downarrow}(\mathcal{H}) \oplus \widetilde{\ell}_2(\mathcal{H}), \quad i \in \mathbf{Z},$$

In order that $\begin{pmatrix} A & B_- + B_+ \\ C & D \end{pmatrix}$ has a strictly contractive extension, it is necessary that

$$\sup_{i=-\infty,\dots,\infty} \|K_i\| < 1. \tag{3.1}$$

This condition is also sufficient as we shall see in the next theorem. To state the results we need the following operators:

$$L_i = \begin{pmatrix} (A_{jk})_{j=-\infty,\ k=-\infty}^{\infty\quad i} \\ (C_{jk})_{j=-\infty,\ k=-\infty}^{\infty\quad i} \end{pmatrix} : \ell_2{}^{\uparrow}(\mathcal{H}) \to \widetilde{\ell}_2(\mathcal{H}) \oplus \widetilde{\ell}_2(\mathcal{H}),$$

and

$$M_i = \begin{pmatrix} (C_{jk})_{j=i,\ k=-\infty}^{\infty\quad \infty} & (D_{jk})_{j=i,\ k=-\infty}^{\infty\quad \infty} \end{pmatrix} : \widetilde{\ell}_2(\mathcal{H}) \oplus \widetilde{\ell}_2(\mathcal{H}) \to \ell_2^{\downarrow}(\mathcal{H}).$$

THEOREM 3.1. *Let $A = (A_{ij})_{i,j=-\infty}^{\infty}$, $B_- = (B_{ij}^-)_{i,j=-\infty}^{\infty}$, $C = (C_{ij})_{i,j=-\infty}^{\infty}$ and $D = (D_{ij})_{i,j=-\infty}^{\infty}$ in $\mathcal{W}$ be given with $B_{ij}^- = 0$ for $i < j$. Suppose that (3.1) is fulfilled. Put*

$$\begin{pmatrix} ((\hat{\alpha}_{11}^-)_{ij})_{i=j}^{\infty} \\ ((\hat{\alpha}_{21})_{ij})_{i=-\infty}^{\infty} \end{pmatrix} := (I - K_j K_j^*)^{-1} \begin{pmatrix} I \\ 0 \\ \vdots \\ - \\ \vdots \\ 0 \\ 0 \\ \vdots \end{pmatrix},$$

$$((\hat{\alpha}_{22}^-)_{ij})_{i=j}^{\infty} := (I - M_j M_j^*)^{-1} \begin{pmatrix} I \\ 0 \\ \vdots \end{pmatrix},$$

$$((\hat{\delta}^+_{11})_{ij})^j_{i=-\infty} := (I - L_j^* L_j)^{-1} \begin{pmatrix} \vdots \\ 0 \\ I \end{pmatrix},$$

$$\begin{pmatrix} ((\hat{\delta}_{12})_{ij})^\infty_{i=-\infty} \\ \\ ((\hat{\delta}^+_{22})_{ij})^j_{i=-\infty} \end{pmatrix} := (I - K_j^* K_j)^{-1} \begin{pmatrix} \vdots \\ 0 \\ 0 \\ \vdots \\ - \\ \vdots \\ 0 \\ I \end{pmatrix},$$

$$(\alpha^-_{p,p})_{ij} := (\hat{\alpha}^-_{pp})_{ij}((\hat{\alpha}^-_{pp})_{jj})^{-1/2}, \quad p = 1,2, \quad j \le i,$$

$$(\delta^+_{p,p})_{ij} := (\hat{\delta}^+_{pp})_{ij}((\hat{\delta}^+_{pp})_{ij})^{-1/2}, \quad p = 1,2, \quad i \le j,$$

$$(\alpha_{21})_{ij} := (\hat{\alpha}_{21})_{ij}((\hat{\alpha}^-_{11})_{jj})^{-1/2}, \quad i,j \in \mathbb{Z},$$

$$(\delta_{12})_{ij} := (\hat{\delta}_{12})_{ij}((\hat{\delta}^+_{22})_{jj})^{-1/2}, \quad i,j \in \mathbb{Z},$$

$$\begin{pmatrix} ((\beta_{11})_{ij})^\infty_{i=-\infty} \\ \\ ((\beta_{21})_{ij})^\infty_{i=-\infty} \end{pmatrix} = L_j\,(\,((\delta^+_{11})_{ij})^j_{i=-\infty}\,),$$

$$\begin{pmatrix} ((\beta^-_{12})_{ij})^\infty_{i=j} \\ \\ ((\beta_{22})_{ij})^\infty_{i=-\infty} \end{pmatrix} = K_j \begin{pmatrix} ((\delta_{12})_{ij})^\infty_{i=-\infty} \\ \\ ((\delta^+_{22})_{ij})^j_{i=-\infty} \end{pmatrix},$$

$$\begin{pmatrix} ((\gamma_{11})_{ij})^\infty_{i=-\infty} \\ \\ ((\gamma^+_{21})_{ij})^j_{i=-\infty} \end{pmatrix} = K_j^* \begin{pmatrix} ((\alpha^-_{11})_{ij})^\infty_{i=j} \\ \\ ((\alpha_{21})_{ij})^\infty_{i=-\infty} \end{pmatrix},$$

$$\begin{pmatrix} ((\gamma_{12})_{ij})^\infty_{i=-\infty} \\ \\ ((\gamma_{22})_{ij})^\infty_{i=-\infty} \end{pmatrix} = M_j^*\,(\,((\alpha^-_{22})_{ij})^\infty_{i=j}\,),$$

where j runs through all integers $\ldots,-1,0,1,\ldots$. Further, let $(\alpha^-_{11})_{ij} = (\alpha^-_{22})_{ij} = (\beta^-_{12})_{ij} = 0$ for $i < j$, $(\gamma^+_{21})_{ij} = (\delta^+_{11})_{ij} = (\delta^+_{22})_{ij} = 0$ for $i > j$ and

$$(3.2) \qquad \alpha = \begin{pmatrix} \alpha^-_{11} & 0 \\ \alpha_{21} & \alpha^-_{22} \end{pmatrix}, \beta = \begin{pmatrix} \beta_{11} & \beta^-_{12} \\ \beta_{21} & \beta_{22} \end{pmatrix}, \gamma = \begin{pmatrix} \gamma_{11} & \gamma_{12} \\ \gamma^+_{21} & \gamma_{22} \end{pmatrix}, \delta = \begin{pmatrix} \delta^+_{11} & \delta_{12} \\ 0 & \delta^+_{22} \end{pmatrix}.$$

Then the element G in $\mathcal{W}$ defined by

$$G := \beta\delta^{-1} = \alpha^{*^{-1}}\gamma^*$$

is the unique stricly contractive extension of $\begin{pmatrix} A & B_- \\ C & D \end{pmatrix}$ *such that the (2,2) block entry of* $G(I - G^*G)^{-1}$ *is lower triangular in* $\mathcal{W}$*, i.e.,*

$$((G(I - G^*G)^{-1})_{22})_{ij} = 0, \quad i < j.$$

We call the extension G constructed in the above theorem the *triangular extension* of $\begin{pmatrix} A & B_- \\ C & D \end{pmatrix}$. The following theorem gives a linear fractional description for the set of all strictly contractive extensions of a given lower triangular part.

THEOREM 3.2. *Let A, B_-, C and D in $\mathcal{W}$ be given, where $B_- = (B_{ij}^-)_{i,j=-\infty}^{\infty}$ with $B_{ij}^- = 0$ for $i < j$. Then* $\begin{pmatrix} A & B_- \\ C & D \end{pmatrix}$ *has a strictly contractive extension in $\mathcal{W}$ if and only if (3.1) holds. In that case, let α, β, γ and δ be defined as in (3.2). Then each strictly contractive extension* $\mathcal{F} = \begin{pmatrix} A & B_- + B_+ \\ C & D \end{pmatrix}$ *is given by*

$$\mathcal{F} = (\alpha\mathcal{E} + \beta)(\gamma\mathcal{E} + \delta)^{-1}, \tag{3.3}$$

where $\mathcal{E} = \begin{pmatrix} 0 & E_+ \\ 0 & 0 \end{pmatrix} \in \mathcal{W}^{2\times 2}$ *with $(E_+)_{ij} = 0$ for $j \leq i$, and $\|E_+\| < 1$. Furthermore, (3.3) gives a one-one correspondence between all such $\mathcal{E}$ and all strictly contractive extensions $\mathcal{F}$. Alternatively, each strictly contractive extension* $\mathcal{F} = \begin{pmatrix} A & B_- + B_+ \\ C & D \end{pmatrix}$ *is of the form*

$$\mathcal{F} = (\alpha^* + \mathcal{K}\beta^*)^{-1}(\gamma^* + \mathcal{K}\delta^*) \tag{3.4}$$

where $\mathcal{K} = \begin{pmatrix} 0 & 0 \\ K_- & 0 \end{pmatrix} \in \mathcal{W}^{2\times 2}$ *with $(K_-)_{ij} = 0$ for $i \leq j$, and $\|K_-\| < 1$. Also, (3.4) parametrizes all strictly contractive extensions $\mathcal{F}$.*

To the state the maximum entropy principle we need the notion of a (left) multiplicative diagonal of a (positive) element in $\mathcal{W}^{2\times 2}$. Let $\begin{pmatrix} A & B \\ C & D \end{pmatrix} \in \mathcal{W}^{2\times 2}$ be positive as an operator on $\widetilde{\ell}_2(\mathcal{H}) \oplus \widetilde{\ell}_2(\mathcal{H})$. Then (see [11]) this 2×2 block matrix allows a factorization

$$\begin{pmatrix} A & B \\ C & D \end{pmatrix} = \begin{pmatrix} U_{11}^+ & U_{12} \\ 0 & U_{22}^+ \end{pmatrix} \begin{pmatrix} \Delta_1 & 0 \\ 0 & \Delta_2 \end{pmatrix} \begin{pmatrix} (U_{11}^+)^* & 0 \\ U_{12}^* & (U_{22}^+)^* \end{pmatrix},$$

where U_{11}^+ and U_{22}^+ are in $\mathcal{W}$, U_{11}^+ and U_{22}^+ are invertible, U_{11}^+, U_{22}^+ and their inverses are uppertriangular, and Δ_1 and Δ_2 are diagonals. The block matrix $\begin{pmatrix} \Delta_1 & 0 \\ 0 & \Delta_2 \end{pmatrix}$ is called the *(left) multiplicative diagonal* of $\begin{pmatrix} A & B \\ C & D \end{pmatrix}$.

THEOREM 3.3. *Let A, B_-, C and D in $\mathcal{W}$ be given, where $B_- = (B_{ij}^-)_{i,j=-\infty}^{\infty}$ with $B_{ij}^- = 0$ for $i < j$. Suppose that (3.1) holds. Put*

$$Q_1 = \operatorname{diag}([((I - K_iK_i^*)^{-1})_{11}]_{00})_{i=-\infty}^{\infty},$$

$$Q_2 = \operatorname{diag}([(I - M_iM_i^*)^{-1}]_{00})_{i=-\infty}^{\infty}.$$

Then for the (left) multiplicative diagonal $\begin{pmatrix} \Delta_1(\mathcal{F}) & 0 \\ 0 & \Delta_2(\mathcal{F}) \end{pmatrix}$ of $I - \mathcal{F}\mathcal{F}^*$, where $\mathcal{F}$ is a strictly contractive extension of $\begin{pmatrix} A & B_- \\ C & D \end{pmatrix}$, the following inequality holds:

$$\begin{pmatrix} \Delta_1(\mathcal{F}) & 0 \\ 0 & \Delta_2(\mathcal{F}) \end{pmatrix} \leq \begin{pmatrix} Q_1^{-1} & 0 \\ 0 & Q_2^{-1} \end{pmatrix}. \tag{3.5}$$

Moreover, equality holds in (3.5) if and only if $\mathcal{F}$ is the unique triangular of $\begin{pmatrix} A & B_- \\ C & D \end{pmatrix}$.

4. The band method for contractive extension problems

The proofs of the theorems in Sections 2 and 3 are based on a general extension principle for algebras with a band structure which has been developed in the papers [8,9,10]. The general scheme appearing in these papers allows one to deal with a number of different extension problems from one point of view. In the present section we review this method for contractive extension problems.

First we introduce the general setting. In what follows $\mathcal{B}$ is an algebra with unit e and an involution operation *, and we assume that $\mathcal{B}$ admits a direct sum decomposition of the form

$$\mathcal{B} = \mathcal{B}_1 \dot{+} \mathcal{B}_2^0 \dot{+} \mathcal{B}_d \dot{+} \mathcal{B}_3^0 \dot{+} \mathcal{B}_4 \tag{4.1}$$

where $\mathcal{B}_1, \mathcal{B}_2^0, \mathcal{B}_d, \mathcal{B}_3^0$ and $\mathcal{B}_4$ are linear subspaces of $\mathcal{B}$ and the following conditions are satisfied:

(i) $\mathcal{B}_1^* = \mathcal{B}_4$, $\quad (\mathcal{B}_2^0)^* = \mathcal{B}_3^0$, $\quad \mathcal{B}_d^* = \mathcal{B}_d$,
(ii) $e \in \mathcal{B}_d$,
(iii) the following multiplication table describes some additional rules on the multiplication in $\mathcal{B}$:

(4.2)

	$\mathcal{B}_1$	$\mathcal{B}_2^0$	$\mathcal{B}_d$	$\mathcal{B}_3^0$	$\mathcal{B}_4$
$\mathcal{B}_1$	$\mathcal{B}_1$	$\mathcal{B}_1$	$\mathcal{B}_1$	$\mathcal{B}_+^0$	$\mathcal{B}$
$\mathcal{B}_2^0$	$\mathcal{B}_1$	$\mathcal{B}_+^0$	$\mathcal{B}_2^0$	$\mathcal{B}_c$	$\mathcal{B}_-^0$
$\mathcal{B}_d$	$\mathcal{B}_1$	$\mathcal{B}_2^0$	$\mathcal{B}_d$	$\mathcal{B}_3^0$	$\mathcal{B}_4$
$\mathcal{B}_3^0$	$\mathcal{B}_+^0$	$\mathcal{B}_c$	$\mathcal{B}_3^0$	$\mathcal{B}_-^0$	$\mathcal{B}_4$
$\mathcal{B}_4$	$\mathcal{B}$	$\mathcal{B}_-^0$	$\mathcal{B}_4$	$\mathcal{B}_4$	$\mathcal{B}_4$

,

where

$$\mathcal{B}_+^0 := \mathcal{B}_1 \dot{+} \mathcal{B}_2^0, \quad \mathcal{B}_-^0 := \mathcal{B}_3^0 \dot{+} \mathcal{B}_4, \quad \mathcal{B}_c = \mathcal{B}_2^0 \dot{+} \mathcal{B}_d \dot{+} \mathcal{B}_3^0.$$

We shall refer to $\mathcal{B}$ as the *algebra with the band structure* (4.1). We need the following additional notation:

$$\mathcal{B}_+ := \mathcal{B}_+^0 \dot{+} \mathcal{B}_d, \quad \mathcal{B}_- := \mathcal{B}_-^0 \dot{+} \mathcal{B}_d \tag{4.3}$$

$$\mathcal{B}_u := \mathcal{B}_+ \dot{+} \mathcal{B}_3^0, \quad \mathcal{B}_\ell := \mathcal{B}_2^0 \dot{+} \mathcal{B}_- \tag{4.4}$$

For $\alpha = +, -, \ell, u$ we define P_α to be the projection of $\mathcal{B}$ onto $\mathcal{B}_\alpha$ along the natural complement of $\mathcal{B}_\alpha$ in $\mathcal{B}$. For example, P_+ is the projection of $\mathcal{B}$ onto $\mathcal{B}_+$ along $\mathcal{B}_3^0 \dot{+} \mathcal{B}_4$ and P_ℓ is the projection of $\mathcal{B}$ onto $\mathcal{B}_\ell$ along $\mathcal{B}_1$.

For elements of $\mathcal{B}$ we consider two types of factorizations. We say that $b \in \mathcal{B}$ admits a *left spectral factorization* (relative to the decomposition (4.1)) if $b = b_+ b_+^*$ for some b_+ in $\mathcal{B}_+$ with b_+ invertible and $b_+^{-1} \in \mathcal{B}_+$. Analogously, b is said to have a *right spectral factorization* (relative to the decomposition (4.1)) if $b = b_- b_-^*$ for some b_- in $\mathcal{B}_-$ with b_- invertible and $b_-^{-1} \in \mathcal{B}_-$.

An element $b \in \mathcal{B}$ is called *nonnegative definite* (notation $b \geq 0$) if b can be represented in the form

$$b = c^* c \tag{4.5}$$

for some c in $\mathcal{B}$. If in this representation of b the element c is invertible, then b is said to be *positive definite.* We say b that *positive definite in* $\mathcal{B}_d$ if (4.5) holds with $c \in \mathcal{B}_d$ and c is invertible. In the latter case both c^* and c^{-1} are in $\mathcal{B}_d$. Note that b is positive definite whenever b admits a right or left spectral factorization.

We are now ready to state the abstract version of the strictly contractive extension problem. Let $\varphi \in \mathcal{B}_\ell$. We say that ψ is a *strictly contractive extension* of φ if

(i) $\psi - \varphi \in \mathcal{B}_1$,

(ii) $e - \psi^* \psi$ is positive definite.

Given $\varphi \in \mathcal{B}_\ell$ we are interested to find all strictly contractive extensions of φ. It turns out that in order to solve this problem one has to look first for a so-called (strictly contractive) *triangular extension* ψ of φ. The latter means that ψ is a strictly contractive extension of φ with the additional property that $\psi(e - \psi^* \psi)^{-1}$ belongs to $\mathcal{B}_\ell$.

To solve the abstract strictly contractive extension problem we need an additional property. Note that the algebra $\mathcal{B}^{2\times 2}$ of 2×2 matrices with entries from $\mathcal{B}$ is again an algebra with a * operation, namely

$$\begin{pmatrix} a & b \\ c & d \end{pmatrix}^* := \begin{pmatrix} a^* & c^* \\ b^* & d^* \end{pmatrix}. \tag{4.6}$$

Hence for elements in $\mathcal{B}^{2\times 2}$ we can speak about nonnegative definite and positive definite. Our first extension theorem will need the first of the next two axioms; the second will play a role in the abstract maximum entropy theorem.

AXIOM (A0a). If $\begin{pmatrix} a & b \\ c & d \end{pmatrix}$ is positive definite in $\mathcal{B}^{2\times 2}$, then a and d are positve definite in $\mathcal{B}$.

AXIOM (A0b). If $\begin{pmatrix} a & b \\ c & d \end{pmatrix}$ is nonnegative definite in $\mathcal{B}^{2\times 2}$, then a and d are nonnegative definite in $\mathcal{B}$.

THEOREM 4.1. *Let $\mathcal{B}$ be an algebra with band structure (4.1), and assume that Axiom (A0a) is fulfilled. Let $\varphi \in \mathcal{B}_\ell$. The element φ has a triangular extension τ such that $e - \tau\tau^*$ admits a left and $e - \tau^*\tau$ admits a right spectral factorization if and only if the equations*

$$a - P_-\varphi P_u(\varphi^* a) = e, \quad d - P_+\varphi^* P_\ell(\varphi d) = e \tag{4.7}$$

have solutions a and d with the following properties:

(i) $a \in \mathcal{B}_-$ and $d \in \mathcal{B}_+$,
(ii) a and d are invertible, $a^{-1} \in \mathcal{B}_-$, $d^{-1} \in \mathcal{B}_+$
(iii) $P_d a$ and $P_d d$ are positive definite in $\mathcal{B}_d$.

In that case φ has a unique triangular extension τ for which $e - \tau\tau^*$ admits a left and $e - \tau^*\tau$ admits a right spectral factorization, and this τ is given by

$$\tau := bd^{-1} = a^{*-1}c^*,$$

where

$$b = P_\ell(\varphi d), \quad c = P_u(\varphi^* a). \tag{4.8}$$

The spectral factorizations of $e - \tau\tau^*$ and $e - \tau^*\tau$ are given by

$$e - \tau\tau^* = a^{*-1}(P_d a)a^{-1}, \quad e - \tau^*\tau = d^{*-1}(P_d d)d^{-1}. \tag{4.9}$$

If the algebra $\mathcal{B}$ is a Banach algebra, then there are conditions on the maps

$$a \to a - \epsilon P_-\varphi P_u(\varphi^* a), \quad d \to d - \epsilon P_+\varphi^* P_\ell(\varphi d) \quad (0 \le \epsilon \le 1)$$

which guarantee that equation (4.6) has solutions with the desired properties (i)-(iii). (See [8], Theorem II.1.1 and [9], Theorem I.2.2 for further details.)

To describe the set of all strictly contractive extensions of a given $\varphi \in \mathcal{B}_\ell$, we need extra requirements on $\mathcal{B}$. We shall assume that $\mathcal{B}$ is a *-subalgebra of a C^*-algebra $\mathcal{R}$ with norm $\|\cdot\|_\mathcal{R}$, and $\mathcal{R}$ has a unit e which belongs to $\mathcal{B}$. Further, we assume that the following axiomas hold:

AXIOM (A1). If $f \in \mathcal{B}$ is invertible in $\mathcal{R}$, then $f^{-1} \in \mathcal{B}$;

AXIOM (A2). If $f_n \in \mathcal{B}_+$, $f \in \mathcal{B}$ and $\lim_{n\to\infty} \|f_n - f\|_{\mathcal{R}} = 0$, then $f \in \mathcal{B}_+$.

If $e - f^*f$ is positive definite in $\mathcal{B}$, then $e - f^*f$ is positive definite as an element of $\mathcal{R}$, and hence $\|f\|_{\mathcal{R}} < 1$.

The following theorem describes the set of all strictly contractive extensions.

THEOREM 4.2. *Let $\mathcal{B}$ be an algebra with band structure (4.1), and assume that Axiom (A0a) is fulfilled. Furthermore, assume that $\mathcal{B}$ is a *-subalgebra of C^*-algebra $\mathcal{R}$ such that the unit e of $\mathcal{B}$ is in $\mathcal{R}$ and Axioms (A1) and (A2) hold true. Let $\varphi \in \mathcal{B}_\ell$, and suppose that φ has a triangular extension τ such that $e - \tau\tau^*$ admits a left and $e - \tau^*\tau$ admits a right spectral factorization. Let α, δ be invertible elements in $\mathcal{B}$ such that $\alpha^{\pm 1} \in \mathcal{B}_-$, $\delta^{\pm 1} \in \mathcal{B}_+$ and*

$$(e - \tau\tau^*)^{-1} = \alpha\alpha^*, \quad (e - \tau^*\tau)^{-1} = \delta\delta^*,$$

and put

$$\beta = P_\ell(\varphi\delta), \qquad \gamma = P_u(\varphi^*\alpha).$$

Then each strictly contractive extension $\psi \in \mathcal{B}$ of ϕ is of the form

$$\psi = (\alpha h + \beta)(\gamma h + \delta)^{-1}, \tag{4.10}$$

where h is an element in $\mathcal{B}_1$ such that $e - h^*h$ is positive definite in $\mathcal{B}$. Furthermore, equation (4.10) gives a one-one correspondence between all such h and all strictly contractive extensions ψ of ϕ. Alternatively, each strictly contractive extension $\psi \in \mathcal{B}$ of ϕ is of the form

$$\psi = (\alpha^* + f^*\beta^*)^{-1}(\gamma^* + f^*\delta^*), \tag{4.11}$$

where f is an element in $\mathcal{B}_1$ such that $e - ff^*$ is positive definite in $\mathcal{B}$. Furthermore, equation (4.11) gives a one-one correspondence between all such f and all strictly contractive extensions ψ of ϕ.

Note that one can find α and δ in Theorem 4.2 from the elements a and d in Theorem 4.1 by using (4.8). Indeed, write $P_d a$ and $P_d d$ in (4.8) as $P_d a = r^*r$ and $P_d d = s^*s$, where r and s are invertible in $\mathcal{B}_d$, and put $\alpha = ar^{-1}$ and $\delta = ds^{-1}$.

To state that the maximum entropy principle in the present abstract setting we need the notion of a multiplicative diagonal and two additional axioms.

Let b be a positive definite element of $\mathcal{B}$, and assume that b admits a left spectral factorization $b = b_+b_+^*$, where $b_+^{\pm 1} \in \mathcal{B}_+$. By definition the *left multiplicative diagonal* $\Delta_{left}(b)$ of b is the element

$$\Delta_{left}(b) := P_d(b_+)P_d(b_+)^*. \tag{4.12}$$

The right hand side of (4.12) does not depend on the particular choice of the spectral factorization, and hence $\Delta_{left}(b)$ is well-defined (see [10], Section I.1 for further details). A right multiplicative diagonal is defined in a similar way starting from a right spectral factorization.

The two additional axioms that are needed concern nonnegative elements in $\mathcal{B}$. They read as follows:

AXIOM (A3). The element $P_d(aa^* + bb^*)$ is nonnegative definite for all a and b in $\mathcal{B}$.

AXIOM (A4). If $P_d(aa^* + bb^*) = 0$, then a and b are both equal to zero.

THEOREM 4.3. *Let $\mathcal{B}$ be an algebra with band structure* (4.1), *and assume that the Axioms* (A0a), (A0b), (A3), *and* (A4) *are fulfilled. Let $\varphi \in \mathcal{B}_\ell$, and suppose that φ has a triangular extension τ such that $e - \tau\tau^*$ admits a left spectral factorization. Then for any strictly contractive extension ψ of φ such that $e - \psi\psi^*$ admits a left spectral factorization we have*

$$\Delta_{left}(e - \tau\tau^*) \geq \Delta_{left}(e - \psi\psi^*),$$

and equality holds if and only if $\psi = \tau$.

5. Sketch of proofs

To derive the results in Sections 2 en 3 from those in Section 4 we have to specify the algebra $\mathcal{B}$ and its band structure (4.1). To do this let us return to the algebra $\mathcal{W}$ introduced in Section 2. In $\mathcal{W}$ we consider the following subspaces:

$$\mathcal{W}_+^0 = \{V = (V_{jk})_{j,k=-\infty}^{\infty} \in \mathcal{W} \mid V_{jk} = 0 \text{ for } j \leq k\},$$

$$\mathcal{W}_-^0 = \{V = (V_{jk})_{j,k=-\infty}^{\infty} \in \mathcal{W} \mid V_{jk} = 0 \text{ for } j \geq k\},$$

$$\mathcal{W}_d = \{V = (V_{jk})_{j,k=-\infty}^{\infty} \in \mathcal{W} \mid V_{jk} = 0 \text{ for } j \neq k\}.$$

Now take $\mathcal{B} = \mathcal{W}$. Then $\mathcal{B}$ has a unit, $E = (\delta_{jk}I)_{j,k=-\infty}^{\infty}$, where δ_{jk} is the Kronecker delta, and $\mathcal{B}$ has an involution, namely

$$\left((V_{jk})_{j,k=-\infty}^{\infty}\right)^* := (V_{kj}^*)_{j,k=-\infty}^{\infty}$$

Put

$$\mathcal{B}_1 = \mathcal{W}_+^0, \quad \mathcal{B}_2^0 = \{0\}, \quad \mathcal{B}_d = \mathcal{W}_d, \quad \mathcal{B}_3^0 = \{0\}, \quad \mathcal{B}_4 = \mathcal{W}_-^0.$$

With $\mathcal{B}_1$, $\mathcal{B}_2^0$, $\mathcal{B}_d$, $\mathcal{B}_3^0$ and $\mathcal{B}_4$ defined in this way, (4.1) holds and the conditions (i), (ii) and (iii) in the second paragraph of Section 4 are fulfilled. One can show that an element $V \in \mathcal{W}$ is positive definite in $\mathcal{W}$ if and only if V is positive definite as an operator on $\widetilde{\ell}_2(H)$ (cf., Lemma II.3.2 in [9]), and hence $I - V^*V$ is positive definite if and only if $\|V\| < 1$. Note that in this case $\mathcal{B}_\ell$ consists of the lower triangular elements in $\mathcal{W}$, and hence for $\mathcal{B} = \mathcal{W}$ the abstract problem considered in Section 4 is precisely the contractive extension problem of Section 2. It is a simple matter to check that AXIOMS (A0a), (A0b), (A3) and

(A4) are fulfilled for $\mathcal{B} = \mathcal{W}$. One can show (cf., Section II.3 of [9]) that with $\mathcal{R}$ equal to the C^*-algebra of all bounded linear operators on $\widetilde{\ell}_2(H)$, Axioms (A1) and (A2) hold for $\mathcal{B} = \mathcal{W}$. Hence we can apply the abstract results of Section 4 in the concrete setting of Section 2. In this way one may prove the theorems in Section 2.

To prove the theorems in Section 3 one has to apply the results of Section 4 to the algebra $\mathcal{B} = \mathcal{W}^{2\times 2}$. Since $\mathcal{W}$ is a *-algebra with a unit, the same is true for $\mathcal{W}^{2\times 2}$. Now put

$$\mathcal{B}_1 = \begin{pmatrix} 0 & \mathcal{W}_+^0 \\ 0 & 0 \end{pmatrix} = \{\begin{pmatrix} 0 & V \\ 0 & 0 \end{pmatrix} \mid V \in \mathcal{W}_+^0\},$$

$$\mathcal{B}_2^0 = \begin{pmatrix} \mathcal{W}_+^0 & \mathcal{W}_- \\ 0 & \mathcal{W}_+^0 \end{pmatrix} = \{\begin{pmatrix} A & B \\ 0 & D \end{pmatrix} \mid A, D \in \mathcal{W}_+^0, B \in \mathcal{W}_-\},$$

$$\mathcal{B}_d = \begin{pmatrix} \mathcal{W}_d & 0 \\ 0 & \mathcal{W}_d \end{pmatrix} = \{\begin{pmatrix} A & 0 \\ 0 & D \end{pmatrix} \mid A, D \in \mathcal{W}_d\},$$

$$\mathcal{B}_3^0 = \begin{pmatrix} \mathcal{W}_-^0 & 0 \\ \mathcal{W}_+ & \mathcal{W}_-^0 \end{pmatrix} = \{\begin{pmatrix} A & 0 \\ C & D \end{pmatrix} \mid A, D \in \mathcal{W}_-^0, C \in \mathcal{W}_1\},$$

$$\mathcal{B}_4 = \begin{pmatrix} 0 & 0 \\ \mathcal{W}_-^0 & 0 \end{pmatrix} = \{\begin{pmatrix} 0 & 0 \\ V & 0 \end{pmatrix} \mid V \in \mathcal{W}_-^0\}.$$

Here $\mathcal{W}_+ = \mathcal{W}_+^0 + \mathcal{W}_d$ and $\mathcal{W}_- = \mathcal{W}_-^0 + \mathcal{W}_d$. Next, one checks that with $\mathcal{B}_1$, $\mathcal{B}_2^0$, $\mathcal{B}_d$, $\mathcal{B}_3^0$ and $\mathcal{B}_4$ as above the algebra $\mathcal{B} = \mathcal{W}^{2\times 2}$ is an algebra with a band structure. Note that in this case

$$\mathcal{B}_\ell = \{\begin{pmatrix} A & B \\ C & D \end{pmatrix} \mid A, B, C \in \mathcal{W}, \quad B \in \mathcal{W}_-\},$$

and hence for $\mathcal{B} = \mathcal{W}^{2\times 2}$ the abstract contractive extension problem in Section 4 is precisely the extension problem considered in Section 3. To get the results of Section 3 one now follows the same line of reasoning as sketched above for the case $\mathcal{B} = \mathcal{W}$.

For semi-infinite operator matrices full proofs of Theorems 2.1 and 2.2 are given in [9], Section II.3 and of Theorem 2.3 in [10], Section II.4. All details concerning Theorems 3.1 - 3.3 will appear in a future publication [11].

References

[1] V.M. Adamjan, D.Z. Arov and M.G. Krein, Infinite block Hankel matrices and their connection with the interpolation problem, *Izv. Akad. Nauk. Arm. SSR Ser. Mat.* 6 (1971), 87-112; English transl. in *Amer. Math. Soc. Transl.* (2) 111 (1978), 133-156.

[2] J.C. Doyle, Synthesis of robust controllers, *Proc. IEEE conf. Dec. Control*, 1983.

[3] J.C. Doyle, Lecture notes in advances in multivariable control,*ONR/Honeywell Workshop*, Minneapolis, 1984.

[4] H. Dym and I. Gohberg, A maximum entropy principle for contractive interpolants, *J. Funct. Anal.* 65 (1986), 83-125.

[5] H. Dym and I. Gohberg, A new class of contractive interpolants and maximum entropy principles, in: *Topics in operator theory and interpolation* (Ed. I. Gohberg), Operator Theory: Advances and Applications, Vol. 29, Birkhäuser Verlag, Basel, 1988, pp. 117-150.

[6] A. Feintuch, On asymptotic Toeplitz and Hankel operators, in: *The Gohberg Anniversary Collection*, Vol. 2 (Ed. H. Dym, S. Goldberg, M.A. Kaashoek and P. Lancaster), Operator Theory: Advances and Applications, Vol. 41, Birkhäuser Verlag, Basel, 1989, pp. 241-254.

[7] A. Feintuch and B.A. Francis, Distance formulas for operator algebras arising in optimal control problems, in: *Topics in operator theory and interpolation* (Ed. I. Gohberg), Operator Theory: Advances and Applications, Vol. 29, Birkhäuser Verlag, Basel, 1988, pp. 151-170.

[8] I. Gohberg, M.A. Kaashoek and H.J. Woerdeman, The band method for positive and contractive extension problems, *J. Operator Theory* 22 (1989), 109-155.

[9] I. Gohberg, M.A. Kaashoek and H.J. Woerdeman, The band method for positive and contractive extension problems: an alternative version and new applications, *Integral Equations Operator Theory* 12 (1989), 343-382.

[10] I. Gohberg, M.A. Kaashoek and H.J. Woerdeman, A maximum entropy principle in the general framework of the band method, *J. Funct. Anal.* 95 (1991), 231-254.

[11] I. Gohberg, M.A. Kaashoek and H.J. Woerdeman, The band method for multiblock problems, in preparation.

C.I.M.E. Session on "Recent Developments in H^∞ Control Theory"

List of Participants

G. ARSENE, INCREST, Bd. Pacii 220, 79622 Bucharest

M. BAKONYI, INCREST, Bd. Pacii 220, 70622 Bucharest

S. BATTILOTTI, Dip. di Inf. e Sist., Via Eudossiana 18, 00184 Roma

D. BENSOUSSAN, Ecole de Techn. Sup., Univ. du Québec, 4750 Av. Henri-Julien, Montréal, Québec H2T 2CB

R. M. BIANCHINI, Dip. di Mat., Viale Morgagni 67/A, 50134 Firenze

F. BLANCHINI, Dip. di Mat., Univ. Udine, Via Zanon 6, 33100 Udine

C.A. BONATTI, c/o Aermacchi S.p.A., Via S.Silvestro 80, 21100 Varese

A. BÖTTCHER, Techn. Univ., Sekt. Math., PSF 964, Karl-Marx-Stadt 9010

F. CALLIER, Dept. of Math., Fac. Univ. de Namur, Rempart de la Vierge 8, B-5000 Namur

M. CAMPI, Dip. di Elettr., Pol. di Milano, Piazza Leonardo da Vinci 32, 20133 Milano

R. CAPONETTO, Ist. di Elettrot. ed Elettr., Viale A. Doria 6, 95125 Catania

A. CASAVOLA, Dip. di Sist. e Inf., Via di S. Marta 3, 50139 Firenze

B. CASTILLO-TOLEDO, Dip. di Inf. e Sist., Via Eudossiana 18, 00184 Roma

A. CAVALLO, Dip. di Inf. e Sist., Via Claudio 21, 80125 Napoli

L. CHISCI, Dip. di Sist. e Inf., Via di S. Marta 3, 50139 Firenze

P. COLANERI, Dip. di Elettr., Pol. di Milano, P.za Leonardo da Vinci 32, 20133 Milano

G. DE MARIA, Dip. di Inf. e Sist., Via Claudio 21, 80125 Napoli

G. DE NICOLAO, Centro Teoria dei Sistemi, CNR, Dip. di Elettr., Pol. di Milano, Piazza Leonardo da Vinci 32, 20133 Milano

P. DORATO, EECE Dept., Univ. New Mexico, Albuquerque, NM 87131

P. FUHRMANN, Dept. of Math., Ben-Gurion Univ., Beer-Sheva

R. GADIDOV, INCREST, Bd. Pacii 220, 79622 Bucharest

M.T. GALIZIA, Dip. di Mat., Pol. di Torino, Corso Duca degli Abruzzi 24, 10129 Torino

L. GIARRE', Dip. di Sist. e Inf., Via di S. Marta 3, 50139 Firenze

I. GOHBERG, Dept. of Math., Tel-Aviv Univ., Ramat-Aviv, 69978 Tel-Aviv

A. GOMBANI, LADSEB-CNR, Corso Stati Uniti 4, 35020 Padova

Andrei HALANAY, Dept. of Math., Univ. of Bucharest, Bucharest

G. HEINIG, Karl-Marx-Univ. Leipzig, Sekt. Math., Hauptgebaude, Leipzig

S. HUI, Dept. of Math. Sci.,Coll. of Sci., San Diego State Univ., San Diego, CA 92182

A. IONESCU, INCREST, Bd. Pacii 220, 79622 Bucharest

V. IONESCU, Str. Emile Zola nr. 3, 71272 Bucharest, Sector 1.

A. ISIDORI, Dip. di Inf. e Sist., Via Eudossiana 18, 00184 Roma

M.A. KAASHOEK, Fac. Wisk. en Inf., Vrije Univ., De Boelelaan 1081, 1081 HV Amsterdam

M. KINNAERT, Lab. d'Autom., CP 165, Univ. Libre de Bruxelles, 50 Av. F. D. Roosevelt, B-1050 Bruxelles

L. LANARI, Dip. di Inf. e Sist., Via Eudossiana 18, 00184 Roma

L. LERER, Fac. Wisk. en Inf., Vrije Univ., Postbus 7161, 1007 MC Amsterdam

S. LONGHI, Dip. di Elettr. e Autom., Via Brecce Bianche, 60131 Ancona

S. MALAN, Dip. di Aut. e Inf., Pol. di Torino, Corso Duca degli Abruzzi 24, 10129 Torino

G. MARRO, D.E.I.S., Univ. di Bologna, Viale Risorgimento 2, 40136 Bologna

G. MEINSMA, Fac. of Appl. Math., Univ. of Twente, P.O.Box 217, 7500 AE Enschede

G. MIRABELLA, Ist. di Elettrot. ed Elettr., Viale A. Doria 6, 90125 Catania

G. NAEVDAL, Div. of Math. Sci., The Norwegian Inst. of Techn., N-7034 Trondheim

P. NISTRI, Dip. di Sist. e Inf., Via di S. Marta 3, 50139 Firenze

G. NUNNARI, Ist. di Elettrot. ed Elettr., Viale A. Doria 6, 95125 Catania

P.-O. NYMAN, Dept. of Appl. Math., Univ. of Twente, P.O.Box 217, 7500 AE Enschede

G. ORIOLO, Dip. di Inf. e Sist., Via Eudossiana 18, 00184 Roma

A. PIAZZI, D.E.I.S., Univ. di Bologna, Viale Risorgimento 2, 40136 Bologna

S. PINZONI, LADSEB-CNR, Corso Stati Uniti 4, 35020 Padova

J. RAISCH, Inst. fur Systemdynamik und Regelungstechnik, Univ. Stuttgart, Pfaffenwaldring 9, D-7000 Stuttgart 80

V. RASVAN, Craiova Univ., Dept. of Control Engng & Comp. Sci., A.I.Cuza Str. 13, RO-1100 Craiova

M. ŠEBEK, Systems and Control Group, Dept. of Appl. Math., Univ. of Twente, P.O. Box 217, 7500 AE Enschede

B. TERRENI, Dip. di Mat., Univ. di Milano, Via C. Saldini 50, 20133 Milano

D. TIMOTIN, INCREST, Bd. Pacii 220, 79622 Bucharest

A.-J. VAN DER VEEN, Delft Univ. of Techn., Network Theory Sect., Dept. EE, Mekelweg 4, P.O. Box 5031, 2600 GA Delft

M. VERHAEGEN, Delft Univ. of Techn., Network Theory Sect., Dept. EE, Mekelweg 4, P.O. Box 5031, 2600 GA Delft

E. WEGERT, Mining Acad. Freiberg, Markusstr. 34, Karl-Marx-Stadt, DDR-9072

M.G. XIBILIA, Ist. di Elettrot. ed Elettr., Viale A. Doria 6, 95125 Catania

J. ZHANG, Dip. di Sist. e Inf., Via di S. Marta 3, 50139 Firenze

FONDAZIONE C.I.M.E.
CENTRO INTERNAZIONALE MATEMATICO ESTIVO
INTERNATIONAL MATHEMATICAL SUMMER CENTER

"Recent Developments in Geometric Topology and Related Topics"

is the subject of the First 1990 C.I.M.E. Session.

The Session, sponsored by the Consiglio Nazionale delle Ricerche and the Ministero della Pubblica Istruzione, will take place under the scientific direction of Prof. PAOLO DE BARTOLOMEIS (Università di Firenze), and Prof. FRANCO TRICERRI (Università di Firenze) at Villa "La Querceta", Montecatini Terme (Pistoia), from **June 4 to June 12, 1990.**

Courses

a) **Curvature Topology and Critical Points.** (8 lectures in English).
Prof. Jeff CHEEGER (NYU - Courant Institute).

Outline

In about 1976, K. Grove and K. Shiohama made a basic advance in riemannian geometry, by observing that the Isotopy Lemma of Morse Theory could be generalized to distance functions on riemannian manifolds.

In combination with Toponogov's comparison theorem, their observation has a number of very interesting applications which relate curvature and topology. The main goal of our lectures will be to expose three such results.

1) The theorem of Grove-Petersen, asserting that the collection of riemannian-manifolds with diameter $\leqslant$ d, and sectional curvature $\geqslant$ H., contains only finitely many homotopy types.
2) The theorem of Gromov asserting that there is a universal bound on the sum of the Betti numbers for riemannian n-manifolds with diameter $\leqslant$ d and sectional curvature $\geqslant$ H.
3) The theorem of Abresch-Gromoll asserting that a complete manifold with non-negative Ricci curvature, sectional curvature bounded below by H ($>-\infty$) and whose "size at infinity grows slowly", has finite topological type.

The prerequisites for these lectures are chapters 1 and 2 of [CE] (some of this material will be rapidly reviewed).

References

[AG] Abresch, U., Gromoll, D., On Complete Manifolds with Nonnegative Ricci Curvature, Gromoll, D., JAMS (to appear).

[CE] Cheeger, J., Ebin, D., Comparison theorems in Riemannian geometry, North-Holland, N.Y., 1975.

[G] Gromov, M., Curvature, diameter and Betti numbers, Comm. Math. Helv. 56 (1981), 179-195.

[GrP] Grove, K., Petersen, P., Bounding homotopy types by geometry, Annals of Mathematics, 128 (1988), 195-206.

[GrPW] Grove, K., Petersen, P., Wu, J.Y., Controlled Topology in Geometry, BAMS, Volume 20, Number 2, April 1989.

[GrS] Grove, K., Shiohama, K., A generalized sphere theorem, Ann. of Math. 106 (1977), 201-211.

[EH] Eschenberg, J., Heintze, E., An elementary proof of the Cheeger-Gromoll splitting theorem, Ann. Glob. Analysis and Geometry (1984), 141-151.

b) Rigidity Theory of Locally Symmetric Spaces. (8 lectures in English).
Prof. Mikhail GROMOV (IHES)
Prof. Pierre PANSU (Ecole Polytechnique, Palaiseau)

Contents

1. Statement and discussion of basic rigidity theorems of Mostow and Margulis.
2. Behaviour maps at the ideal boundary.
3. Harmonic maps and their applications to rigidity.

Bibliography

G.D. Mostow, Strong Ridigity for locally symmetric spaces, Princeton University Press, 1973.
N. Mok, Metric Rigidity Theorems on Hermitian Locally Symmetric Manifolds, World Scientific, 1989.

c) Instantons, stable bundles and the differential topology of algebraic surfaces. (8 lectures in English)
Prof. Christian OKONEK (Universität Bonn)

Contents

1. Topology of algebraic surfaces
2. Stable vector bundles
3. Instantons
4. Hermitian - Einstein structures
5. Donaldson's invariants
6. C^{∞}-structures of algebraic surfaces

Basic literature (books only)

[B/P/V] Barth, W., Peters, C., Van de Ven, A.: "Compact complex surfaces". Erg. der Math. 3. Folge, Bd. 4. Berlin, Heidelberg, New York, Tokyo, Springer 1984.
[F/U] Freed, D., Uhlenbeck, K.: "Instantons and four-manifolds". M.S.R.I. publ. no. 1. New York, Berlin, Heidelberg, Tokyo, Springer 1984.
[K] Kobayashi, S.: "Differential geometry of complex vector bundles". Iwanami Shoten and Princeton University Press 1987.
[O/S/S] Okonek, C., Schneider, M., Spindler, H.: "Vector bundles over complex projective spaces". Progress in Math. 3. Boston, Basel, Stuttgart, Birkhäuser 1980.

FONDAZIONE C.I.M.E.
CENTRO INTERNAZIONALE MATEMATICO ESTIVO
INTERNATIONAL MATHEMATICAL SUMMER CENTER

"Continua with microstructures"

is the subject of the Third 1990 C.I.M.E. Session.

The Session, sponsored by the Consiglio Nazionale delle Ricerche and the Ministero della Pubblica Istruzione, will take place under the scientific direction of Prof. GIANFRANCO CAPRIZ (Università di Pisa) at Villa "La Querceta", Montecatini Terme (Pistoia), Italy, **from July 2 to July 10, 1990.**

Courses

a) **Invariants in the theory of crystal defects.** (6 lectures in English).
Prof. Cesare DAVINI (Università di Udine, Italy).

The course presents recent results on a continuum theory of defects in crystals. The theory is based on the notion that defects should be measured by descriptors which are additive over the parts of the crystal and which are invariant under elastic deformations. It is shown that there is an infinite list of descriptors with these properties with a finite functional basis. This complete list strictly includes the Burgers' vectors and the dislocation density tensor of the classical theory of dislocations. Connections with the basic mechanisms of crystal plasticity are also discussed.

Outline of the course

1. Crystal lattices. Old molecular theories of elasticity. Defects and their role in the mechanics of materials.
2. Continuous theories of defects: the contributions of Bilby and Kondo. A continuum model for defective crystals. Elastic invariants.
3. Characterization of the first order invariants and their interpretation. Invariants of higher order. Measures of local defectiveness.
4. Neutral deformations. Conjugacy and canonical states.
5. Characterization of canonical states. A complete list of invariants.
6. A connection with a classical theorem of Frobenius. Equidefective states. Slips and rearrangements.

Basic references

[1] Taylor, G.I., The mechanism of plastic deformation of crystals, Part I and II, Proc. Roy. Soc. A 145 (1934), 362-387, 388-404.
[2] Bilby, B.A., Continuous distributions of dislocations. In: "Progress in solids mechanics", Vol. I (I.E. Sneddon, ed.), North-Holland Publishing Co., Amsterdam, 1960.

[3] Kroner, E., Allgemeine Kontinuumstheorie der Versetzungen und Eigenspannungen, Arch. Rational Mech. Anal. 4 (1960), 273-334.
[4] Davini, C., A proposal for a continuum theory of defective crystals, Arch. Rational Mech. Anal. 96 (1986), 295-317.
[5] Davini, C., Elastic invariants in crystal theory, In: "Material instabilities in continuum mechanics and related mathematical problems" (J.M. Ball, ed.), Clarendon Press, Oxford, 1988.
[6] Davini, C. and B.P. Parry, On defect-preserving deformations in crystals, Int. J. Plasticity 5 (1989), 337-369.
[7] Davini, C. and G.P. Parry, A complete list of invariants for defective crystals, (to appear).

b) **Microstructural theories for granular materials.** (6 lectures in English).
Prof. James T. JENKINS (Cornell University).

Outline

We outline the derivation of continuum theories for granular materials that are appropriate in the two extremes of their behavior: rapid flows involving particle collisions, and quasi-static deformations with enduring, frictional, interparticle contacts. In each extreme the microstructural variable that is important is a symmetric second rank tensor. For rapid flows this tensor is the second moment of the velocity fluctuations. In quasi-static situations it is a measure of the orientational distribution of the contact area. In each case we discuss the determination of the microstructure from the appropriate field equations and assess its influence on the stresses necessary to maintain a given flow or deformation. For rapid flows, boundary conditions may be derived using methods similar to those employed to obtain the field equations and constitutive relations. We indicate how the boundary conditions influence flows and illustrate this by employing the results of the theory in a simple hydraulic model for a rock debris slide.

References

Jenkins, J.T., Cundall, P.A. and Ishibashi, I., Micromechanical modeling of granular materials with the assistance of experiments and numerical simulations, in "Powders and Grains" (J. Biarez and R. Gourves, eds.), pp. 257-264, Balkema: Rotterdam, 1989.
Jenkins, J.T., Balance laws and constitutive relations for rapid flows of granular materials, in "Constitutive Models of Deformation" (J. Chandra and R.P. Srivastav, eds.), pp. 109-119, SIAM, Philadelphia, 1987.
Jenkins, J.T., Rapid flows of granular materials, in "Non-classical Continuum Mechanics" (R.J. Knops and A.A. Lacey, eds.), pp. 213-225, University Press, Cambridge, 1987.

c) **Defects and textures in liquid crystals.** (6 lectures in English).
Prof. Maurice KLEMAN (Université Paris-Sud).

Outline

1. Microstructure, the director $\vec{n}$, equations of equilibrium.
2. Layered phases: the equation $\mathrm{curl}\,\vec{n} = 0$ and the geometry of focal conics; topology of defects at the Sm A - Sm C transition, analogy with monopoles.
3. Columnar phases; the equation $\mathrm{div}\,\vec{n} = 0$ and the geometry of developable domains.
4. Double helical patterns in sinectics and cholesterics; presence of minimal surfaces, frustration, a model for the chromosome of dinoflagellate.
5. Cubic phases and minimal surfaces.
6. Some related aspects in ferromagnets.

d) **The topological theory of defects in ordered media.** (6 lectures in English).
Prof. David MERMIN (Cornell University).

Outline

Example of ordered media and their associated spaces of internal states.

2. Defects and their physical importance, classes of mutually homotopic loops in the state space, and the relation between the two.
3. The fundamental group of the state space and its relation to the combination laws for defects; media with non-abelian fundamental groups.
4. Some simple topological properties of continuous groups; group theoretic characterization of the state space in terms of broken symmetry.
5. How to deduce the fundamental group directly from the symmetry of the uniform medium.
6. The second homotopy group, its relation to point defects in 3 dimensions, the conversion of point defects by moving them around line defects, and how to deduce all this directly from the symmetry of the uniform medium.

FONDAZIONE C.I.M.E.
CENTRO INTERNAZIONALE MATEMATICO ESTIVO
INTERNATIONAL MATHEMATICAL SUMMER CENTER

"Mathematical Modelling of Industrial Processes"

is the subject of the Fourth 1990 C.I.M.E. Session.

The Session, sponsored by the Consiglio Nazionale delle Ricerche and the Ministero della Pubblica Istruzione, will be under the auspices of ECMI (European Consortium for Mathematics in Industry) and in collaboration with SASIAM (School for Advanced Studies in Industrial and Applied Mathematics).

It will take place under the Scientific direction of Prof. VINCENZO CAPASSO (Director of SASIAM) and Prof. ANTONIO FASANO (Università di Firenze), in TECNOPOLIS (Valenzano, Bari), from **September 24 to September 29, 1990.**

Courses

a) **Case studies of Industrial Mathematics Projects.** (7 lectures in English).
Prof. Stavros BUSENBERG (Harvey Mudd College, Claremont, USA).

Outline

These lectures will describe several industrial projects in which I have been involved over the past twenty years. A number of these projects originated in the Claremont Mathematics Clinic Program where small teams of students and faculty study problems sponsored and funded by industrial concerns. The other projects originated in consulting activities or in different University-Industry mathematics programs.

The first lecture will give an overview of a variety of Industrial Mathematics problems and of the settings in which they arose. Each of the remaining lectures will be organized about a particular mathematical area which has been useful in specific projects. However, it is the nature of Industrial Mathematics that it cannot be easily encapsulated in tidy mathematical fields which are defined for the convenience of academics, and we will end up touching upon a variety of techniques and theories.

- Case Studies of Industrial Mathematics Problems
- Semiconductor Contact Resistivity: Inverse Elliptic Problems
- Inverse Problems: Examples, Theory, and Computation
- Adaptive Pattern Recognition via Neural Networks: Optimization
- Static and Dynamic Optimization Problems
- Agricultural and Animal Resource Management: Dynamical Systems.

General References

. H.T. Banks and K. Kunisch, Estimation Techniques for Distributed Parameter Systems, Birkhauser, Boston, 1989.
. D.P. Bertsekas, Constrained Optimization and Lagrange Multiplier Methods, Academic Press, New York, 1982.
. Tarun Khanna, Foundations of Neural Networks, Addison-Wesley, Reading, Massachusetts, 1990.
. C. Castillo-Chavez, S.A. Levin and C.A. Shoemaker (Eds.) - Mathematical Approaches to Problems in Resource Management and Epidemiology, Lecture Notes in Biomathematics 81, Springer Verlag, New York, 1989.

b) **Inverse Problems in Mathematics for Industry.** (7 lectures in English).
Prof. Bruno FORTE (University of Waterloo, Ontario, Canada).

Outline

The process of deriving a deterministic mathematical model from the knowledge of particular solutions(s) and/or global properties of solutions will be analyzed. Examples of inverse problems related to some typical industrial process will be presented. Mainly we will be dealing with: inverse problems in classical mechanics (dynamical systems), inverse problems in diffusion processes.

References

1. A.S. Galiullin, Inverse problems of dynamics, Mir Publisher, Moscow, 1984
2. Frederic Y.M. Wan, Mathematical models and their analysis, Harper and Row, New York, 1989.

c) **Mathematical Aspects of Some Industrial Problems.** (7 lectures in English).
Prof. Hendrik K. KUIKEN (Philips Research Lab., Eindhoven).

Lectures 1 and 2: The determination of surface tension and contact angle from the shape of a sessile drop.

Literature

C.A. Smolders and E.M. Duyvis, Contact angles and de-wetting of mercury. Recueil 80 (1961), 635-649.
C.J. Lyons, E. Elbing and I.R. Wilson, A general selected plane method for measuring interfacial tensions from the shape of pendant and sessile drops. J. Chem. Soc. Farad. Trans. 81 (1985), 327-339.
Y. Rotenberg, L. Buruvka and A.W. Newman, Determination of surface tension and contact angle from the shape of axisymmetric fluid interfaces. J. Coll. Interf. Sci. 93 (1983), 169-183.
H.K. Kuiken, The determination of surface tension and contact angle from shape of a sessile drop revisited. To be published.

Lecture 3 and 4: The mathematical modelling of viscous sintering processes

Literature

H.E. Exnor, Principles of single phase sintering. Revs. Powder Metall. Phys. Chem. 1 (1979), 7-251
H.K. Kuiken, Viscous sintering: the surface-tension-driven flow of a liquid form under the influence of curvature gradients at its surface. To appear in J. Fluid Mech.
H.K. Kuiken, Deforming surfaces and viscous sintering. To appear in Proc. Conf. on the Math. and Comp. of Deforming Surfaces. Cambridge 1988. Oxford U. Press 1990.

Lectures 5, 6 and 7: Mathematical modelling of etching processes

Literature

H.K. Kuiken, Etching: a two-dimensional mathematical approach. Proc. R. Soc. London A392 (1984), 199-225.
H.K. Kuiken, Etching through a slit. Proc. R. Soc. London A396 (1984), 95-117.
H.K. Kuiken, J.J. Kelly and P.H.L. Notten, Etching at resist edges. J. Elchem. Soc. 133 (1986), 1217-1226 (part 1), 1227-1232 (part 2).
H.K. Kuiken, Mathematical modelling of etching processes. Proc. 1987 Irsee Conf. on Free and Moving Boundaries. Pitman 1990.

LIST OF C.I.M.E. SEMINARS

		Publisher
1954 -	1. Analisi funzionale	C.I.M.E.
	2. Quadratura delle superficie e questioni connesse	"
	3. Equazioni differenziali non lineari	"
1955 -	4. Teorema di Riemann-Roch e questioni connesse	"
	5. Teoria dei numeri	"
	6. Topologia	"
	7. Teorie non linearizzate in elasticità, idrodinamica,aerodinamica	"
	8. Geometria proiettivo-differenziale	"
1956 -	9. Equazioni alle derivate parziali a caratteristiche reali	"
	10. Propagazione delle onde elettromagnetiche	"
	11. Teoria della funzioni di più variabili complesse e delle funzioni automorfe	"
1957 -	12. Geometria aritmetica e algebrica (2 vol.)	"
	13. Integrali singolari e questioni connesse	"
	14. Teoria della turbolenza (2 vol.)	"
1958 -	15. Vedute e problemi attuali in relatività generale	"
	16. Problemi di geometria differenziale in grande	"
	17. Il principio di minimo e le sue applicazioni alle equazioni funzionali	"
1959 -	18. Induzione e statistica	"
	19. Teoria algebrica dei meccanismi automatici (2 vol.)	"
	20. Gruppi, anelli di Lie e teoria della coomologia	"
1960 -	21. Sistemi dinamici e teoremi ergodici	"
	22. Forme differenziali e loro integrali	"
1961 -	23. Geometria del calcolo delle variazioni (2 vol.)	"
	24. Teoria delle distribuzioni	"
	25. Onde superficiali	"
1962 -	26. Topologia differenziale	"
	27. Autovalori e autosoluzioni	"
	28. Magnetofluidodinamica	"

1963 - 29. Equazioni differenziali astratte "
30. Funzioni e varietà complesse "
31. Proprietà di media e teoremi di confronto in Fisica Matematica "

1964 - 32. Relatività generale "
33. Dinamica dei gas rarefatti "
34. Alcune questioni di analisi numerica "
35. Equazioni differenziali non lineari "

1965 - 36. Non-linear continuum theories "
37. Some aspects of ring theory "
38. Mathematical optimization in economics "

1966 - 39. Calculus of variations Ed. Cremonese, Firenze
40. Economia matematica "
41. Classi caratteristiche e questioni connesse "
42. Some aspects of diffusion theory "

1967 - 43. Modern questions of celestial mechanics "
44. Numerical analysis of partial differential equations "
45. Geometry of homogeneous bounded domains "

1968 - 46. Controllability and observability "
47. Pseudo-differential operators "
48. Aspects of mathematical logic "

1969 - 49. Potential theory "
50. Non-linear continuum theories in mechanics and physics and their applications "
51. Questions of algebraic varieties "

1970 - 52. Relativistic fluid dynamics "
53. Theory of group representations and Fourier analysis "
54. Functional equations and inequalities "
55. Problems in non-linear analysis "

1971 - 56. Stereodynamics "
57. Constructive aspects of functional analysis (2 vol.) "
58. Categories and commutative algebra "

1972 - 59.	Non-linear mechanics		"
60.	Finite geometric structures and their applications		"
61.	Geometric measure theory and minimal surfaces		"
1973 - 62.	Complex analysis		"
63.	New variational techniques in mathematical physics		"
64.	Spectral analysis		"
1974 - 65.	Stability problems		"
66.	Singularities of analytic spaces		"
67.	Eigenvalues of non linear problems		"
1975 - 68.	Theoretical computer sciences		"
69.	Model theory and applications		"
70.	Differential operators and manifolds		"
1976 - 71.	Statistical Mechanics		Ed Liguori, Napoli
72.	Hyperbolicity		"
73.	Differential topology		"
1977 - 74.	Materials with memory		"
75.	Pseudodifferential operators with applications		"
76.	Algebraic surfaces		"
1978 - 77.	Stochastic differential equations		"
78.	Dynamical systems		Ed Liguori, Napoli and Birhäuser Verlag
1979 - 79.	Recursion theory and computational complexity		"
80.	Mathematics of biology		"
1980 - 81.	Wave propagation		"
82.	Harmonic analysis and group representations		"
83.	Matroid theory and its applications		"
1981 - 84.	Kinetic Theories and the Boltzmann Equation	(LNM 1048)	Springer-Verlag
85.	Algebraic Threefolds	(LNM 947)	"
86.	Nonlinear Filtering and Stochastic Control	(LNM 972)	"
1982 - 87.	Invariant Theory	(LNM 996)	"
88.	Thermodynamics and Constitutive Equations	(LN Physics 228)	"
89.	Fluid Dynamics	(LNM 1047)	"

1983 -	90.	Complete Intersections	(LNM 1092)	"
	91.	Bifurcation Theory and Applications	(LNM 1057)	"
	92.	Numerical Methods in Fluid Dynamics	(LNM 1127)	"
1984 -	93.	Harmonic Mappings and Minimal Immersions	(LNM 1161)	"
	94.	Schrödinger Operators	(LNM 1159)	"
	95.	Buildings and the Geometry of Diagrams	(LNM 1181)	"
1985 -	96.	Probability and Analysis	(LNM 1206)	"
	97.	Some Problems in Nonlinear Diffusion	(LNM 1224)	"
	98.	Theory of Moduli	(LNM 1337)	"
1986 -	99.	Inverse Problems	(LNM 1225)	"
	100.	Mathematical Economics	(LNM 1330)	"
	101.	Combinatorial Optimization	(LNM 1403)	"
1987 -	102.	Relativistic Fluid Dynamics	(LNM 1385)	"
	103.	Topics in Calculus of Variations	(LNM 1365)	"
1988 -	104.	Logic and Computer Science	(LNM 1429)	"
	105.	Global Geometry and Mathematical Physics	(LNM 1451)	"
1989 -	106.	Methods of nonconvex analysis	(LNM 1446)	"
	107.	Microlocal Analysis and Applications	(LNM 1495)	"
1990 -	108.	Geoemtric Topology: Recent Developments	(LNM 1504)	
	109.	H_∞ Control Theory	(LNM 1496)	"
	110.	Continua with microstructures	to appear	"
	111.	Mathematical Modelling of Industrial Processes	to appear	"
1991 -	112.	Topological Methods in the Theory of Ordinary Differential Equations in Finite and Infinite Dimensions	to appear	

Lecture Notes in Mathematics

For information about Vols. 1–1312
please contact your bookseller or Springer-Verlag

Vol. 1313: F. Colonius. Optimal Periodic Control. VI, 177 pages. 1988.

Vol. 1314: A. Futaki. Kähler-Einstein Metrics and Integral Invariants. IV, 140 pages. 1988.

Vol. 1315: R.A. McCoy, I. Ntantu, Topological Properties of Spaces of Continuous, Functions. IV, 124 pages. 1988.

Vol. 1316: H. Korezlioglu, A.S. Ustunel (Eds.), Stochastic Analysis and Related Topics. Proceedings, 1986. V, 371 pages. 1988.

Vol. 1317: J. Lindenstrauss, V.D. Milman (Eds.), Geometric Aspects of Functional Analysis. Seminar, 1986–87. VII, 289 pages. 1988.

Vol. 1318: Y. Felix (Ed.), Algebraic Topology – Rational Homotopy. Proceedings, 1986. VIII, 245 pages. 1988.

Vol. 1319: M. Vuorinen, Conformal Geometry and Quasiregular Mappings. XIX, 209 pages. 1988.

Vol. 1320: H. Jürgensen, G. Lallement, H.J. Weinert (Eds.), Semigroups, Theory and Applications. Proceedings, 1986. X, 416 pages. 1988.

Vol. 1321: J. Azéma, P.A. Meyer, M. Yor (Eds.), Séminaire de Probabilités XXII. Proceedings. IV, 600 pages. 1988.

Vol. 1322: M. Métivier, S. Watanabe (Eds.), Stochastic Analysis. Proceedings, 1987. VII, 197 pages. 1988.

Vol. 1323: D.R. Anderson, H.J. Munkholm, Boundedly Controlled Topology. XII, 309 pages. 1988.

Vol. 1324: F. Cardoso, D.G. de Figueiredo, R. Iório, O. Lopes (Eds.), Partial Differential Equations. Proceedings, 1986. VIII, 433 pages. 1988.

Vol. 1325: A. Truman, I.M. Davies (Eds.), Stochastic Mechanics and Stochastic Processes. Proceedings, 1986. V, 220 pages. 1988.

Vol. 1326: P.S. Landweber (Ed.), Elliptic Curves and Modular Forms in Algebraic Topology. Proceedings, 1986. V, 224 pages. 1988.

Vol. 1327: W. Bruns, U. Vetter, Determinantal Rings. VII, 236 pages. 1988.

Vol. 1328: J.L. Bueso, P. Jara, B. Torrecillas (Eds.), Ring Theory. Proceedings, 1986. IX, 331 pages. 1988.

Vol. 1329: M. Alfaro, J.S. Dehesa, F.J. Marcellan, J.L. Rubio de Francia, J. Vinuesa (Eds.): Orthogonal Polynomials and their Applications. Proceedings, 1986. XV, 334 pages. 1988.

Vol. 1330: A. Ambrosetti, F. Gori, R. Lucchetti (Eds.), Mathematical Economics. Montecatini Terme 1986. Seminar. VII, 137 pages. 1988.

Vol. 1331: R. Bamón, R. Labarca, J. Palis Jr. (Eds.), Dynamical Systems, Valparaiso 1986. Proceedings. VI, 250 pages. 1988.

Vol. 1332: E. Odell, H. Rosenthal (Eds.), Functional Analysis. Proceedings. 1986–87. V, 202 pages. 1988.

Vol. 1333: A.S. Kechris, D.A. Martin, J.R. Steel (Eds.), Cabal Seminar 81–85. Proceedings, 1981–85. V, 224 pages. 1988.

Vol. 1334: Yu.G. Borisovich, Yu.E. Gliklikh (Eds.), Global Analysis – Studies and Applications III. V, 331 pages. 1988.

Vol. 1335: F. Guillén, V. Navarro Aznar, P. Pascual-Gainza, F. Puerta, Hyperrésolutions cubiques et descente cohomologique. XII, 192 pages. 1988.

Vol. 1336: B. Helffer, Semi-Classical Analysis for the Schrödinger Operator and Applications. V, 107 pages. 1988.

Vol. 1337: E. Sernesi (Ed.), Theory of Moduli. Seminar, 1985. VIII, 232 pages. 1988.

Vol. 1338: A.B. Mingarelli, S.G. Halvorsen. Non-Oscillation Domains of Differential Equations with Two Parameters. XI, 109 pages. 1988.

Vol. 1339: T. Sunada (Ed.), Geometry and Analysis of Manifolds. Proceedings, 1987. IX, 277 pages. 1988.

Vol. 1340: S. Hildebrandt, D.S. Kinderlehrer, M. Miranda (Eds.), Calculus of Variations and Partial Differential Equations. Proceedings, 1986. IX, 301 pages. 1988.

Vol. 1341: M. Dauge, Elliptic Boundary Value Problems on Corner Domains. VIII, 259 pages. 1988.

Vol. 1342: J.C. Alexander (Ed.), Dynamical Systems. Proceedings, 1986–87. VIII, 726 pages. 1988.

Vol. 1343: H. Ulrich, Fixed Point Theory of Parametrized Equivariant Maps. VII, 147 pages. 1988.

Vol. 1344: J. Král, J. Lukes, J. Netuka, J. Vesely´ (Eds.), Potential Theory – Surveys and Problems. Proceedings, 1987. VIII, 271 pages. 1988.

Vol. 1345: X. Gomez-Mont, J. Seade, A. Verjovski (Eds.), Holomorphic Dynamics. Proceedings, 1986. VII. 321 pages. 1988.

Vol. 1346: O.Ya. Viro (Ed.), Topology and Geometry – Rohlin Seminar. XI, 581 pages. 1988.

Vol. 1347: C. Preston, Iterates of Piecewise Monotone Mappings on an Interval. V, 166 pages. 1988.

Vol. 1348: F. Borceux (Ed.), Categorical Algebra and its Applications. Proceedings, 1987. VIII, 375 pages. 1988.

Vol. 1349: E. Novak, Deterministic and Stochastic Error Bounds in Numerical Analysis. V, 113 pages. 1988.

Vol. 1350: U. Koschorke (Ed.), Differential Topology Proceedings, 1987, VI, 269 pages. 1988.

Vol. 1351: I. Laine, S. Rickman, T. Sorvali (Eds.), Complex Analysis, Joensuu 1987. Proceedings. XV, 378 pages. 1988.

Vol. 1352: L.L. Avramov, K.B. Tchakerian (Eds.), Algebra – Some Current Trends. Proceedings. 1986. IX, 240 Seiten. 1988.

Vol. 1353: R.S. Palais, Ch.-l. Teng, Critical Point Theory and Submanifold Geometry. X, 272 pages. 1988.

Vol. 1354: A. Gómez, F. Guerra, M.A. Jiménez, G. López (Eds.), Approximation and Optimization. Proceedings, 1987. VI, 280 pages. 1988.

Vol. 1355: J. Bokowski, B. Sturmfels, Computational Synthetic Geometry. V, 168 pages. 1989.

Vol. 1356: H. Volkmer, Multiparameter Eigenvalue Problems and Expansion Theorems. VI, 157 pages. 1988.

Vol. 1357: S. Hildebrandt, R. Leis (Eds.), Partial Differential Equations and Calculus of Variations. VI, 423 pages. 1988.

Vol. 1358: D. Mumford, The Red Book of Varieties and Schemes. V, 309 pages. 1988.

Vol. 1359: P. Eymard, J.-P. Pier (Eds.) Harmonic Analysis. Proceedings, 1987. VIII, 287 pages. 1988.

Vol. 1360: G. Anderson, C. Greengard (Eds.), Vortex Methods. Proceedings, 1987. V, 141 pages. 1988.

Vol. 1361: T. tom Dieck (Ed.), Algebraic Topology and Transformation Groups. Proceedings. 1987. VI, 298 pages. 1988.

Vol. 1362: P. Diaconis, D. Elworthy, H. Föllmer, E. Nelson, G.C. Papanicolaou, S.R.S. Varadhan. École d´ Été de Probabilités de Saint-Flour XV–XVII. 1985–87 Editor: P.L. Hennequin. V, 459 pages. 1988.

Vol. 1363: P.G. Casazza, T.J. Shura, Tsirelson´s Space. VIII, 204 pages. 1988.

Vol. 1364: R.R. Phelps, Convex Functions, Monotone Operators and Differentiability. IX, 115 pages. 1989.

Vol. 1365: M. Giaquinta (Ed.), Topics in Calculus of Variations. Seminar, 1987. X, 196 pages. 1989.

Vol. 1366: N. Levitt, Grassmannians and Gauss Maps in PL-Topology. V, 203 pages. 1989.

Vol. 1367: M. Knebusch, Weakly Semialgebraic Spaces. XX, 376 pages. 1989.

Vol. 1368: R. Hübl, Traces of Differential Forms and Hochschild Homology. III, 111 pages. 1989.

Vol. 1369: B. Jiang, Ch.-K. Peng, Z. Hou (Eds.), Differential Geometry and Topology. Proceedings, 1986–87. VI, 366 pages. 1989.

Vol. 1370: G. Carlsson, R.L. Cohen, H.R. Miller, D.C. Ravenel (Eds.), Algebraic Topology. Proceedings, 1986. IX, 456 pages. 1989.

Vol. 1371: S. Glaz, Commutative Coherent Rings. XI, 347 pages. 1989.

Vol. 1372: J. Azéma, P.A. Meyer, M. Yor (Eds.), Séminaire de Probabilités XXIII. Proceedings. IV, 583 pages. 1989.

Vol. 1373: G. Benkart, J.M. Osborn (Eds.), Lie Algebras. Madison 1987. Proceedings. V, 145 pages. 1989.

Vol. 1374: R.C. Kirby, The Topology of 4-Manifolds. VI, 108 pages. 1989.

Vol. 1375: K. Kawakubo (Ed.), Transformation Groups. Proceedings, 1987. VIII, 394 pages, 1989.

Vol. 1376: J. Lindenstrauss, V.D. Milman (Eds.), Geometric Aspects of Functional Analysis. Seminar (GAFA) 1987–88. VII, 288 pages. 1989.

Vol. 1377: J.F. Pierce, Singularity Theory, Rod Theory, and Symmetry-Breaking Loads. IV, 177 pages. 1989.

Vol. 1378: R.S. Rumely, Capacity Theory on Algebraic Curves. III, 437 pages. 1989.

Vol. 1379: H. Heyer (Ed.), Probability Measures on Groups IX. Proceedings, 1988. VIII, 437 pages. 1989.

Vol. 1380: H.P. Schlickewei, E. Wirsing (Eds.), Number Theory, Ulm 1987. Proceedings. V, 266 pages. 1989.

Vol. 1381: J.-O. Strömberg, A. Torchinsky, Weighted Hardy Spaces. V, 193 pages. 1989.

Vol. 1382: H. Reiter, Metaplectic Groups and Segal Algebras. XI, 128 pages. 1989.

Vol. 1383: D.V. Chudnovsky, G.V. Chudnovsky, H. Cohn, M.B. Nathanson (Eds.), Number Theory, New York 1985–88. Seminar. V, 256 pages. 1989.

Vol. 1384: J. Garcia-Cuerva (Ed.), Harmonic Analysis and Partial Differential Equations. Proceedings, 1987. VII, 213 pages. 1989.

Vol. 1385: A.M. Anile, Y. Choquet-Bruhat (Eds.), Relativistic Fluid Dynamics. Seminar, 1987. V, 308 pages. 1989.

Vol. 1386: A. Bellen, C.W. Gear, E. Russo (Eds.), Numerical Methods for Ordinary Differential Equations. Proceedings, 1987. VII, 136 pages. 1989.

Vol. 1387: M. Petkovi´c, Iterative Methods for Simultaneous Inclusion of Polynomial Zeros. X, 263 pages. 1989.

Vol. 1388: J. Shinoda, T.A. Slaman, T. Tugué (Eds.), Mathematical Logic and Applications. Proceedings, 1987. V, 223 pages. 1989.

Vol. 1000: Second Edition. H. Hopf, Differential Geometry in the Large. VII, 184 pages. 1989.

Vol. 1389: E. Ballico, C. Ciliberto (Eds.), Algebraic Curves and Projective Geometry. Proceedings, 1988. V, 288 pages. 1989.

Vol. 1390: G. Da Prato, L. Tubaro (Eds.), Stochastic Partial Differential Equations and Applications II. Proceedings, 1988. VI, 258 pages. 1989.

Vol. 1391: S. Cambanis, A. Weron (Eds.), Probability Theory on Vector Spaces IV. Proceedings, 1987. VIII, 424 pages. 1989.

Vol. 1392: R. Silhol, Real Algebraic Surfaces. X, 215 pages. 1989.

Vol. 1393: N. Bouleau, D. Feyel, F. Hirsch, G. Mokobodzki (Eds.), Séminaire de Théorie du Potentiel Paris, No. 9. Proceedings. VI, 265 pages. 1989.

Vol. 1394: T.L. Gill, W.W. Zachary (Eds.), Nonlinear Semigroups, Partial Differential Equations and Attractors. Proceedings, 1987. IX, 233 pages. 1989.

Vol. 1395: K. Alladi (Ed.), Number Theory, Madras 1987. Proceedings. VII, 234 pages. 1989.

Vol. 1396: L. Accardi, W. von Waldenfels (Eds.), Quantum Probability and Applications IV. Proceedings, 1987. VI, 355 pages. 1989.

Vol. 1397: P.R. Turner (Ed.), Numerical Analysis and Parallel Processing. Seminar, 1987. VI, 264 pages. 1989.

Vol. 1398: A.C. Kim, B.H. Neumann (Eds.), Groups – Korea 1988. Proceedings. V, 189 pages. 1989.

Vol. 1399: W.-P. Barth, H. Lange (Eds.), Arithmetic of Complex Manifolds. Proceedings, 1988. V, 171 pages. 1989.

Vol. 1400: U. Jannsen. Mixed Motives and Algebraic K-Theory. XIII, 246 pages. 1990.

Vol. 1401: J. Steprans, S. Watson (Eds.), Set Theory and its Applications. Proceedings, 1987. V, 227 pages. 1989.

Vol. 1402: C. Carasso, P. Charrier, B. Hanouzet, J.-L. Joly (Eds.), Nonlinear Hyperbolic Problems. Proceedings, 1988. V, 249 pages. 1989.

Vol. 1403: B. Simeone (Ed.), Combinatorial Optimization. Seminar, 1986. V, 314 pages. 1989.

Vol. 1404: M.-P. Malliavin (Ed.), Séminaire d´Algèbre Paul Dubreil et Marie-Paul Malliavin. Proceedings, 1987–1988. IV, 410 pages. 1989.

Vol. 1405: S. Dolecki (Ed.), Optimization. Proceedings, 1988. V, 223 pages. 1989. Vol. 1406: L. Jacobsen (Ed.), Analytic Theory of Continued Fractions III. Proceedings, 1988. VI, 142 pages. 1989.

Vol. 1407: W. Pohlers, Proof Theory. VI, 213 pages. 1989.

Vol. 1408: W. Lück, Transformation Groups and Algebraic K-Theory. XII, 443 pages. 1989.

Vol. 1409: E. Hairer, Ch. Lubich, M. Roche. The Numerical Solution of Differential-Algebraic Systems by Runge-Kutta Methods. VII, 139 pages. 1989.

Vol. 1410: F.J. Carreras, O. Gil-Medrano, A.M. Naveira (Eds.), Differential Geometry. Proceedings, 1988. V, 308 pages. 1989.

Vol. 1411: B. Jiang (Ed.), Topological Fixed Point Theory and Applications. Proceedings. 1988. VI, 203 pages. 1989.

Vol. 1412: V.V. Kalashnikov, V.M. Zolotarev (Eds.), Stability Problems for Stochastic Models. Proceedings, 1987. X, 380 pages. 1989.

Vol. 1413: S. Wright, Uniqueness of the Injective III$_1$ Factor. III, 108 pages. 1989.

Vol. 1414: E. Ramirez de Arellano (Ed.), Algebraic Geometry and Complex Analysis. Proceedings, 1987. VI, 180 pages. 1989.

Vol. 1415: M. Langevin, M. Waldschmidt (Eds.), Cinquante Ans de Polynômes. Fifty Years of Polynomials. Proceedings, 1988. IX, 235 pages.1990.

Vol. 1416: C. Albert (Ed.), Géométrie Symplectique et Mécanique. Proceedings, 1988. V, 289 pages. 1990.

Vol. 1417: A.J. Sommese, A. Biancofiore, E.L. Livorni (Eds.), Algebraic Geometry. Proceedings, 1988. V, 320 pages. 1990.

Vol. 1418: M. Mimura (Ed.), Homotopy Theory and Related Topics. Proceedings, 1988. V, 241 pages. 1990.

Vol. 1419: P.S. Bullen, P.Y. Lee, J.L. Mawhin, P. Muldowney, W.F. Pfeffer (Eds.), New Integrals. Proceedings, 1988. V, 202 pages. 1990.

Vol. 1420: M. Galbiati, A. Tognoli (Eds.), Real Analytic Geometry. Proceedings, 1988. IV, 366 pages. 1990.

Vol. 1421: H.A. Biagioni, A Nonlinear Theory of Generalized Functions, XII, 214 pages. 1990.

Vol. 1422: V. Villani (Ed.), Complex Geometry and Analysis. Proceedings, 1988. V, 109 pages. 1990.

Vol. 1423: S.O. Kochman, Stable Homotopy Groups of Spheres: A Computer-Assisted Approach. VIII, 330 pages. 1990.

Vol. 1424: F.E. Burstall, J.H. Rawnsley, Twistor Theory for Riemannian Symmetric Spaces. III, 112 pages. 1990.

Vol. 1425: R.A. Piccinini (Ed.), Groups of Self-Equivalences and Related Topics. Proceedings, 1988. V, 214 pages. 1990.

Vol. 1426: J. Azéma, P.A. Meyer, M. Yor (Eds.), Séminaire de Probabilités XXIV, 1988/89. V, 490 pages. 1990.

Vol. 1427: A. Ancona, D. Geman, N. Ikeda, École d'Eté de Probabilités de Saint Flour XVIII, 1988. Ed.: P.L. Hennequin. VII, 330 pages. 1990.

Vol. 1428: K. Erdmann, Blocks of Tame Representation Type and Related Algebras. XV. 312 pages. 1990.

Vol. 1429: S. Homer, A. Nerode, R.A. Platek, G.E. Sacks, A. Scedrov, Logic and Computer Science. Seminar, 1988. Editor: P. Odifreddi. V, 162 pages. 1990.

Vol. 1430: W. Bruns, A. Simis (Eds.), Commutative Algebra. Proceedings. 1988. V, 160 pages. 1990.

Vol. 1431: J.G. Heywood, K. Masuda, R. Rautmann, V.A. Solonnikov (Eds.), The Navier-Stokes Equations – Theory and Numerical Methods. Proceedings, 1988. VII, 238 pages. 1990.

Vol. 1432: K. Ambos-Spies, G.H. Müller, G.E. Sacks (Eds.), Recursion Theory Week. Proceedings, 1989. VI, 393 pages. 1990.

Vol. 1433: S. Lang, W. Cherry, Topics in Nevanlinna Theory. I, 174 pages.1990.

Vol. 1434: K. Nagasaka, E. Fouvry (Eds.), Analytic Number Theory. Proceedings, 1988. VI, 218 pages. 1990.

Vol. 1435: St. Ruscheweyh, E.B. Saff, L.C. Salinas, R.S. Varga (Eds.), Computational Methods and Function Theory. Proceedings, 1989. VI, 211 pages. 1990.

Vol. 1436: S. Xambó-Descamps (Ed.), Enumerative Geometry. Proceedings, 1987. V, 303 pages. 1990.

Vol. 1437: H. Inassaridze (Ed.), K-theory and Homological Algebra. Seminar, 1987–88. V, 313 pages. 1990.

Vol. 1438: P.G. Lemarié (Ed.) Les Ondelettes en 1989. Seminar. IV, 212 pages. 1990.

Vol. 1439: E. Bujalance, J.J. Etayo, J.M. Gamboa, G. Gromadzki. Automorphism Groups of Compact Bordered Klein Surfaces: A Combinatorial Approach. XIII, 201 pages. 1990.

Vol. 1440: P. Latiolais (Ed.), Topology and Combinatorial Groups Theory. Seminar, 1985–1988. VI, 207 pages. 1990.

Vol. 1441: M. Coornaert, T. Delzant, A. Papadopoulos. Géométrie et théorie des groupes. X, 165 pages. 1990.

Vol. 1442: L. Accardi, M. von Waldenfels (Eds.), Quantum Probability and Applications V. Proceedings, 1988. VI, 413 pages. 1990.

Vol. 1443: K.H. Dovermann, R. Schultz, Equivariant Surgery Theories and Their Periodicity Properties. VI, 227 pages. 1990.

Vol. 1444: H. Korezlioglu, A.S. Ustunel (Eds.), Stochastic Analysis and Related Topics VI. Proceedings, 1988. V, 268 pages. 1990.

Vol. 1445: F. Schulz, Regularity Theory for Quasilinear Elliptic Systems and – Monge Ampère Equations in Two Dimensions. XV, 123 pages. 1990.

Vol. 1446: Methods of Nonconvex Analysis. Seminar, 1989. Editor: A. Cellina. V, 206 pages. 1990.

Vol. 1447: J.-G. Labesse, J. Schwermer (Eds), Cohomology of Arithmetic Groups and Automorphic Forms. Proceedings, 1989. V, 358 pages. 1990.

Vol. 1448: S.K. Jain, S.R. López-Permouth (Eds.), Non-Commutative Ring Theory. Proceedings, 1989. V, 166 pages. 1990.

Vol. 1449: W. Odyniec, G. Lewicki, Minimal Projections in Banach Spaces. VIII, 168 pages. 1990.

Vol. 1450: H. Fujita, T. Ikebe, S.T. Kuroda (Eds.), Functional-Analytic Methods for Partial Differential Equations. Proceedings, 1989. VII, 252 pages. 1990.

Vol. 1451: L. Alvarez-Gaumé, E. Arbarello, C. De Concini, N.J. Hitchin, Global Geometry and Mathematical Physics. Montecatini Terme 1988. Seminar. Editors: M. Francaviglia, F. Gherardelli. IX, 197 pages. 1990.

Vol. 1452: E. Hlawka, R.F. Tichy (Eds.), Number-Theoretic Analysis. Seminar, 1988–89. V, 220 pages. 1990.

Vol. 1453: Yu.G. Borisovich, Yu.E. Gliklikh (Eds.), Global Analysis – Studies and Applications IV. V, 320 pages. 1990.

Vol. 1454: F. Baldassari, S. Bosch, B. Dwork (Eds.), p-adic Analysis. Proceedings, 1989. V, 382 pages. 1990.

Vol. 1455: J.-P. Françoise, R. Roussarie (Eds.), Bifurcations of Planar Vector Fields. Proceedings, 1989. VI, 396 pages. 1990.

Vol. 1456: L.G. Kovács (Ed.), Groups – Canberra 1989. Proceedings. XII, 198 pages. 1990.

Vol. 1457: O. Axelsson, L.Yu. Kolotilina (Eds.), Preconditioned Conjugate Gradient Methods. Proceedings, 1989. V, 196 pages. 1990.

Vol. 1458: R. Schaaf, Global Solution Branches of Two Point Boundary Value Problems. XIX, 141 pages. 1990.

Vol. 1459: D. Tiba, Optimal Control of Nonsmooth Distributed Parameter Systems. VII, 159 pages. 1990.

Vol. 1460: G. Toscani, V. Boffi, S. Rionero (Eds.), Mathematical Aspects of Fluid Plasma Dynamics. Proceedings, 1988. V, 221 pages. 1991.

Vol. 1461: R. Gorenflo, S. Vessella, Abel Integral Equations. VII, 215 pages. 1991.

Vol. 1462: D. Mond, J. Montaldi (Eds.), Singularity Theory and its Applications. Warwick 1989, Part I. VIII, 405 pages. 1991.

Vol. 1463: R. Roberts, I. Stewart (Eds.), Singularity Theory and its Applications. Warwick 1989, Part II. VIII, 322 pages. 1991.

Vol. 1464: D. L. Burkholder, E. Pardoux, A. Sznitman, Ecole d'Eté de Probabilités de Saint- Flour XIX-1989. Editor: P. L. Hennequin. VI, 256 pages. 1991.

Vol. 1465: G. David, Wavelets and Singular Integrals on Curves and Surfaces. X, 107 pages. 1991.

Vol. 1466: W. Banaszczyk, Additive Subgroups of Topological Vector Spaces. VII, 178 pages. 1991.

Vol. 1467: W. M. Schmidt, Diophantine Approximations and Diophantine Equations. VIII, 217 pages. 1991.

Vol. 1468: J. Noguchi, T. Ohsawa (Eds.), Prospects in Complex Geometry. Proceedings, 1989. VII, 421 pages. 1991.

Vol. 1469: J. Lindenstrauss, V. D. Milman (Eds.), Geometric Aspects of Functional Analysis. Seminar 1989-90. XI, 191 pages. 1991.

Vol. 1470: E. Odell, H. Rosenthal (Eds.), Functional Analysis. Proceedings, 1987-89. VII, 199 pages. 1991.

Vol. 1471: A. A. Panchishkin, Non-Archimedean L-Functions of Siegel and Hilbert Modular Forms. VII, 157 pages. 1991.

Vol. 1472: T. T. Nielsen, Bose Algebras: The Complex and Real Wave Representations. V, 132 pages. 1991.

Vol. 1473: Y. Hino, S. Murakami, T. Naito, Functional Differential Equations with Infinite Delay. X, 317 pages. 1991.

Vol. 1474: S. Jackowski, B. Oliver, K. Pawałowski (Eds.), Algebraic Topology, Poznań 1989. Proceedings. VIII, 397 pages. 1991.

Vol. 1475: S. Busenberg, M. Martelli (Eds.), Delay Differential Equations and Dynamical Systems. Proceedings, 1990. VIII, 249 pages. 1991.

Vol. 1476: M. Bekkali, Topics in Set Theory. VII, 120 pages. 1991.

Vol. 1477: R. Jajte, Strong Limit Theorems in Noncommutative L_2-Spaces. X, 113 pages. 1991.

Vol. 1478: M.-P. Malliavin (Ed.), Topics in Invariant Theory. Seminar 1989-1990. VI, 272 pages. 1991.

Vol. 1479: S. Bloch, I. Dolgachev, W. Fulton (Eds.), Algebraic Geometry. Proceedings, 1989. VII, 300 pages. 1991.

Vol. 1480: F. Dumortier, R. Roussarie, J. Sotomayor, H. Żoładek, Bifurcations of Planar Vector Fields: Nilpotent Singularities and Abelian Integrals. VIII, 226 pages. 1991.

Vol. 1481: D. Ferus, U. Pinkall, U. Simon, B. Wegner (Eds.), Global Differential Geometry and Global Analysis. Proceedings, 1991. VIII, 283 pages. 1991.

Vol. 1482: J. Chabrowski, The Dirichlet Problem with L^2 Boundary Data for Elliptic Linear Equations. VI, 173 pages. 1991.

Vol. 1483: E. Reithmeier, Periodic Solutions of Nonlinear Dynamical Systems. VI, 171 pages. 1991.

Vol. 1484: H. Delfs, Homology of Locally Semialgebraic Spaces. IX, 136 pages. 1991.

Vol. 1485: J. Azéma, P. A. Meyer, M. Yor (Eds.), Séminaire de Probabilités XXV. VIII, 440 pages. 1991.

Vol. 1486: L. Arnold, H. Crauel, J.-P. Eckmann (Eds.), Lyapunov Exponents. Proceedings, 1990. VIII, 365 pages. 1991.

Vol. 1487: E. Freitag, Singular Modular Forms and Theta Relations. VI, 172 pages. 1991.

Vol. 1488: A. Carboni, M. C. Pedicchio, G. Rosolini (Eds.), Category Theory. Proceedings, 1990. VII, 494 pages. 1991.

Vol. 1490: K. Metsch, Linear Spaces with Few Lines. XIII, 196 pages. 1991.

Vol. 1491: H. Gillet, E. Lluis- Puebla, J.-L. Loday, V. Snaith, C. Soulé, Higher Algebraic K-theory: an overview. VI, 164 pages. 1991.

Vol. 1492: K. R. Wicks, Fractals and Hyperspaces. VIII, 168 pages. 1991.

Vol. 1493: E. Benoît (Ed.), Dynamic Bifurcations. Proceedings, Luminy 1990. VII, 219 pages. 1991.

Vol. 1494: M.-T. Cheng, X.-W. Zhou, D.-G. Deng (Eds.), Harmonic Analysis. Proceedings, 1988. IX, 226 pages. 1991.

Vol. 1495: J. M. Bony, G. Grubb, L. Hörmander, H. Komatsu, J. Sjöstrand, Microlocal Analysis and Applications. Montecatini Terme, 1989. Editors: L. Cattabriga, L. Rodino. VII, 349 pages. 1991.

Vol. 1496: C. Foias, B. Francis, J. W. Helton, H. Kwakernaak, J. B. Pearson, H_∞-Control Theory. Como, 1990. Editors: E. Mosca, L. Pandolfi. VII, 336 pages. 1991.